UNNATURAL

Philip Ball is a writer and contributor to *Nature*, where he previously worked as an editor for physical sciences. He writes regularly in the scientific and popular media, often combining the arenas of science and art, and delivers lectures with equal success at NASA and the V&A Museum. His many books include *The Self-Made Tapestry*, *H_2O: A Biography of Water*, *The Devil's Doctor*, *Critical Mass* (winner of the 2005 Aventis Prize for Science Books), *Universe of Stone*, *Nature's Patterns* and, most recently, the acclaimed *The Music Instinct*. He obtained a PhD in physics from the University of Bristol.

T0174494

PHILIP BALL

Unnatural

The Heretical Idea of Making People

VINTAGE BOOKS
London

Published by Vintage 2012

2 4 6 8 10 9 7 5 3

Copyright © Philip Ball 2011

Philip Ball has asserted his right under the Copyright, Designs
and Patents Act 1988 to be identified as the author of this work

First published in Great Britain in 2011 by
The Bodley Head

Vintage
Random House, 20 Vauxhall Bridge Road,
London SW1V 2SA

www.vintage-books.co.uk

Addresses for companies within
The Penguin Random House Group can be found at:
global.penguinrandomhouse.com

The Random House Group Limited Reg. No. 954009

A CIP catalogue record for this book
is available from the British Library

ISBN 9780099551836

Penguin Random House is committed to a sustainable future for
our business, our readers and our planet. This book is made from
Forest Stewardship Council® certified paper.

Printed and bound in Great Britain by Clays Ltd, Elcograf S.p.A.

Contents

Introduction
It's Not Natural

We know, of course, how to make people. Millions do it every day, and if they don't always succeed, my impression is that on the whole they enjoy trying. Needless to say, this book is not about *that* method – or rather, it is about where that method ends, and about what lies beyond it. And how, or whether, we can know where the boundary lies.

We can say with some confidence that the historical alternatives for making people – fashioning them from clay, from putrescent matter, from scavenged body parts – do not work. What is intriguing is that a consensus about the futility of these methods is relatively recent. That we did not notice their failure, or at least were not sure of it, for a long time tells us something interesting, and, I contend, something important.

First, it tells us that a belief in the possibility of making artificial humans, like all beliefs that lack a basis in objective fact – astrology, predestination, a heaven-and-hell afterlife – expresses deep-seated desires and anxieties. Second, the belief is wrong not trivially but in rather complex ways that are not just the result of ignorance – and in a particular sense it may not be wrong at all. Third, the mere fact that it is mistaken to think people can be made by these means does not banish the influence of that notion on our customs, assumptions, actions and judgements, and might sometimes perversely amplify it.

The idea of artificial procreation has never been more relevant than it is today. An allusion to the 'old' myths – to the alchemical homunculus, Faust, Frankenstein, *Brave New World* – is almost de rigueur in public discussions of assisted conception, 'designer babies', genetic modification, embryo research and cloning. Usually this is nothing more than a knee-jerk reflex, made with scant regard to what precisely those words are supposed to invoke. Yet however lazy and unconsidered such journalistic references to Frankenstein are, they convey *something*. We think we know what they're getting at. And it is not something good.

Scientists engaged in new ways of 'making people' – what I here call 'anthropoeia', which condenses that phrase into its Greek equivalent – in modern times, such as those researching *in vitro* fertilization and cloning, often resent and lament these intrusions of myth and legend into their field of work. Here we are, the scientists will say, trying to improve medicine and to relieve man's estate – trying to *do good* – and all the rest of the world can see are Gothic ghouls and mad inventors. 'Whatever today's embryologists may do, Frankenstein or Faust or Jekyll will have foreshadowed, looming over every biological debate', said Robert Edwards, a pioneer of IVF, in 1989 at the height of the debate about research on the human embryos that IVF had suddenly made available. Edwards was impatient with the way, in his view, science-fiction narratives were shaping the discussion: 'The necessity or otherwise for experiments on human embryos sparks the most intense argument, as fears arise about tailor-made babies, or clones, or cyborgs, or some other nightmarish fancy.'

'The trouble really started way back in the 1930s, by courtesy of the brilliant Aldous Huxley', Edwards asserted. But he was wrong about that. Aldous Huxley did not conceive a tale that subsequently shaped thinking about embryo research, any more than did Mary Shelley, Robert Louis Stevenson or Goethe. Rather, they and other writers gave particular embodiments to pre-existing myths and legends that would have exerted their influence come what may. Edwards might well have wished that *Brave New World* had never been written, but as we shall see, Huxley's authorship of that novel was almost incidental; the ideas were firmly bedded down before he put pen to paper.

Edwards also failed to perceive the true role of fictional tropes of anthropoeia. It is not simply the case that there happen to be stories and legends that create inconvenient and misleading stereotypes. In the stories we tell about artificial people – how they are made, and what we assume they are like – we reveal some of our most profound feelings about what is natural and what is not, and about what this distinction implies in moral terms. For making people has always been cause for moral judgement, which is at root a judgement about naturalness. As molecular biologist Lee Silver puts it, 'Nearly every literate person perceives *natural* as a synonym for good, whereas the opposite idea – unnatural, artificial, or synthetic – evokes a reflexive negative reaction.' As a result, he says, 'all naturalistic arguments against biotechnology are actually spiritual arguments in disguise'.

This connotation of the natural is, as we shall see, a historical construction. The German prefix 'un-' was attached in the early modern period to connote acts that were deemed reprehensible because they were *contra naturam*, against nature. As historian Helmut Puff says:

Un-natural is not simply non-natural, the opposite of natural. By sheer weight of the rhetorical tradition and frequent usage in moralizing contexts, 'un-' words take on additional connotations, the other side of the norm. From the point of view of the speaker, 'un-natural' articulates a polemical stance. 'Un-' enunciations condemn that which is expressed, declare it as dangerous, treacherous ground . . . It is a word that polices the dangerous boundary between the normative and the non-normative, the pure and the impure . . . Un-natural connotes a wretched state that ought to bring about the most vocal condemnation. It is meant to activate, though the precise nature of the implied action remains undefined. Yet the emotional response solicited from the listener / reader by means of this wording is clear: horror.

Old stories and brave new worlds

Robert Edwards had ample reason to complain about the way his work was framed for public debate. When the UK government's provisional framework for regulating embryo research was released in late 1987, the *Today* newspaper covered it under the headline 'Clamp on Frankenstein scientists', while the *Sun* accompanied its report with a still from a Frankenstein movie. *The Times* spoke of 'creating babies from the dead', and even the *Independent on Sunday*, which was broadly supportive of embryo research, could not resist reporting on the Human Fertilization and Embryology Act of 1990 with a piece called 'Brave new embryos'. Opponents of the research worried that, to quote one British Member of Parliament, 'The ultimate goal may be to produce a child entirely *in vitro* or to produce genetically identical individuals by cloning.'

But the real story behind these headlines is complicated. The temptation for pro-research lobbies to present it as a case of benighted superstition and paranoia pitched against rational attempts to understand nature and improve the human lot – a rerun, if you will, of Galileo's persecution – has been undermined many times by scientists themselves. Indeed, in pronouncements of this sort, even the scientists were unconsciously drawing on old myths.

And it is quite wrong, perhaps dangerously so, to suppose as Edwards did that the public and the media have simply been hoodwinked by the stories of Huxley and Shelley. These, after all, are not the sources that induced Pope John Paul II to condemn IVF in 2004 by calling it 'a technology that wants to substitute true paternity and maternity and therefore that does harm to the dignity of parents and children alike', and to complain that the conjugal act 'cannot be substituted by a mere technological procedure which is devoid of human value and subject to the dictates of science and technology'. Yet in order to understand what truly motivates objections like these, we need to appreciate that they stem from the same tradition that created the science-fiction tales of artificial procreation – a tradition that predates the Christian Church itself.

So for the sake of a clear, honest and humane debate about how we allow, enable and assist people to come into being, it is time to bring these myths out into the open, and examine what they tell us about our fears, fantasies and fetishes concerning the idea of making people.

1 Art Versus Nature?

So, over that art
Which you say adds to nature, is an art
That nature makes.
　　William Shakespeare, *The Winter's Tale* (c.1611)

All Nature is but Art, unknown to thee,
All Chance, Direction, which thou canst not see.
　　Alexander Pope, *Essay on Man* (1733–4)

In our dealings with nature we cannot avoid thinking in
metaphors of religious origin.
　　J. B. S. Haldane, *What Is Life?* (1949)

Making people artificially is a form of technology.

And that, some will say, is precisely what is wrong with it. For isn't technology, for all its undoubted conveniences, a cold, impersonal thing which should have nothing to do with the wonder of human procreation? Indeed, some would argue that to make reproduction a technological process is to divorce it from humanity entirely. 'Would not the laboratory production of human beings no longer be *human* procreation?' asked bioethicist Leon Kass in the wake of early reports on successful *in vitro* fertilization.

Quite what Kass meant by 'laboratory production' was not clear, but if, as one might suppose, a reasonable definition embraces the mixing of human ova and sperm in a Petri dish until they unite, or the injection of a spermatozoon into the ovum through a tiny syringe, then his implication that people alive today as a result of such procedures are not the result of human procreation seems absurd and positively insulting. But of course, the matter of technological intervention reaches far beyond the techniques of IVF, for there are few pregnancies and births in the developed world that do not involve it to some degree. While that is not to the unalloyed advantage of mother and child, it does mean that the death or impairment of either of them is far less likely than in the days when procreation was a more 'natural' affair.

But why would Kass then say such a thing? Clearly, it has something to do with how he feels about the character of *technical* or *artificial* intervention in 'nature'. Now, my opening statement above is arguably little more than a tautology, for the root of 'art', *ars*, is only the Latin equivalent of the Greek root of 'technology', *techne*. Our instinctive aversion to the image supplied by this phrase focuses on the single concept embodied in those two words, artifice and technology: the concept of *art*, of human-made productions.

Yet what could possibly be wrong with *art*? The contemporary meaning we assign to this word embodies all that we deem noble in the human spirit: its creativity, imagination, capacity for wonder. Isn't the link between the works of Bach, Rembrandt and Goethe and the shuddering, oil-mired engines of industrial technology just an etymological curiosity?

Maybe that's how it seems to us now, but it is not at all how it appeared in the ancient and medieval world. For what literature and painting have in common with machines and synthetic materials is that they are all human-made: they are things that *do not exist in nature*. They are, in the literal sense, unnatural.

There we have it. What provokes us about the idea of 'making people artificially', of 'the laboratory production of human beings', is that it is unnatural. The logical absurdity of Kass's statement (and its illogic goes deeper, as we will see later) is illustrative of how instinctive feelings about what is and is not 'natural' trump any attempt to think the matter through.

I aim in this chapter to lay the ground for thinking the matter through. The distinctions that have been drawn in earlier ages between the natural and the artificial were in many ways subtler and more sophisticated than those we colloquially recognize today. We can show now beyond question that the positions of the atoms in insulin or vitamin C made artificially – by industrial chemical synthesis – are identical to those in the same substances harvested from natural sources. Yet some people still prefer their drugs and food ingredients to come from the latter direction. This is not attributable merely to ignorance (although let's not deny that this may play a part), but stems unwittingly from old associations of the natural and the artificial that pervade the cultural atmosphere. In the ancient and early modern worlds these preconceptions were often more explicit, and therefore more productively explored.

Man the maker

Even the most ardent technophobe has to concede that making things – the skill of *techne* – is a part of what makes us human: we are as much *Homo faciens* as we are *Homo sapiens*. The debate about whether all technology should therefore be deemed 'natural' is, however, apt quickly to become arid. The better question, long discussed by the ancients, is how exactly our artefacts differ from the substances and objects we find in nature.

If one insists that there *must* be a difference, there is little left to discuss: art and nature simply yield distinct classes of things. But such a distinction was challenged by the discipline of alchemy. Historians of science are widely agreed now that alchemy was never primarily an esoteric, mystical quest to make gold, but was instead a practical technology used to create all the artificial substances that early cultures needed or desired: dyes, drugs, metals, soap, glass. All the same, alchemists *did* claim to be making gold – and that is what brought the art/nature distinction into focus.

The first alchemists, working in Egypt and Hellenistic Greece before about 200 BC, admitted that they were just mimicking gold, generally for decorative purposes such as gilding and painting. The methods and materials of this tradition are revealed in a pair of manuscripts called the Leiden and Stockholm papyri, thought to be fragments of a workshop manual written by an Egyptian artisan around the third century AD but compiling recipes from earlier sources. Among them are prescriptions for imitating gold, such as 'Giving objects of copper the appearance of gold'. There are hints that these procedures may have had deceptive intent: this particular recipe, for example, assures the reader that metalsmiths will find it hard to detect the difference, while another set of instructions involving the alloying of gold with iron is called 'Fraud of gold'. But that is precisely the point: the author recognizes that he is making a kind of fake gold, which mimics the real thing in appearance only. Even when he suggests that the result might be better than what it mimics, that is no different from the sort of sales pitch made for modern synthetic replicas of natural materials.

Around 100 BC, this began to change. Now alchemists started to suggest they were actually replicating, and not just simulating, the precious metal. It was no coincidence that at the same time alchemy began to acquire a theoretical and mystical content, for the bold claim to be changing the fundamental nature of matter demanded philosophical justification. In the view of Zosimos of Panopolis, an early Christian Egyptian who lived around AD 300 and whose alchemical writings imply that the sole aim of this art is now trans- mutation to gold, metals consist of two parts: a 'body' (*soma*) and a 'spirit' (*pneuma*). It is in the *pneuma* that the defining characteristics of a particular metal reside, while the *soma* is apparently the same for all metals (Zosimos hints that it may be equivalent to the metal mercury). The *pneuma* can be evaporated from one metal – the 'freeing' of the metal's spirit – distilled, and united with another. Transmutation thus becomes equated with processes – the 'death' of a metal and its reanimation by transmigration of spirit – that were more commonly associated with living organisms, and especially with humankind. The notion of a spirit that can be liberated from one kind of matter and united with another also lay behind the ritualistic animation of statues, an aspect of the discipline known as theurgy. Such practices were deemed to give the animated matter special powers, so that statues might become oracles. They were also likely to be seen as dabbling with phenomena that once had been the preserve of the gods.

So efforts to justify and explain the alchemical claim to be *repro-ducing* nature, rather than simply simulating it, led very quickly to suggestions that the alchemists could manipulate and perhaps even induce life. According to the great thirteenth-century German scholar Albertus Magnus, alchemists act towards metals as doctors do towards their patients. The idea that both are guiding their subjects towards a more 'perfect' state clearly extended beyond the realm of mere metaphor.

Makers or deceivers?

In the ancient and medieval Western world, Christian, Muslim and pagan alike deferred to the learning of the Greek philosophers, particularly Aristotle and Plato, on matters concerning the natural world. What did those writers have to say about the distinctions between art and nature, and the possibility of transcending them?

The central issue is deftly summarized by historians of science Bernadette Bensaude-Vincent and William Newman: 'Is *techne* a continuation of nature's activity, a rebellion against nature, or a challenge to nature?' In the first of these interpretations the technologist is simply helping along a process that occurs in nature anyway, and towards the same goal. In the second, he is performing something that cannot occur in nature, while in the third he is claiming to be able to make something superior to that which nature may produce. In either of these latter two views, one can say that the activity is or results in something that is *unnatural*.

In his *Republic*, Plato deplores the imitative feats of poetry and painting because of their deceptive character. Anyone, he scoffs, can replicate 'the sun and the heavens, and the earth and yourself, and other animals and plants, and all the other things', simply by holding up a mirror to them. But that would be to make the 'appearances only': a feeble feat of mimicry. There are three types of makers, says Plato. The highest is the divine Creator, who makes not actual things but the archetypes of things. Take a bed, for example. The Creator has made only one bed, what we might call the essence of a bed or, as we'd now say, the Platonic ideal of 'bed'. Then there is the carpenter, who makes individual instances of this ideal: an inferior sort of making, but nevertheless worthy and important, for this is after all an actual bed on which one might rest. The third 'maker' is the painter, who, however, does not truly make anything at all, but only the appearance of something. What good is that when you need a bed to lie on? Besides, it is only a single appearance: the bed made by a carpenter can be seen in the round, from any angle, but the painter can show us only one view of a bed. He does not even have to know that it is a bed, or what beds are for. 'A painter', says Plato, 'will paint a cobbler, carpenter, or any other artist, though he knows nothing of their arts.' In short, the painter is a wizard, a shallow deceiver whose productions are not to be trusted. We venerate Homer, Plato says, and yet for what? All he gave us was a story, an illusion of the world, lacking in truth, wisdom and virtue:

Had he in his lifetime friends who loved to associate with him, and who handed down to posterity an Homeric way of life, such as was established by Pythagoras who was so greatly beloved for his wisdom, and whose followers are to this day quite celebrated for the order which was named after him? Nothing of the kind is recorded of him . . . But can you imagine that if Homer had really been able to educate and improve mankind – if he had possessed knowledge and not been a mere imitator – can you imagine, I say, that he would not have had many followers, and been honoured and loved by them?

This astonishing attack on imitative art – 'an inferior who marries an inferior, and has inferior offspring', as Plato puts it – seems to reflect a common attitude in the ancient world, whereby mimesis was seen as a form of deception.

So much the worse for the fine arts. But some technological arts were also deemed to possess an unnatural character. In a book called *Mechanica* (*Mechanical Problems*), falsely attributed to Aristotle but probably written by another ancient Greek scholar,* mechanical machines are said to act 'against nature' because they can, for example, oppose the natural tendency of heavy objects to fall to the ground, instead lifting them aloft and thereby coercing them into a new, 'contra-natural' behaviour. Today, 'against nature' has a pejorative ring to it, and that connotation was not absent in ancient times. According to William Newman, 'the very term *mechanomai*, the verbal form of the Greek word for machine, was often used in a negative sense, to mean the act of deceitful contrivance'. Among the most notorious of such deceits were the moving statues and automata attributed to Daedalus (see p. 33).

---

* Some think the author was the Pythagorean scholar Archytas of Tarentum, active in the fourth century BC, who was an acquaintance of Plato.

Not everyone took such a dim view of art. Indeed, on the whole the Greeks tended to regard *techne* as something intrinsically beneficial, and even Pseudo-Aristotle's *Mechanical Problems*, for all its talk of contravention, celebrates the triumph over nature that machines afford. To Aeschylus, the bestowing of technologies on humankind by Prometheus (examined in the next chapter) was an act of liberation. The Athenian poet Antiphon says that art enables humans to supersede nature where otherwise we would be subservient to it. From around the time that he expressed these sentiments in the late fifth century BC, Greek philosophers began to portray *techne* as a means of coercing nature, by force or even by 'torture', so as to gain mastery over it and transgress its boundaries. This was seen as a praiseworthy goal, or at least a useful one.

Aristotle himself was silent on the propriety of such coercion. He did, however, agree that art had its benefits. He said that it could act as a kind of handmaid to nature, correcting its mistakes as one artisan might correct those of another, and helping to bring it to a state of greater perfection. Here, nature and art necessarily act both in the same direction and towards the same end, because in Aristotle's teleological world nature was itself guided as if by intelligent agency towards making things as well designed as they could be: 'nature is a cause,' he said, 'a cause that operates for a purpose'. And so:

if a house, e.g. had been a thing made by nature, it would have been made in the same way as it is now by art; and if things made by nature were made also by art, they would come to be in the same way as by nature. Each step then in the series is for the sake of the next; and generally art partly completes what nature cannot bring to a finish, and partly imitates her. If, therefore, artificial products are for the sake of an end, so clearly also are natural products. The relation of the later to the earlier terms of the series is the same in both.

This teleological aspect of nature, Aristotle said, was revealed by the 'design' of non-human artefacts and forms, such as the way animals seemed shaped and disposed for their natural role – an expression of 'purpose' that continued to baffle and confuse scientists until Darwin furnished an explanation of how it might arise in the absence of a designer. As Aristotle wrote:

This is most obvious in the animals other than man: they make things neither by art nor after inquiry or deliberation. Wherefore people discuss whether it is by intelligence or by some other faculty that these creatures work, spiders, ants, and the like. By gradual advance in this direction we come to see clearly that in plants too that is produced which is conducive to the end – leaves, e.g. grow to provide shade for the fruit. If then it is both by nature and for an end that the swallow makes its nest and the spider its web, and plants grow leaves for the sake of the fruit and send their roots down (not up) for the sake of nourishment, it is plain that this kind of cause is operative in things which come to be and are by nature.

This is, then, why art must either imitate nature or supplement it, and in either case it acts in parallel. For Aristotle, and for many others who came after, *techne* was therefore not only a mimic of nature but actually provided a model for understanding how nature works: as in 'art', so in nature. For the Alexandrian anatomist Erastistratus in the third century BC, we might comprehend the heart's mechanism by reference to the bellows pump. By the same token, nature had many tricks to teach the technologist – a belief now very much back in vogue in the scientific field known as biomimetics. The Greek historian Diodorus Siculus claimed that Talos, the nephew of the great inventor Daedalus, invented the saw by copying in iron the shape of a snake's jawbone, which he found would cut through wood.* And echoing Aristotle's comment about 'houses' built in nature, the Roman architect and technologist Marcus Vitruvius Pollio asserted that the first houses built by men imitated the nests of swallows. In this view, how could technology be anything other than benign?

This correspondence between art and nature justified the way Aristotle used the terminology of technology to describe processes in nature. In his *Meteorology* he speaks of 'boiling' and 'roasting' – terms derived from cookery – in the context of natural phenomena, and equates such processes in nature explicitly with the methods of art:

---

* In his *Natural History*, Pliny attributes Daedalus himself with the invention of carpentry. Daedalus is said to have murdered Talos, partly out of jealousy because his nephew's skills exceeded his own. For this crime he was exiled to Knossos, where he entered into the service of King Minos of Crete – a story that I pick up in the next chapter.

This, then, is what is called concoction by boiling: and it makes no difference whether it takes place in an artificial or a natural vessel, for the cause is the same in all cases.

Here, art and nature are placed on equal footing: artisans imitate nature when they carry out these processes in the kitchen or the workshop, and their vocabulary can be imported back into the natural world. This correspondence became a central theme in medieval alchemy, whose practitioners considered that their laboratory procedures represented in microcosm the transactions of macrocosmic nature, to the extent that alchemical transformation became the explicatory principle of all natural events: the circulation of water between sky and ocean, say, was a form of alchemical evaporation and condensation.

Inevitably, this two-way exchange blurred the boundary itself: art and nature began to look seamless. Even an apparently synthetic procedure such as glass-making might be called 'natural', because it uses natural materials – sand and soda, the latter for example as the mineral called natron – and processes them using extreme heat, which also occurs in nature – for example in volcanic regions. How, then, could the product be anything other than natural itself? After all, a kind of natural glass *did* sometimes turn up at volcanic sites where the earth's heat had melted sand.

Aristotle therefore allows the possibility that, by mimicking nature's processes, the artist might produce not merely an imitation of its products but something that genuinely warrants the description 'natural'. In his *Theorica et practica*, the thirteenth-century Italian writer Paul of Taranto offered an ambitious attempt to defend alchemy by aligning it with the natural philosophy of medieval scholasticism, and with Aristotelianism in particular. According to Aristotle, all substances are comprised of matter and form, and Paul remarked that art does nothing more or less than impress a new form into a natural substance. Yet he drew the Aristotelian distinction between incidental, superficial form – appearance, one might say – and intrinsic, substantial form. Sometimes art manipulates only the former, as in painting and sculpture, but other kinds of art bring about a fundamental change, working on what Paul calls the 'form of nature'. So there are certainly fraudulent and ignorant alchemists who merely reproduce the appearance of gold – they are 'painters of metals' – but there are also genuine ones who transform a metal's essence.

There is something of Aristotle's viewpoint in the comments of Francis Bacon on the art–nature distinction in the seventeenth century. Bacon insisted that the emerging scientific enterprise needed experiments that led to practical outcomes: to artefacts, new substances and forms, novel technologies. For Bacon, true scientists are like bees, which extract the goodness from nature and use it to make valuable things. These included medicines, 'artificial metals', fabrics and machines – and also engineered life. In Bacon's utopian fable *The New Atlantis* (1627), the scientist-priests who rule the fictional land of Bensalem say that:

We make, by art . . . trees and flowers to come earlier or later than their seasons; and to come up and bear more speedily than by their natural course they do . . . We have also means to make divers plants rise by mixtures of earths without seeds; and likewise to make divers new plants, differing from the vulgar; and to make one tree or plant turn into another. We have also parts and enclosures of all sorts of beasts and birds . . . By art likewise, we make them greater or taller than their kind is; and contrariwise dwarf them, and stay their growth: we make them more fruitful and bearing than their kind is; and contrariwise barren and not generative. Also we make them differ in colour, shape, activity, many ways . . . Neither do we this by chance, but by what we know beforehand, of what matter and commixture what kind of those creatures will arise.

Bacon has been credited with overthrowing entirely the old distinctions between art and nature. But he didn't completely deny that there was a difference, saying only that art may compel nature 'to do that which without art would not be done'. Like Aristotle, he says that art can help natural processes to go further than they would otherwise. Yet he insists that a production may be deemed artificial only because of the way it was made, and not because of what it is. Lamenting 'the fashion to talk as if art were something different from nature, so that things artificial should be separated from things natural, as differing totally in kind', he insisted that 'men ought on the contrary to have a settled conviction, that things artificial differ from things natural not in form or essence, but only in the efficient' – that is, in the way they have been made.*

Until the seventeenth century, these issues were debated with the greatest heat and contention in regard to alchemy. Many historians of science have been tempted to pose the central battle of alchemy as an argument between those who believed that base metals can be transmuted to gold and those who did not. Recent scholarship has introduced more subtlety by showing that even renowned 'doubters', such as the seventeenth-century Anglo-Irish scientist Robert Boyle, were generally more concerned to draw a distinction between those who knew the 'true art' and the charlatans and bunglers who merely claimed to. But now we see that there is another facet to the disputes too: it wasn't necessarily a matter of whether or not you believed that gold could be made, but whether you thought the alchemists' gold was as good as that in nature. After all, it was almost universally thought that metals form in the earth by a process of maturation in which the lesser passes gradually to the greater: lead to iron to gold. The alchemist simply claimed to be inducing this 'natural' process at a faster rate in the laboratory. To the fourteenth-century Muslim polymath Ibn Khaldun, that was a crucial distinction: he argued that alchemical metals cannot be as good as natural ones because they are made too quickly. The maturation of metals cannot be speeded up without compromising it, he said, because 'Nature always takes the shortest way in what it does.' The falsehood of alchemists then lay not in passing one thing off as another, but in trafficking inferior goods.

---

* Robert Hooke, examining natural and artificial objects through a microscope later in the century, did adduce a more concrete distinction relating to the issue of fabrication. He found that natural objects seemed more finely wrought: thorns, for example, were always sharper and smoother than pins. Compared to the fabulous intricacy of the fly's eye, human art seemed paltry and clumsy at the microscopic scale, no matter how elegant to the naked eye. 'So unaccurate is [Art]', he wrote in his *Micrographia* (1665), 'in all its productions, even in those which seem most neat, that if we examin'd with an organ more acute then that by which they were made, the more we see of their *shape*, the less appearance will there be of their *beauty*: whereas in the works of *Nature*, the deepest Discoveries shew us the greatest Excellencies.' This, however, is a far more materialistic distinction than that which could be applied to the 'essences' of substances natural and artificial, such as gold and minerals. It merely showed that God was a superior craftsman, which no pious observer – and Hooke was certainly that – would ever have doubted.

Alchemists challenged that prejudice. An anonymous work called the *Book of Hermes* dating from around the late thirteenth century says that 'human works are variously the same as natural ones', pointing out that several salts and minerals such as 'green salt' (probably verdigris), vitriol (metal sulphate salts), tutia (zinc oxide) and sal ammoniac (ammonium chloride) 'are both artificial and natural'. Indeed, the author claims, 'the artificial are even better than the natural' – which might have been true in at least one sense, given that the laboratory-made materials might be more pure. He adds (anticipating Francis Bacon):

Nor does art make all these things, rather it helps nature to make them. Therefore the assistance of this art does not alter the nature of things. Hence the works of man can be both natural with regard to essence and artificial with regard to mode of production.

Even the 'artificiality' of alchemical methods is open to question, given that they mimic those of nature. In the 1330s, the Italian physician and alchemist Petrus Bonus argued that bricks, for instance, are not 'unnatural', for they are not fundamentally different to clay baked in the warmth of the sun. Petrus repeats the claim that substances and objects made by human art can be even better than their 'natural' analogues: nature can be improved. This is certainly one way to interpret Aristotle's comment that art may 'partly complete what nature cannot bring to a finish'.

Spirit of life

I have dwelt at some length on the debate over the alchemical transmutation of metals and minerals because many of the same ideas carry over to early theories of life. However peculiar it might seem to us to generalize from inert to organic matter – from metals to men – this was a natural extension within a world view that was in some sense animistic, in which all of creation was imbued with vital spirits. From ancient Greece to Renaissance Europe, it was widely held that there was a steady, progressive quickening of matter from the lowliest rock to the stuff of humankind. Metals were already considered to possess in some sense more 'animation' than stones –

they were known, for example, to be able to grow in curious branching forms that resembled plants. Plants themselves were still higher on the scale, having powers of growth, nutrition and to a limited extent movement. Animals were higher still, and finally humans.

This belief was the corollary of an Aristotelian view of matter in which a fundamental, undifferentiated substance (what was called by some Greek philosophers *prote hyle*) was given specific qualities by a 'spirit', a volatile ingredient located somewhere between the physical and the metaphysical. If all organisms are composed of the same basic matter, and if the spirits that define a particular substance can be sublimed and transferred, then it is only reasonable to suppose that matter lower on the scale can acquire a greater degree of animation – that it can be, as it were, brought to life. This was the rationale for the universal belief in spontaneous generation, in which simple animals such as insects and worms, even rodents, can spring from apparently lifeless matter. It was common enough, of course, to find such life thronging in putrid organic matter such as rotten food or carcasses. That spontaneous generation must to be preceded by decay had a parallel in the alchemical idea that metals must 'die' before being 'reborn' in a higher state.

Spontaneous generation from a warm, moist, fetid matrix was the canonical scheme in Greek thought for the origin of (non-human) life. Ovid's *Metamorphoses* puts it this way:

> All other forms of life the earth brought forth,
> In diverse species, of her own accord,
> When the sun's radiance warmed the pristine moisture
> And slime and oozy marshlands swelled with heat,
> And in that pregnant soil the seeds of things,
> Nourished as in a mother's womb, gained life
> And grew and gradually assumed a shape.

This transformation even furnished a kind of primitive evolutionary theory: Anaximander of Miletus in the sixth century BC said that fish grew from warm mud, and that from them humans eventually evolved.

Aristotle concurred that spontaneous generation is a real phenomenon, saying, for example, that lice appear this way in irritated skin. Each type of spontaneously generated creature was considered to spring from its own specific matrix: bees from the carcasses of cattle, wasps from horses or donkeys, scorpions from crabs, snakes from decomposing spinal cords. There is some sense here that like begets like, a reflection of the emphasis on *form*: the Islamic philosopher Avicenna (Ibn Sina, *c*.980–1037) claimed, for example, that snakes may be made from the hairs of women, kept in a warm, moist place.

And just as alchemists purported to recreate the maturation of metals, so they began to assert that they could summon life out of lifeless stuff by an artificial type of spontaneous generation. There was a respectable pedigree for such claims in holy scripture, as Thomas Aquinas pointed out: Aaron turned his wooden staff into a snake in front of the Pharaoh, and the Pharaoh's magicians did the same trick using their 'secret arts'.* There was not necessarily anything esoteric about such knowledge; for the Roman writer Virgil, spontaneous generation of bees was simply a useful form of biotechnology for the bee-keeper – a craft known as *bougonia*. He supplies a recipe, making the creation of new life sound no different from the art of making wine or bread:

> They find a two-year calf with sprouting horns
> Whose nostrils and whose mouth they stop,
> Despite his struggles; beat and pulverize
> The carcass, while they leave the skin intact.
> Here enclosed they leave him, laying sticks
> And sprays of thyme and new-cut cinnamon
> Beneath his flanks . . .
> Meanwhile, within the corpse the fluids heat, the soft bones tepefy
>   [become warm],
> And creatures fashioned wonderfully appear;
> First void of limbs, but soon awhir with wings.

It is understandable that this artificial generation of life should be regarded almost casually: in contrast to the transmutation of metals, almost any fool, trickster or careless housemaid could 'generate'

---

* The Pharaoh's magicians also copied Aaron's feat of conjuring up frogs, but it is not clear whether this was an act of spontaneous generation or simply of summoning. See Exodus 7:8–12.

maggots in rotten meat. But again the persistent question about artifice arises: are these creatures the *same* as 'natural' ones? Not according
to Avicenna, who wrote that 'Art is weaker than nature and does not
overtake it', so that art cannot make products identical to those in
nature. These words appear in a document known to the Latin-speaking
medieval world as *Sciant artifices* ('Let the artificers know'), part of a
work on geology and minerals, *Liber de congelatione et conglutinatione
lapidum* (*Book on the Congealment and Concretion of Stones*), which was
for some time believed to be the work of Aristotle himself (and thus
to carry his full authority). Avicenna was here speaking only of
alchemy in relation to metals, but his argument was later considered
to apply to any form of art.

The Muslim scholar Averroes (Ibn Rushd, 1126–98), whose writings
exerted a strong influence on the natural philosophy of the European
Middle Ages, took much the same position but addressed explicitly
the artificial production of life. Commenting on Aristotle's *De gener-
atione animalium* (*On the Generation of Animals*), he says that creatures
such as insects or mice made by spontaneous generation (whether by
art or by chance) in rotten matter are quite different from those that
result from sexual reproduction, even though they look identical,
because they originate from different types of matter. In particular,
he says, they are *sterile*. Remember that.

The response of the medieval world to the idea of 'artificial life'
was thus quite different from the horror it now typically engenders,
which must suggest to us that feelings of revulsion about these 'unnatural' creations are by no means inevitable. Medieval people saw nothing
intrinsically *distasteful* in creating humans and other forms of life –
the problem was rather that, as Averroes said, organisms made by art
were, like alchemists' gold, a kind of fake. In defending their art,
alchemists were therefore compelled to make counterclaims about the
status of artificial beings.

But wasn't it hubristic to imagine that God's creations could be
equalled, perhaps even bettered? That accusation gained strength as
the Renaissance brought fresh vigour to the art–nature debate and
stimulated interest in strange transmutations. The sixteenth-century
Sienese engineer Vannoccio Biringuccio argued that if alchemists could
really make the philosopher's stone then they could claim to 'hold
prisoner in a bottle that God which is the creator of all these things'.

Later that century, the ceramicist and philosopher Bernard Palissy said of alchemical transmutation that it is 'a rash undertaking against the glory of God to wish to usurp that which is of his estate'.

Petrus Bonus had previously rejected accusations of impiety, asserting that alchemy was given to man by God so that we might improve on the raw substance of nature. Yet this line of defence did little to prevent Church prohibitions against the practising of alchemy by clerics, beginning in the late thirteenth century. One of the most important, a bull issued by Pope John XXII in the early fourteenth century, appears to label alchemists as charlatans and forgers who 'deceive the ignorant populace as to the alchemical fire of the furnace', thereby condemning them in secular terms as deceivers. But John XXII was also concerned about the alleged links between alchemy and sorcery (he had been the intended victim of an assassination plot using black magic), and his prohibition argued that transmutation is 'not in the nature of things': that it is improper for more fundamental reasons than mere trickery. And his accusation of deception does not itself necessarily imply intentional fraud for financial gain: it was often the mere *claim* to reproduce the substances of nature that led alchemists to be called fraudulent.

Nature's errors

Questions about whether nature can be rivalled or surpassed by art must confront the matter of how perfect nature is in the first place. There was good reason to believe that not everything in the natural world was wrought with skill, foresight and precision. Here is what the English traveller John Mandeville claimed to see in the fourteenth century on the islands around a region he called Dondun, probably in the East Indies:

In one of these isles be folk of great stature, as giants. And they be hideous for to look upon. And they have but one eye, and that is in the middle of the front . . . And in another isle toward the south dwelt folk of foul stature and of cursed kind that have no heads. And their eyes be in their shoulders. And in another isle be folk that have the face all flat, all plain, without nose and without mouth. But they have two small holes, all round, instead of their eyes, and their mouth is flat also without lips . . .

His catalogue of monstrosities continues: people with horses' feet, with ears 'that hang down to their knees', with feathers. The monsters encountered in voyages to faraway lands are a familiar trope in the Middle Ages, and all are comparably incredible. But they *were* believed, as often as not, because folk knew that the world did spawn 'monsters' of one sort or another. The Florentine writer Luca Landucci recorded one that visited his home city in 1513:

A Spaniard came to Florence, who had with him a boy of about thirteen, a kind of monstrosity, whom he went round showing everywhere, gaining much money. He had another creature coming out of his body, who had his head inside the boy's body, with his legs and his genitals and part of his body hanging outside.

Aristotle's close analogy between nature and art made it natural to regard 'monsters' as mere mistakes, like those even the best craftsmen would occasionally make. As he put it:

Now mistakes come to pass even in the operations of art: the grammarian makes a mistake in writing and the doctor pours out the wrong dose. Hence clearly mistakes are possible in the operations of nature also. If then in art there are cases in which what is rightly produced serves a purpose, and if where mistakes occur there was a purpose in what was attempted, only it was not attained, so must it be also in natural products, and monstrosities will be failures in the purposive effort.

To Aristotle, monsters were contrary to nature only insofar 'as it holds for the most part': they deviate from the normal course of nature, but deviations are themselves intrinsic to the way nature works. One might say that monsters are thus 'contra-natural' without being exactly unnatural.

Aristotle was not burdened by the obligation to insist on God's perfection. But Christian writers could hardly assert that monsters were divine errors, and so they decided these beings must be intentional – that they signified God's purpose. To St Augustine, writing in the fifth century, monsters serve as divine warnings about the consequences of vice and folly. This significative role of monsters is reflected in the word itself, derived either (or equally) from the Latin *monstrare*, to show, and *monere*, to warn.* In the Middle Ages, *monstrum* (monstrous) was almost cognate with *portentum* and *ostentum* as words connoting deviations from nature with portentous import. According to Isidore of Seville in the early seventh century, monsters 'predict future things', and 'come by the divine will'. As portents, they therefore have moral connotations: in Augustine's view, the monster delivers an admonition about degeneracy.

Mandeville's bizarre races were not viewed in quite the same way as deformed individuals. Wondrous species suggested that nature was strange, even perverse, but they were considered a part of the natural order all the same. Only monsters that differed in their deformity from others of their kind were *contra naturam*, and they therefore inspired horror where dog-headed men were merely cause for amazement. Isidore explained that monstrous babies did not live long because they did not need to: they had fulfilled their role as an augury once they were born. But were they, then, human, and did this mean they should be baptised? If they were conjoined twins, was one baptism needed, or two? No one knew the answers to these questions;† more surprisingly, as we shall see later, they are questions that have never really gone away.

It fell to Thomas Aquinas to reconcile Aristotle's view of monsters as nature's errors with Augustine's insistence on monsters being portents made by divine intervention. Yet for all his ingenuity in giving Aristotelianism a Christian interpretation, Aquinas could not do much more here than force an awkward marriage: violations of the natural order, he said, can sometimes be accidental (preternatural) and sometimes divinely arranged. 'The order imposed on things by God is based on what usually occurs, in most cases, in things', he wrote, 'but not on what is always done.' The sixteenth-century French surgeon Ambroise Paré hedged his bets to an even greater degree, listing a dozen causes of monsters that included the glory, or the wrath, of God, too much or too little seed involved in the generation, hereditary or accidental illness, too narrow a womb, and the intervention of

---

* The Greek equivalent is *teras*, meaning portent. The modern study of anomalies in physical development – growth defects – is called teratology, and agents known to cause such defects are known as teratogens.

† A common rule of thumb relied on a head count: the number of souls was equal to the number of heads.

witchcraft and demons. In any event, he stressed, 'monsters are things that appear outside the course of Nature'.

That was the point: in the Middle Ages and the early modern periods, monsters might not necessarily be portents but they were aberrations loaded with moral baggage. Often they were caused by deviant acts or thoughts. Birth deformities could be the fruit of warped imagination in the mother during conception and gestation, or of unnatural unions: a 'dog-headed' child indicated that the mother had coupled with a dog. As biologist François Jacob expresses it:

Each monster is the result of iniquity and bears witness to a certain disorder: an act (or even an intention) not in conformity with the order of the world. Physical or moral, each divergence from nature produces an unnatural fruit. Nature, too, has its morality.

Although Francis Bacon was not the atheist that some have tried to make him (he considered atheism 'in all respects hateful'), he had little time for a God who was constantly intervening in the world. He considered that nature ran by its own accord, and that monsters need be given no religious interpretation. But rather than dismissing them, *pace* Aristotle, as mere useless mistakes, he regarded them as a source of creative inspiration: rather than destroying nature's order, they suggested a new way in which things might be contrived. In this view, monsters and prodigies seemed to display in nature an innate creative potential, a capacity to evade its own rules. Later in the seventeenth century, Gottfried Leibniz argued that in its inventiveness nature displayed an intelligent autonomy, which in turn implied that nature herself had a kind of soul. Any form of novelty and curiosity of shape in nature was considered to be evidence of this soul at work. This was what led John Beaumont, describing 'stone-plants' (the mineral growths now called dendrites that look tree-like and organic) to the Royal Society in the 1680s, to award minerals a 'vegetative soul' like that postulated by Aristotle.

Although Leibniz cautioned against regarding nature's soul as akin to the rational soul of humankind, nonetheless views like his seem to personify nature as an artisan, making strange objects that had about them something of the 'artificial'. In other words, not only did art mimic nature, but vice versa. Yet if nature was an intelligent agency, where did God feature? Might nature go its own way, heedless of God's design? As historians Lorraine Daston and Katharine Park remark, 'If philosophers deprived nature of skill and autonomy, it was out of openly voiced fear that she might usurp the praise due to God.'

This reawakened troubling questions that had emerged four centuries earlier in the wake of enthusiasm for Aristotle's mechanistic, rule-bound nature. Was nature God's servant, a contrivance that unfolded of its own accord so that God was spared the indignity of keeping it daily in motion? If so, could God nonetheless alter the laws, or was nature autonomous? You could argue it either way. To the seventeenth-century English physician Walter Charleton it was demeaning to suppose that God needed servants; he was in constant personal attendance in the world. Robert Boyle also rejected God's dependency on servants, but derived the opposite conclusion that the universe must therefore be entirely mechanistic, a complex device that operated as precisely and as regularly as the famously intricate clock made for Strasbourg cathedral, with its automata that observed the hours: 'all things', he wrote,

proceed, according to the artificer's first design, and the motions of the little statues, that at such hours perform these or those things, [and] do not require, like those of puppets, the peculiar interposing of the artificer, or any intelligent agent employed by him.

Natural law

The occurrence of 'monsters' thus led to a rather subtle considera-tion of what 'natural' means. One view was that nature was a fallible craftsman; another was that these aberrations were ordained by God for specific reasons. In either case, one was left with no cause to attribute 'perfection' to nature.

The wresting of purpose from nature that has characterized the scientific enterprise for the past 200 years, particularly in Darwinian biology and in cosmology, has paradoxically forced a greater insistence on an idealized nature. The theory of evolution by natural selection seems at face value to provide a mechanism for the inexorable improve-ment and optimization of nature, and the modern desire to mimic biological shape and form in the applied sciences sometimes invokes and fosters the notion that 'nature knows best' – that natural products are as good as they could possibly be. Evolutionary biologists know that this is an illusion. For one thing, there is no absolute 'optimum' towards which evolution can aim: the target is constantly shifting, because both the non-living and the living environments of any organism are always in flux. But in any case, evolution need seek only a solution that is 'good enough', providing the recipient with a marginal reproductive advantage. It generates ad hoc solutions with the materials at hand, and cannot easily go back to restructure the platform on which natural selection tinkers. So the natural world is full of botched designs such as the back-to-front neural wiring of the human eye, locked into place by 'accidents' of evolutionary history. Yet no matter how often biologists point out such things, or explain some of the (to our eyes) vile and pernicious strategies that predatory organisms use on their prey, we appear now to be saddled with a reified nature.

Let's be clear: humans do have a sad aptitude for fouling up the natural world. Nature often works in a delicately balanced fashion which human activities can easily send awry. But that is not because nature is good and *techne* is bad – it is because we are frequently clumsy, misinformed or stupid. Eradicating pathogenic viruses and microbes does not seem in principle like a terrible intervention, even if in practice we may sometimes do it blunderingly. Extinction is a natural process (which has killed off just about every species that ever existed) – we are bringing it about at an alarming rate now, but it is far from clear that there is any moral obligation to preserve species that would otherwise go extinct without any push from us. Nature doesn't know best; it is just that we usually don't understand or appreciate it sufficiently to know any better.

Today nature is seen as something amenable to objective scientific study: it is what happens out there in the world when we humans do not interfere. But the 'goodness' that is popularly attributed to this nature comes from another source, which is theological and philosophical. Thus the contemporary reification of nature arises from a confusion of terms: 'nature' as a physical and biological entity, and 'nature' as a predisposition. Because in Judaic, Christian and Islamic tradition nature was under God's jurisdiction and guidance, what happened 'naturally' was the result of God's will and therefore was good by definition.

This principle is embodied in the ancient notion of 'natural law', an inherently theistic idea. For Thomas Aquinas, natural law stemmed from a faith in a rationalistic yet teleological universe in which everything has a part to play. He invoked an 'eternal law' by which all creation is ordered, and considered natural law to be the way humans participate in this plan. Whereas non-rational creatures are simply a part of eternal law, humans act freely within it. Yet our rationality, Aquinas said, predisposes us to behave in certain ways: when we do so, we observe natural law, and by doing so we are guided towards the 'good'. When we act rationally, we act naturally: 'The first rule of reason', Aquinas wrote, 'is the law of nature'.

One of the most important arenas in which natural law was deemed to operate was human procreation. In the early Christian era, the accusation of *contra naturam* was most commonly directed at acts of a 'deviant' sexual nature, such as homosexuality, masturbation and bestiality. These were considered serious vices precisely because they resisted nature's perceived call for procreation: they spent human seed for the 'wrong' purposes. Sodomy was in particular considered to be the 'sin against nature'. Heretics, witches and minorities were frequently accused of having committed sexual depravities: a sure sign of their wickedness. And within the persecutory medieval society, the charge of having acted against nature acquired a deadly, unanswerable force.

Natural law, then, is a human-centred concept that associates certain actions with moral rectitude.* But clearly we do not all do the same thing in all circumstances, so how do we know which is the behaviour dictated by natural law and which is not? That's a difficult problem, as Aquinas acknowledged. Although he believed that we have an innate sense of which *ends* are good, he did not think it would be possible to state general principles that, in all situations, determine the right choice of action. He considered some things, such as lying, murder of the innocent, adultery and blasphemy, to be self-evidently contrary to natural law, but accepted that in other cases the right course was less evident.

---

* Aristotle's comment that 'moral virtues are not in us by nature' posed some problems for this view among medieval Aristotelians. The resulting tension is discussed by historian Joan Cadden in *The Moral Authority of Nature*, L. Daston & F. Vidal (eds) (University of Chicago Press, 2004).

This view of natural law has been debated by many subsequent philosophers, and still has adherents. Even if the particulars of a situation leave the most 'proper' action open to question, some feel that it should be possible to draw up a list of what 'ends' are to be considered good. For example, we might look at the natural tendencies of most human beings – what 'comes naturally', such as preservation of life and the seeking of knowledge, friendship and love. If God's plan is usually observed, these at least must be natural goods. But it is not at all clear how to compile an exhaustive list – do power and prestige qualify, for example? And besides, there is a danger in equating what is 'normal' with what is good and moral. (We see the remnants of this position in the moralistic overtones of the complaint that 'it's not normal'.) This attitude was criticized by David Hume, who argued that one cannot get an *ought* from an *is* – a philosophical solecism known as the naturalistic fallacy. His comment could apply as much to the natural world as to human nature.

Although modern bioethicists are generally alert now to the danger of reifying nature in the view that 'nature knows best', some are reluctant to part with the moral overtones of perceived 'naturalness'. Leon Kass states that 'It may be as much a mistake to claim that "the natural" has *no* moral force as it is to suggest that the natural way is best, because natural.' Kass seemed to feel no compulsion to explain *why* this might be a mistake, however, presumably because he has the weight of such a strong tradition behind him.

The distinction between nature as a description of the world and nature as a moral concept is central to what follows. Appeals to a concept of 'naturalness' are not directed at any objective view of an 'order of things', but stem from a theological and teleological vision of how things happen. They also tend to build on a fundamental distinction between 'nature' and 'art' that makes assumptions about the capabilities and the values of each of them. And these proclivities are most evident, and also most emotive, when they deal with the issue of making people.

2 Work of the Gods

A Greek it was who first opposing dared
Raise mortal eyes that terror to withstand,
Whom nor the fame of Gods nor lightning's stroke
Nor threatening thunder of the ominous sky
Abashed; but rather chafed to angry zest
His dauntless heart to be the first to rend
The crossbars at the gates of Nature old.
    Lucretius, *On the Nature of Things* (*c.*50 BC)

A mighty lesson we inherit:
Thou art a symbol and a sign
To Mortals of their fate and force;
Like thee, Man is in part divine
    Lord Byron, 'Prometheus' (1816)

The engineer is usually a humble fellow, narrowly goal-oriented, content to tinker with stolid diligence until he (it is usually he) gets the bridge built or the machine running. But when he abandons his humility, when he attempts to soar, to exceed the limits that his skill and judgement ought properly to impose, then he becomes mythical. Then he becomes a Prometheus.

To be Promethean is not, as some dictionaries imply, merely to be bold, original and creative, but also to be transgressive and hubristic. That is undoubtedly what Pope John Paul II had in mind when, alluding to the goals (as he perceived them) of modern biomedical research, he spoke of the 'Promethean ambitions' of science. But others find something noble and inspiring in the mythical figure's defiance of the gods, and abhor the idea that the modern Prometheus should desist and submit. As the British biologist J. B. S. Haldane wrote in 1924:

The chemical or physical inventor is always a Prometheus. There is no great invention, from fire to flying, which has not been hailed as an insult to some god. But if every physical and chemical invention is a blasphemy, every biological invention is a perversion. There is hardly one which, on first being brought to the notice of an observer from any nation which had not previously heard of their existence, would not appear to him as indecent and unnatural.

Haldane's proclamation is paraphrased, surely unintentionally, in a letter written by a scientist to his son in a 1939 movie: 'If you, like me, burn with the irresistible desire to penetrate the unknown, carry on . . . Like every seeker after truth, you will be hated, blasphemed, and condemned.' The author here is 'Baron' Frankenstein, the movie *Son of Frankenstein*. In the same defiant spirit, the title character (played by Basil Rathbone, best known for portraying that arch-rationalist Sherlock Holmes), alters the graffiti on his father's tomb from 'Maker of monsters' to the Promethean 'Maker of men'.

Scientists have a tendency to allude to myth as a mere rhetorical flourish, but Haldane's intent went deeper. The book in which his comment appeared was called *Daedalus: Science and the Future*.* And Daedalus, the master craftsman of antiquity, is responsible for one of the earliest stories of human-like creatures crafted by artificial means. The tales of Prometheus and Daedalus offer a fitting overture to the history of anthropoeia, for they illustrate how from the outset this enterprise was intimately linked to the morality of human artifice and its problematic relationship to 'nature'. Prometheus works with divine resources, while Daedalus, a mere man, has only his manual skill and ingenuity. But both are in essence engaged in forms of *techne*, and both pay a bitter price for their presumption.

---

* Titles taken from the characters of Greek mythology were the leitmotif of the 'To-day and To-morrow' series, published in the 1920s by Kegan Paul, of which Haldane's was the first volume. Another, by American geneticist Herbert Spencer Jennings, was called *Prometheus, or Biology and the Advancement of Man* (1925). We will hear more from this auspicious series in Chapter 8.

Friend of man

Who made humans? That is usually regarded as the work of the gods. Many creation myths insist that the first people were fashioned from clay, mud, dirt: primal matter given a life-soul by supernatural means. The Egyptian god Khnum made humankind this way from the clay of the Nile basin; the goddess Nuwa formed the Chinese people from yellow earth. In the Book of Genesis, the Judaeo-Christian God 'created man of the dust of the ground'.

These creation myths insist on a continuum, or at least a negotiable boundary, between animate and inanimate matter. That continuity is also demanded by any purely scientific theory of life's origins on earth, but in myth this usually takes the form of a kind of animism in which life is immanent in all matter. Ovid's *Metamorphoses* describes all manner of such transitions between life and lifeless substance. He says that Prometheus' son Deucalion, the Greek equivalent of Noah, and his wife Pyrrha repopulated the world after the Deluge by making people from rocks thrown over their shoulders. The rocks are called 'the bones of your mother' by the oracle at Themis – meaning the 'bones' of the earth goddess Gaia, but imputing to them already a kind of latent vitality.

Plato says that, while humans were made by the gods 'out of earth and fire', Prometheus was instructed to equip them for the world:

He found that the other animals were suitably furnished, but that man alone was naked and shoeless, and had neither bed nor arms of defence. The appointed hour was approaching when man in his turn was to go forth into the light of day; and Prometheus, not knowing how he could devise his salvation, stole the mechanical arts of Hephaestus and Athene, and fire with them (they could neither have been acquired nor used without fire), and gave them to man. And in this way man was supplied with the means of life.

Prometheus is usually identified as one of the Titans, the ancient race of deities whose battle with the Olympian gods is most probably a mythical retelling of the struggle between rival theologies in the ancient world. Prometheus saw which way that battle was turning and threw in his lot with Zeus and the Olympians. He acquired architecture, medicine, metallurgy and other practical arts from Athene – whether by theft or study is not clear – and passed them on to humankind, making Zeus increasingly uneasy about the growing powers of men. In one account, Zeus decides to withhold fire from humanity because he is angry at how Prometheus tricks him into choosing the least appetizing parts of a sacrificial bull that is to be divided between the gods and men. But Prometheus steals fire and gives it to humankind, and in retribution Zeus condemns Prometheus to his terrible punishment in chains in the Caucasian mountains, eviscerated forever by a great eagle.

Later versions of the Prometheus myth by Aesop and Ovid make his role even more profound: he becomes not only the agent of humankind's ability to use fire, but the maker of humanity itself, whom he fashions from earth and water. This amplifies both the nature of Prometheus' defiance of the Olympian's authority and his role as friend of the human race. His very name makes him an omen of future knowledge: it comes from *promathein*, loosely meaning to think ahead or to understand in advance.

Fire and earth

The view of life as a union of base matter and spirit or soul – of life breathed into earth – finds echoes in Greek natural philosophy. There was no single 'theory of life' in ancient Greece, but many – and they are not always easy to distinguish or even to understand, because the philosophers were no more agreed than we are today on what life is or where it resides. Is it a kind of substance, or a property of matter? If the latter, is that property immanent within matter or imposed upon it?

There are at least two ancient Greek words that can be translated as 'life', reflecting an ambivalence about whether it should be defined empirically – by what we can see – or from first principles. To wit, *zoe* (the root of zoology and protozoa) refers to the sorts of behaviours associated with life (among which self-determined movement was seen to be crucial), while *psyche* refers to the intrinsic property of being alive, what we might call 'life-as-soul'. But underpinning all these ideas is the fact that the Greeks, and the Western world until at least the Renaissance, saw the cosmos in biological terms – and more particularly in human terms – as a kind of purposeful organism imbued with life-giving agency.

Some early Greek thinkers dwelt on the materialistic aspects of life: on the issue of its composition. To Xenophanes in the sixth century BC, 'all creatures that come into being are earth and water'. For Empedocles, the various tissues of the body are made up of his four classical elements in different proportions: blood contains equal amounts of all four, bone is four parts fire to two of water and earth, and so on. For Anaxagoras, the universe is filled with formative 'seeds' of various sorts of matter, intimately mixed. Thus food can become tissue because it already has some portion of 'tissue-ness' within it.

Yet what animated this stuff? Early theories generally speak of a 'soul' in some form or another, but it usually has little of the transcendental character we associate with the word today. Anaxagoras said that life is latent in all his formative seeds but becomes manifest only under the action of an ordering principle called *nous*, which we might equate with 'mind'. Democritus, widely credited with originating atomistic theory, considered living matter to be a mixture of somatic, tangible material (earth and water) and a psychic substance, composed of fire atoms but also associated with soul. In this sense, soul is more or less explicitly a substance, and indeed it supplied Democritus with a materialistic vision of how life was sustained: by breathing in air, which contains fire atoms among other things. Here we can see an early instance of the long-standing association between soul or 'spirit' and some volatile substance associated with air or breath. Animation was often induced in myth by the action of 'breathing life' into inert matter.

The most explicitly spiritual tradition in ancient Greece, and the one that exerted the greatest influence over early Christian thought, was due to Plato. In his view, all living things have a soul bonded to their constituent particles: a kind of intangible agent that imparts motion. This soul is supplied by the Creator of the cosmos. There is, Plato said, a cosmic soul that is responsible for the motions of the celestial bodies, while the human soul is something similar but less grand, less pure. Thus humans and other organisms are essentially receptacles for these primordial souls. For Aristotle, the soul (*psyche*) is what differentiates living from non-living things. But he considers that there are in fact three different types of soul, which distinguish plants, animals and humans. Plants, he says, have only a *nutritive* soul, which enables them to be self-sustaining and self-replicating, and gives them a limited capacity to move. Animals also possess a *sensitive* soul, allowing them to feel and respond to sensation. And humans have in addition a *rational* soul, which makes us creatures that apply reason and logic. In each case, these souls are responsible for (Aristotle seems to imply that they are virtually equated with) the *form* of the organism: they are that which is responsible for the *organization* evident in living things.

In Aristotelian science, life arises when *psyche* imprints form on receptive clay. In Aristotle's biology, the 'clay' of humankind is menstrual blood, which receives animation and organization from semen in the same way that wood takes on shape and function under the ministrations of the carpenter. On whether this process could be carried out by humans themselves, Aristotle is silent; but because his biology doesn't obviously rule it out, the question was permissible, even inevitable. Incorporeal souls are a primal creation, Aristotle says, but they may transmigrate between material bodies. Might humankind direct that flow? Perhaps all that was needed to induce this process was an artificer sufficiently ingenious.

A monster made by art

Daedalus may have been that person. He was a cunning craftsman who could allegedly animate statues, or at least give them an appearance of life. The comic poet Philippus, son of Aristophanes, claimed that this art was mechanically contrived: Daedalus' wooden statue of Aphrodite, the poet said, was moved by flowing liquid mercury (as other automata in antiquity may have been). But others believed that Daedalus could transfer a human soul into a machine. And besides, if movement constituted one of the soul's chief functions, as Aristotle said, then the distinction between lifeless machine and living automaton wasn't clear in any case.

Daedalus' most fateful invention was an explicitly mechanical contrivance: an artificial cow made of wood. He built it at the behest of Pasiphaë, wife of King Minos of Crete, so that she might have sexual congress with a bull. This magnificent beast had been given to

Minos by Poseidon, who had commanded the king to sacrifice it. When Minos, enraptured by the bull's beauty, could not bring himself to destroy it, Poseidon punished him by filling Pasiphaë with desire for the creature. To consummate her lust, she sought help from Daedalus, who had previously carved wonderful lifelike wooden dolls for the royal couple. Daedalus made the wooden cow, covered it in hide, and wheeled it into the meadow where the bull was at pasture, with Pasiphaë hidden inside. And, you might say, unnature took its course.

But such transgressive acts were never without consequences, and the progeny of this union was a monster, the Minotaur. To avoid scandal, Minos fled to Knossos, where he instructed Daedalus to build the Labyrinth, in the centre of which the Minotaur and his mother were imprisoned – guarded, so Aristotle says, by moving statues at the entrance. Minos is said to have incarcerated Daedalus in the Labyrinth too, according to Ovid, so that he might tell no one else about what Pasiphaë had done, or perhaps simply because of the king's fury at the craftsman's complicity in the deed. Escape from Crete was impossible by sea, since Minos kept a close watch on all ships coming and going. But Daedalus made for himself and his son a pair of great feathered wings, and they escaped by flight. Everyone knows what happened next: some feathers secured with wax were lost from Icarus' wings when he flew too close to the sun, and he plunged to his death. Minos pursued Daedalus, but to his own cost, for the clever inventor contrived to pour boiling water, or perhaps hot pitch, into Minos' bath, killing the king.

Little is settled, and little is simple, in Greek legends, which seem at times to burst at the seams from holding too much symbolic weight. And so it is that Talos, Daedalus' talented nephew whose murder was the cause of his exile, was also the name of a servant made of bronze, with a bull's head, given to Minos by Zeus. Some say that this Talos was made by Hephaestus, the son of the goddess Hera, who was said to have been born asexually by parthenogenesis. Hephaestus was the divine smith, the Greek equivalent of the Roman Vulcan, god of blacksmiths, sculptors, artisans and technologists in general. His brazen Talos was but one of a series of metal automata, and Hephaestus is also credited with making Pandora out of clay, into whom the Four Winds breathed life. Pandora, or more specifically the evils contained in her 'box', wreaked Zeus' revenge on humankind for the misdemeanour of Prometheus.*

To confuse matters further, Hephaestus has been identified with the first Talos, Daedalus' nephew, or perhaps with this Talos' son. In any event, the bronze Talos may have been a kind of artificial being devised by the wider and hazily differentiated family of master craftsmen centred on Daedalus, and was said to be animated by blood held in a single vein from his neck to his ankles. In one version of the legend he is killed by Medea, who pulls out the stopper that plugs the vein at the ankle. This, says Robert Graves, is an allusion to the technique of bronze casting of figurines and statues used by Cretan craftsmen at that time, in which a master figure carved from beeswax was encased in clay and baked, and then a hole was pricked in the ankle of the figure to let the molten wax run out and create a mould for the metal. This linked the closely guarded methods of the metalsmith with the notion of animating inert matter. (It is possibly no coincidence that the theme of molten wax reappears in the Icarus legend.) There is thus a cluster of ideas here that conjoin the mysterious, hidden techniques of 'art' with the practice of making life.

C. S. Lewis expounds the long-standing distrust of applied science and technology in his essay 'The Abolition of Man', at the same time revealing (though it was not exactly his intention) how the lingering suspicion of witchcraft is associated with an immoral impulse to pervert the boundaries of life and death:

There is something which unites magic and applied science while separating both from the wisdom of earlier ages. For the wise men of old the cardinal problem had been how to conform the soul to reality, and the solution had been knowledge, self-discipline, and virtue. For magic and applied science alike the problem is how to subdue reality to the wishes of men: the solution is a technique; and both, in the practice of this technique, are ready to do things hitherto regarded as disgusting and impious – such as digging up and mutilating the dead.

---

*This fateful vessel was not exactly hers. It was kept in the house of Prometheus' brother Epimetheus, who became entranced with Pandora's marvellous beauty despite Prometheus' warning that he should accept no gift from Zeus, and who made her his wife.

This is the legacy yoked to the shoulders of cunning Daedalus. Haldane calls him 'the first modern man', who was unjustly 'exposed to the universal and agelong reprobation of a humanity to whom biological inventions are abhorrent'. In the legend of this great craftsman are the germs of all the themes we will encounter repeatedly in the creation of artificial humans. Here is artistry skilful enough to erase the boundary between the animate and the lifeless. Here this ingenuity is put to unnatural ends, bringing calamity on the perpetrator – a calamity precipitated by quite literally striving to rise too high. And here unnatural couplings, linked to mechanical devices, result in a violation of nature and a monstrous birth. Here, in short, is all that can be expected to flow from tampering with living nature when we make people.

3 Recipes for a Little Man

There were once recipes for making people – or rather, for making 'little men', called homunculi. Accounts of these procedures began to appear during the Middle Ages, a time, according to William Newman, when 'tinkering with natural human generation was a widespread topic of discussion'. The making of a homunculus was seen as simply another kind of art, another technology if you will. Medieval scholars had no reason to suppose that it shouldn't be possible, and like the transmutation of gold, it was occasionally said to have been performed. The real question was: what manner of being was created in this way?

The first recorded homunculus was produced as a straightforward cure for childlessness. But the technique was, needless to say, markedly different from modern IVF, which serves the same purpose. The being was made from sperm kept in a vessel shaped like a mandrake root, an object that was itself widely believed capable of taking on miniature human form. The operation was conducted by Qaliqulas, adviser to a king named Harmanus who ruled a Middle Eastern realm around

AD 300. Harmanus had no heir, not because of infertility but because he found intercourse repugnant. The artificial child was named Salaman, and when he grew to puberty he eloped with his former wet nurse Absal.

This tale, recounted by Avicenna, is atypical of the homunculus tradition, in that the artificial being seems in all respects to have been a normal human – except, notably, that Harmanus, furious at the liaison of Salaman and Absal, made them both sterile by magical means. For there is usually something strange and special about the homunculus which marks him apart from humankind. Moreover, there is nothing particularly Promethean about Qaliqulas' deed, which is carried out as a direct alternative to sexual procreation. And Salaman's 'normality' is reinforced by his later fate. He and Absal attempt to drown themselves in despair at their infertility, but he alone survives and is persuaded by Qaliqulas to forsake carnality and embark on a spiritual life. The story is thus a morality tale, but not one about the morality of making people.

All the same, the 'laboratory' generation of humans is rooted in early Arabic science. In a process described in the works attributed to the scholar Jabir ibn Hayyan* around the start of the tenth century, the artificial being may be superhuman in various ways: a girl with a boy's face, or an adolescent with precocious intelligence. It might even have entirely non-human attributes: the mixed sperm of a man and a bird may generate a winged man.

In the generation of homunculi we find the first fully developed theory and practice of anthropoeia. And because it was closely linked to alchemy, this tradition became embroiled in the art–nature debate, forcing the issue of how well human artifice could mimic nature. But in contrast to the making of gold, the making of people raised a new and problematic issue, because one had to consider the matter of the product's soul.

Decay and rebirth

Artificial beings first appear within the alchemical tradition in the works of Zosimos around the start of the fourth century. Zosimos

---

* Like several figures in the history of alchemy, Jabir may or may not have been an actual person – the biography accorded him in the body of 'Jabirian' writings is fictional.

typically framed his alchemical observations in the form of dreams whose content was highly allegorical, described in arcane symbolic language. In one, he says that he dreamed of an 'altar in the form of a flask', inside which a sacrificed priest is trapped. The priest says he has been 'perfected as *pneuma*' – the term used in Greek philosophy for the breath of life. This figure has been dreadfully mutilated and his flesh and bones burnt, and yet he is suddenly and grotesquely transformed: 'His eyes became like blood and he vomited forth his flesh. And I saw him change into a mutilated homunculus,* biting himself and wounding himself with his own teeth.'

Whatever this bizarre description means, it probably should not be taken literally: the 'being' here is not actually a creature but a substance, whose mutilations and burnings refer to laboratory processes that change its appearance. We have already seen how alchemical transformation was often depicted in anthropomorphic terms, as a form of death and liberation of the spirit. Alchemical illustrations often deploy this sort of imagery: kings, lions and hermaphrodites are burnt, drowned and putrefied, all of them symbolic codifications of the manipulation of inert materials. The allusions to Christian mythology of death and purification are not coincidental, although we should not make the Jungian mistake of supposing that a purely spiritual transformation was alchemy's primary goal.

In the Hebrew tradition, the 'wizard of Samaria' Simon Magus – the original Faust figure who challenged St Peter to a magical duel – is said to have made a kind of homunculus from *pneuma* alone. Allegedly he turned this 'air' first into water, then blood, and from this he made flesh. 'When the flesh had become firm', says one account, 'he had produced a man, not from earth but from air, so convincing himself that he could make a new man. He also claimed that he had returned him to the air by undoing the transformations.' The description sounds very much like an allegory of alchemical transformation, in which a vapour is condensed, solidified, and then melted and evaporated again.

The key source for medieval homunculus recipes was a book apocryphally attributed to Plato but probably based on Islamic texts, called the *Book of the Cow*. This tells first how to create an artificial

---

* The word used in the text is the Greek equivalent *anthroparion*: little man.

Alchemical images in which allegorical figures represent chemical reagents, subjected to various procedures and transformations. From Johann Daniel Mylius, *Anatomia auri* (1626) and Michael Maier, *Atalanta fugiens* (1618).

cow in an operation that is not for the faint-hearted. The initial stages resemble Virgil's recipe for making bees by spontaneous generation. A cow is killed and beheaded, the head reattached, and the orifices sewn shut. Then the body is beaten with a large dog's penis until all the bones are broken (the improbability of this operation raises the suspicion of mistranslation). A 'marrow-like' substance is extracted from the body, ground up with herbs, and left to putrefy while being occasionally sprinkled with powdered bees. If this process is conducted in reverse, a living cow can be regenerated.

But then the *Book of the Cow* moves on to a considerably more ambitious feat: the creation of a 'rational animal', an artificial human, in comparison to which making a cow sounds easy. First, you take human sperm and mix it while still warm with 'the stone that is called stone of the sun' – this seems to refer to a phosphorescent material, for it is 'a stone that shines at night like a lamp until the place in which it is found is illuminated'. The mixture is gestated, as it were, in the womb of a cow which has been cleaned with medicines and anointed with ewe's blood. The cow's vulva is plugged with the stone of the sun, and the animal is kept in a dark place and fed on more ewe's blood. A fetus will grow in the womb, and when this is born, it must be put into a powder made from stone of the sun, sulphur, a magnetic material, and a mineral called 'green tutia' ground up with willow sap. The newborn creature will 'at once be clothed in human skin'.

That's not the end of it. This being must be kept inside a glass or lead vessel, and after three days it will begin to move. It should then be fed for seven days with the 'blood that has gone forth from the mother'. How the artificial being looks and behaves is not revealed, but it seems to be considered not exactly human, and has miraculous abilities. Fed with milk and rainwater for a year, it will develop the power of oracular prophecy. If it is killed, its body fluids will enable a person to walk on water.

The use of a homunculus's body fluids was recommended else-where too. The Arabic physician Al Rhazes allegedly says that its blood has medicinal power. This puts us in mind now of 'saviour siblings' selected by embryo screening after IVF to ensure a genetic match, whose cord blood or stem cells might enable future treatment for their sibling's condition. Or rather, it reminds us of the spectre that looms over this controversial topic: an artificial being harvested for 'spare parts', as depicted in Kazuo Ishiguro's 2005 novel *Never Let Me Go*, where pitiable organ donors are bred by unspecified, artificial means.*

In the Middle Ages, the justification for dismembering a homunculus was that it lacks a 'rational soul' and is thereby less than (as well as perhaps more than) human. But that matter wasn't fully clear, which is partly why the whole enterprise bore a disreputable, even heretical odour. The thirteenth-century Catalan physician and alchemist Arnald of Villanova is said to have begun to make a homunculus from semen treated with drugs, but to have smashed it before it was fully formed for fear that he would be seen to compel God to infuse a soul into it.†

Jabir's homunculus is made by a procedure altogether less messy and more explicitly alchemical. The starting material is probably again human sperm – it is variously called 'matter', 'essence' and 'body', but not definitively identified – which is putrefied within a mould that is itself held in a perforated spherical vessel immersed in a warm water bath. Intriguingly, the vessel is rotated to mimic the motions of the celestial bodies, and it is said that a series of concentric vessels, which

---

* As some reviewers pointed out, it would be a mistake to read Ishiguro's novel literally as a dystopian moral fable about cloning or organ-harvesting. Taken at face value, it is hard to credit the story's scenario, which takes place in a version of contemporary England. This society's casual acceptance of the breeding and mutilation of the organ donors seems implausible: there is no authoritarian Big Brother that commands such inhumanity, and the culture is otherwise no more brutalized or decadent than it is today. Science-fiction writer M. John Harrison considered that the book is instead about 'repressing what you know, which is that in this life people fail one another, grow old and fall to pieces. [It] isn't about cloning, or being a clone, at all.' In this respect, it has an ambivalent position within the corpus of modern anthropoeian fables. Yet if one is determined on a literalist reading, just about the only way to rationalize the acceptance of human body-farming is to assume that, for the society in which these sad beings are created, they have no genuine human status at all – no soul. That, indeed, is almost the impression elicited by the narrator's flat, dispassionate style, an amalgam of platitudes and clichés borrowed from 'real' people. As Harrison says, '*Never Let Me Go* makes you want to have sex, take drugs, run a marathon, dance – anything to convince yourself that you're more alive, more determined, more conscious, more dangerous than any of these characters.' Notice that: more alive.

† As with much of the alchemical literature, it's not clear how much of the writings attributed to Arnald were actually written by him – there is certainly a significant body of 'his' work written pseudonymously. It may even be that Arnald wrote nothing genuinely alchemical at all.

further enhances the resemblance to the heavens, is better still. This is a frank indication that the process is considered to be reproducing in the microcosm of art a phenomenon that transpires in the macrocosm of nature.

The persistence of the homunculus in Western culture owes most to a prescription attributed to the early-sixteenth-century Swiss physician and alchemist Paracelsus. In a Paracelsian book called *De rerum natura*, which may not have been written by Paracelsus himself but is thought genuinely to reflect his views, there is a recipe seemingly indebted to both the *Book of the Cow* and to the Jabirian corpus. Again it proceeds from the putrefaction of semen, which is kept in a round glass vessel and warmed with horse manure for forty days. After this time a creature with a transparent, human-like body will have formed and begun to move. Kept in the flask for forty weeks and nourished with a tincture prepared from human blood, it will become 'a true and living infant, having all the members of a child that is born from a woman, but much smaller'. This homunculus may be educated to become an intelligent creature.

The author prefaced the homunculus recipe with an account of how to conjure a live bird from the putrefied ashes of one that has been killed and burnt: a reframing of the legend of the phoenix. This 'resurrection' of dead matter, called palingenesis, is a recurrent theme in alchemical literature. Alchemists and other practitioners of natural magic commonly claimed to be able to perform it on plants, especially flowers (it is conducted by Paracelsus in Jorge Luis Borges's short story 'The Rose of Paracelsus'). Even Robert Boyle concurred that palingenesis is possible, describing how a Polish doctor had once managed to reconstitute a plant from its component atoms. 'Bodys', he wrote, 'Appear & are often thought Absolutely destroy'd, when they are not', so that 'the Seminall Essence of a burnd Plant, may be preserv'd in its incombustible Parts'. Here Boyle's comments are tinged with vitalism, the belief in an intrinsic life-giving agency in organic matter. It is as though the substance, however transformed, retains a memory of what it is 'meant' to be. Boyle claims that the human body retains a kind of life force if it is, for example, eaten by wild animals: 'its Atoms [are] preserv'd in all their Digestions, & kept capable of being reunited'. He even advanced this as a 'scientific' explanation for the Resurrection of Christ.

In a variant on this theme which invokes something like spontaneous generation, the author of *De rerum natura* says that the flesh
of a snake, chopped up and putrefied, can spawn worms that may
grow into 'a hundred snakes where each is as big as the one that
caused the putrefaction'. The text alleges that this regeneration is
rumoured to work for humans too, and claims that Virgil, widely
regarded in the Middle Ages as a magician, hoped to be reborn by
dismemberment of his dead body. Other early-sixteenth-century
accounts allege that the Roman writer was successful, and that a naked
child was briefly seen running around the barrel in which Virgil's
dismembered corpse was placed.

Here again there is a conflation of the symbolic language of alchemy
with its literal interpretation: the burning and dismemberment of
beings and their reincarnation were images often used to represent
the transformation of inorganic materials by heat. In *The Chymical
Wedding of Christian Rosencreutz* (1616), one of the anonymous* founding
texts of the occult movement known as Rosicrucianism, the symbolism
is woven into a fantastical tale of a king and queen who are married
and then beheaded, after which their blood is subjected to various
alchemical operations from which emerge 'two beautiful bright and
almost *Transparent little* Images': two homunculi four inches tall.

The Flemish physician Gerhard Dorn, one of the most active advocates of Paracelsian chemical philosophy in the late sixteenth century,
suggested that the homunculus recipe is actually a coded prescription
for making a medicine from wheat and wine. Paracelsus's detractors
in later times (of which there were many) would have none of this.
The Cambridge philosopher Henry More asserted that the homunculus
recipe was just another example of Paracelsus's boastful, empty claims,
and he dismissed Dorn's interpretation as a whitewash, accusing Dorn
of being 'ashamed of the grosse sense of it'.

More's scepticism is ironic in view of the fact that he otherwise
plays the part of quasi-mystic adversary to the arch-sceptic of the
seventeenth century, Robert Boyle, probably the most influential
scientist in England at that time. Boyle, whose most famous book *The
Sceptical Chymist* (1661) challenged the obfuscation and posturing of

---

* The author is generally considered to be the Paracelsian alchemist Johann Valentin
Andreae.

the alchemists, dismissed More's claim that a 'spirit of nature' mediated between God and the natural world, instead preferring a mechanical (although ultimately devout) explanation of all phenomena in terms of particles in motion. And yet Boyle himself serves to expunge any simplistic division of seventeenth-century natural philosophers into hard-nosed rationalists and credulous traditionalists. For it was this Anglo-Irish paragon of the cautious new 'experimental philosophy' who was apparently taken in by the report of a French con man called Georges Pierre, who described the production in 1678 of a homunculus by a Chinese 'adept' of the alchemical – and probably fictitious – secret society known as the Asterism. If Boyle was ready to credit homunculi in the midst of the so-called Scientific Revolution, it was unlikely that anyone else in England (at least) would have denied them.

The ambiguity between literal and symbolic meaning pertains also to a recipe in *De rerum natura* for making a basilisk, the mythical reptile that could kill with a glance. Prescriptions for creating or using basilisks crop up in the otherwise sober medieval literature of artisans' workshops, where they seem to have been imported at face value from alchemical texts. It's surely no coincidence that the creature's name is derived from the Greek *basiliskos*, 'little king', linking it to the use of regal terminology for substances in alchemy. In the Paracelsian text, a basilisk is made in a similar procedure to the homunculus by using as the raw material menstrual blood, putrefied in a horse's womb, rather than semen. In this respect, says William Newman, the basilisk is a misogynistic representation of the essence of femininity, concocted from the 'greatest impurity' of women – the evil things attributed to menstruation and menstrual blood are a familiar, insidious aspect of folklore.

The homunculus, in contrast, being created from male seed, is considered to be a noble, exalted life form which, because it is the product of art, has access to all the occult knowledge of *techne*. There is nothing inferior about it, despite its stature: rather, its alchemical origins imbue it with supernatural abilities.

The theology of the homunculus

It is hardly surprising that concerns were raised about the impiety of making homunculi. One of the criticisms of this 'technology' was apt to be directed at any form of natural magic: it was suspected of involving the assistance of demons. All claims of alchemical transmutation were susceptible to such accusations, but those that involved the manipulation of life seemed particularly dubious. In response to the rumour that female witches can change into animals, a tenth-century Church document called the *Canon episcopi*, probably the work of the Benedictine abbot Regino of Prüm, says that:

Whoever therefore believes any created thing to be able to be made or to be changed into better or worse or transmuted into another shape or likeness, except by the Creator, who made all things, and by whom all are made, is without doubt an infidel and worse than a pagan.

Many natural magicians denied this emphatically, saying that they did nothing more than accelerate the normal course of nature.

As well as raising concerns about the techniques used, this form of anthropoeia was deemed transgressive because of what it produced. In the thirteenth century, the Bishop of Paris William of Auvergne alleged that recipes like that in the *Book of the Cow* are liable to generate not human-like beings but monsters. And the *Anthropodemus plutonicus* (1666–7) of the German astronomer Johann Praetorius classed the 'chemical man' among the monsters, freaks and prodigies alluded to in the book's title: an aberration of nature, maybe a portent.

Surprisingly, however, suggestions that making an artificial being posed a hubristic challenge to God's monopoly on creation were rare. The theological worry was rather that, generated *de novo* rather than by coitus, the homunculus seemed to be exempt from original sin. And if the homunculus's lack of a mortal, rational soul absolved humans from any obligation to respect its life, by the same token it left the being without need of Christ's redemption. In this respect the homunculus was free as no human ever was: it was as though, through the art of making people, men could vicariously shake off the bonds of moral corruption.

In the face of such accusations, one remarkable defence of the homunculus linked it with the genesis of Christ. Wasn't Jesus, uniquely among men, made in the 'sealed vessel' of an intact womb? And how then was his body formed, if not from semen? In the fifteenth century the Spanish bishop Alonso Tostado, a critic of alchemy, tried to answer this puzzle. He agreed with Thomas Aquinas that Christ must have appeared fully formed within Mary's womb at the moment of conception, rather than undergoing the normal gestation of a human being; but then what was the formative material of Christ's flesh? Tostado supposed that it was created out of menstrual blood alone. But clearly there cannot have been enough of it to make a full-term baby at a stroke, and so the antenatal Jesus must have appeared first as a tiny but fully formed infant, and simply got bigger. That seems to make him a sort of homunculus.

To practice more than heavenly power permits

What, then, is the homunculus, the artificial man? Although made from human seed, he is either less than or more than human, or both at once. It is questionable whether he has a soul. He is perhaps the spawn of witchcraft and demons, and as such, is something monstrous, even if not in appearance. He has special, hidden knowledge. He may be infertile.

But in the sixteenth century the focus turned more strongly towards the creature's maker. What sort of person dabbles with life and procreation, perhaps enlisting Satan's assistance? The answer has coloured the image of the scientist ever since.

The first homunculus-maker, Simon Magus of Samaria, was a malign wizard and an alchemist who could allegedly turn himself into gold. He was the archetypal Promethean trickster, taunting God by calling himself the Great Power and challenging St Peter to a miracle-working duel. And he called himself the auspicious or favoured one: in Latin, *faustus*.

The Faust legend was revitalized in the sixteenth century when at least one mysterious vagabond plying the roads of the German lands seems to have called himself by that name. One tradition associates him with a German knight called Georg Sabellicus, born around 1480, who wandered in the Rhine valley claiming to have alchemical skills. The late-sixteenth-century German scholar Johann Weyer asserts that the Faust then beginning to be celebrated in popular legend was instead born near Cracow in what is now Poland. This Faust is mentioned by Luther's colleague Philip Melanchthon, who says that he met him and was unimpressed by his bombastic claims and superficial trickery. His life was first recounted at length, to lucrative success, in 1587 by the Frankfurt publisher Johann Spiess, and then – in more obviously fictionalized form, and to greater fame – in Christopher Marlowe's 1592 play *The Tragical History of Doctor Faustus*.

In many of the escapades attributed to Faust in the early accounts, he is more of a clownish scoundrel than a wicked man. He was alleged to have summoned the heroes of the Trojan Wars and the monsters of Greek mythology when he lectured on Homer at Erfurt, and to have made Helen of Troy appear in front of the university faculty at Wittenberg. He once froze immobile the butler of the Archbishop of Salzburg while Faust and his friends made free with the archbishop's wine cellar, finally conjuring the poor man on to a treetop. He won a gold coin from a hapless farmer after betting him that he could eat the farmer's cart of hay, and then proceeding almost to do so.

But soon the Faust stories began to assert that his powers came from a pact with the Devil. Spiess says he was torn limb from limb when Satan came to collect his due, and Weyer also says Faust came to a terrible end. This was the Faust portrayed by Christopher Marlowe, whose story is ultimately moralistic: Faust was doomed by his overweening ambition. As he says in Marlowe's play:

> O what a world of profit and delight,
> Of power, of honour, of omnipotence
> Is promis'd to the studious artisan!
> All things that move between the quiet poles
> Shall be at my command . . .
> A sound magician is a mighty god:
> Here, Faustus, try thy brains to gain a deity.

He is finally dragged to hell, while the Chorus spells out the moral:

> Faustus is gone; regard his hellish fall,
> Whose fiendful fortune may exhort the wise
> Only to wonder at unlawful things,
> Whose deepness doth entice such forward wits
> To practice more than heavenly power permits.

But by the Elizabethan era, the power of the Church to compel humility before God was waning, and the Baconian desire for power over nature, won by scientific investigation, was in the ascendant. No one would question that it was reprehensible to consort with demons (or, for sceptics, even to imagine one could do so); but Faust's thirst for knowledge and his Promethean access to powerful arts no longer seemed to be crimes in themselves. Indeed, in the light of the emerging humanistic philosophy of the early Scientific Revolution, these were noble aspirations.

So by the time Faust's tale was told again by Wolfgang von Goethe at the end of the eighteenth century, he had become a romantic figure, a Job-like pawn manipulated by higher powers and lured away from the righteous path by Mephistopheles in a wager with God. It is true that Goethe's Faust craves knowledge; as Mephistopheles says:

> In madness, half suspected on his part,
> He hankers after heaven's loveliest orbs,
> Demands from earth the choicest joy and art,
> And, far and near, what pleases and absorbs
> Still fails to satisfy his restless heart.

But this obsession is not in itself displeasing to God, who says, 'Though now he serves me in bewildering ways, my light shall lead him soon from his despairing.'

Goethe reunites Faust with the homunculus tradition when his former assistant Wagner, with Mephistopheles in attendance, creates an artificial being in the laboratory. Wagner's motive is telling: both recalling King Harmanus' revulsion of sexual intercourse and prefiguring the debates about sexuality and artificial reproduction in the early twentieth century, Wagner wants by this means to make sex unnecessary:

> That old style we declare
> A poor begetting in a foolish fashion . . .
> What if the beasts still find it their delight,
> In future man, as fits his lofty mind,
> Must have a source more noble and refined.

Here, then, making artificial humans is no longer a matter of usurping God's power, but acquires a kind of prudish moral justification.

Wagner gives only the vaguest intimation of the procedure:

> Now hope may be fulfilled,
> That hundreds of ingredients, mixed, distilled –
> And mixing is the secret – give us power
> The stuff of human nature to compound;
> If in a limbeck [alembic, an alchemical flask] we now seal it
>   round
> And cohobate with final care profound,
> The finished work may crown this silent hour.

Yet it works:

> The substance stirs, is turning clearer! . . .
> What Nature by slow process organized,
> That we have grasped, and crystallized it out.

Goethe's homunculus is no monster but a 'dapper charming lad', a 'lustrous dwarf', and apparently a hermaphrodite. He is a glowing super-being with magical powers, though he must stay confined within his glass, his own private microcosm: 'The cosmos scarce will encompass Nature's kind,/But Man's creations need to be confined.'

The homunculus uses his powers to bring Faust, accompanied by Mephistopheles, to ancient Greece, where they meet the philosophers Thales and Anaxagoras, both of whom speculated on matter and its transformations. Here the homunculus seeks to escape his glass prison by being reborn as a human being. 'Grant me your company and words of worth', he implores the two philosophers, 'For I myself desire to come of birth'. Thales, who believed that elemental water is the fundamental element of all matter ('From moisture all organic living came'), tells him that to attain human shape the fiery homunculus must be joined with water. The group then retires to the Aegean, where they meet the shape-shifting god Proteus. Thales explains to him the predicament of the homunculus:

> Most strangely made, as I have heard him say,
> For birth in his case reached but life's half-way,
> No qualities he lacks of the ideal,
> But sadly lacks the tangible and real.
> Till now the glass alone has given him weight;
> But now he longs for an embodied state.

Proteus encourages the homunculus to enter the waves, and transforms himself into a dolphin so that the artificial being may ride on his back. The homunculus is carried to the throne of Galatea, who Ovid identifies as a sea nymph – but whose name is shared by Pygmalion's animated statue, popularized in a retelling in Goethe's time (see p. 152). Here the homunculus's glass vessel is dashed against the throne, and rather than being incarnated the being is spread in a flash of light across the waves: a union of fire and water with strong alchemical overtones. In this way the homunculus gives himself to the sea, contrasting a submission to nature with Faust's hubristic effort to tame the ocean's might by reclaiming submerged land and thus subvert nature's order: 'To hold the lordly ocean from the shore,/To set the watery waste new boundary lines.'

So Goethe's homunculus is presented as a highly sympathetic, intelligent being: a kind of personification of the liberated human intellect. Yet he is an incomplete being, doomed to remain captive in his glass unless he can acquire a body through mystical alchemical union. And note that he is created not to mock or challenge God, but to make sex obsolete.

Faust, meanwhile, is now permitted redemption at the end of Goethe's play, although it is not altogether clear why. It is true that he has learnt some humility, and his land reclamation project might be deemed a worthy use of his technical prowess (although unknown to Faust, Mephistopheles kills an elderly couple who refuse to relocate from the land on which Faust is working). Goethe's translator Philip Wayne says simply that 'Faust is saved by his will to strive'. But it seems surprising that redemption from a satanic pact can be bought so easily. When a host of angels descends and bears Faust's soul to heaven, they seem intent merely on frustrating Mephistopheles.

It isn't easy, then, to find within Goethe's tale the simple moral about the dangers of Promethean ambition that was central to the original Faust books. In particular, making artificial humanoids is not presented as self-evidently reprehensible; in some ways the synthetic being serves rather to highlight the human shortcomings of his maker. But this complexity in Goethe's narrative was never subsequently developed. Instead, the anthropoetic Faust legend was about to take a new form. It was one more suited to the burgeoning scientific and industrial milieu of the nineteenth century, and it was written by a troubled Englishwoman not yet in her twenties.

4 It Lives!

I ought to be thy Adam, but I am rather the fallen angel.
  Mary Shelley, *Frankenstein* (1818)

You have created a monster, and it will destroy you!
  Dr Waldman in *Frankenstein* (1931, dir. James Whale)

Mary Shelley's *Frankenstein* is with justification called a modern myth, one of the select group of novels that includes *Don Quixote, Robinson Crusoe, The Strange Case of Dr Jekyll and Mr Hyde* and *Dracula*. *Frankenstein* is in some respects the now-familiar story of an arrogant, overreaching Faust/Prometheus figure: a morality tale of anthropoeia in which the Devil's assistance is no longer required to make the artificial being, which is why the creature itself is left to exact the retribution that tradition requires. But if the novel were no more than this, it would not have had such extraordinary resonance and longevity.

In part, Shelley was articulating an ambivalence towards the new sciences of the early nineteenth century, which appeared to be on the verge of explaining life itself. But there are many more layers to the story. It expresses ambivalence towards procreation itself – a process that might understandably be considered inherently treacherous by a woman whose own mother died as a result of her birth. And we can discern the perplexity of a woman in a man's world: a woman who could not even initially publish her novel under her own name, and who was living in the shadow of a famous and domineering father, not to mention a brash, egotistical lover and future husband. Observing men's boundless and perhaps ruthless creative desires, Mary Shelley imagines a man who, by making life itself, manages to usurp even this aspect of a woman's power.

Shelley's biographer Anne Mellor asserts that 'the idea of an entirely man-made monster is Mary Shelley's own', and that she 'created her myth single-handedly'. Not only are these claims emphatically wrong, but to make them is to distort and perhaps even undermine Shelley's achievement. The story did not spring newly minted from her imagination; probably no modern myth can be created that way. Shelley herself proclaimed her debt to the ancient myths with her subtitle: 'The modern Prometheus'. Many of the themes she explores are evident in the older anthropoetic tradition, given as it were new flesh. And insofar as Victor Frankenstein's tragedy is, on the surface, wholly secular and self-inflicted, it completes the romanticization of Faust that Goethe had begun.

Critics are divided between those who feel that *Frankenstein* is a forward-looking work, a prototype of science fiction that is deeply engaged with the scientific themes of the age, and those who regard it as ill-informed and rooted in obsolete notions of science – as critic James Rieger puts it, invoking 'switched-on magic, souped-up alchemy, the electrification of Paracelsus'. Yet it is because *Frankenstein* has elements of both traditions – the modern and the ancient – that it occupies such an important nexus in the evolving myth of making people.* There seems to be little to be gained either by pretending that Mary Shelley was fully au fait with the latest scientific thinking on biology, chemistry and electricity, or by scoffing at her shaky grasp of the 'facts' as they were then known. Hers was inevitably a lay person's perspective, albeit impressive for an eighteen-year-old and all the more so given that young women were not supposed to take an interest in such things. It was the rich mixture of ingredients unwittingly used to cook up *Frankenstein* that gave it such cultural currency: among them, the Romantic equivocation over science and technology,

---

* Science writer Jon Turney argues apropos Rieger that 'it is precisely the electrification of Paracelsus which marks out *Frankenstein* as a pivotal point in the transition from the supernaturally fantastic to the scientifically plausible'. This is true in spirit, but we must remember that the Paracelsian homunculus was also 'scientifically plausible' within the rationalistic natural philosophy of its time, and Paracelsus had (in his own mind) no time for superstition.

Mary Shelley's peculiar family history, and not least the uninhibited, even naive, passion for Gothic narrative that helped prevent the book becoming a self-consciously calculated affair. If Percy Shelley had written it – and heaven knows he tried, heavily editing great swathes of the text – it would probably have been much better written, and far more dull and arid.

The problem of the father

William Godwin (1756–1836), Mary's father, was a radical thinker, an anarchist who considered that government corrupts society. People left to their own devices, he argued in *Enquiry Concerning Political Justice* (1793), will find a natural morality. In 1797 Godwin married Mary Wollstonecraft, another liberal free-thinker and every inch his intellectual match. She was a friend of Tom Paine and William Blake and a former lover of the artist Henry Fuseli (whose famous painting *The Nightmare* epitomizes the Gothic vision of eroticized, subconscious terror), and she had an illegitimate daughter, Fanny, from another previous relationship. Wollstonecraft was already pregnant by Godwin when they married (the couple's previous antipathy to this patriarchal institution occasioned much soul-searching), but she died shortly after giving birth to her daughter Mary. It seems banal to observe that this must have fixed in Mary's mind the intimate connection between birth and death, as well as suggesting to her that ordinary childbirth was fraught with danger; but that does not make these things any the less true. The guilt she felt at her mother's death can only have been heightened by the way her father subsequently idealized his late wife.

Among Godwin's wide-ranging interests was an enthusiasm for the history of the 'occult sciences'. Two years before his death he published *Lives of the Necromancers*, an account of famous wizards, alchemists and seers from antiquity to the seventeenth century, including Paracelsus, Agrippa and Faust. Although Godwin called this a 'delineation of the credulity of the human mind', it is hard to believe that his motivation was purely educational – to demonstrate, as he put it, that 'the wildest extravagances of human fancy, the most deplorable perversion of human faculties, and the most horrible distortions of jurisprudence, may occasionally afford us a salutary lesson'. There is, on the contrary, a hint of approval in the book's opening statement that 'Man is a creature of boundless ambition.' That ambition fascinated Godwin; in his Faustian novel *St Leon*, the protagonist embarks on an ill-fated quest for the philosopher's stone and the elixir of life.

Godwin kept elevated company – among the guests who came to debate politics, philosophy, science and literature at her father's house, Mary would have met William Wordsworth, Samuel Taylor Coleridge, Humphry Davy and William Hazlitt. In early 1812 Godwin received a bold letter from a young poet named Percy Bysshe Shelley, who was eager to introduce himself to the noted philosopher, saying 'I am convinced I could represent myse[lf] to you in such terms as not to be thought wholly unworthy of your friendship.' This young man, who was already married at the age of nineteen, became a regular caller and something of Godwin's protégé. By the summer of 1814 Shelley dined with the Godwins every day, and at the end of June he and Mary declared their love for one another.

It took Godwin barely two weeks to discover the relationship, whereupon he forbade them to meet, despite Percy's assurances that his own marriage was moribund. As a result, on 18 July, Mary and Percy eloped to France, and she was ignored by her father for the next three and a half years. Very soon Mary was pregnant. Her daughter Clara was born prematurely the following February, and died within two weeks. This event provoked fantasies of reanimation: in 1815 Mary described a dream in which she brought her dead child back to life. Her second child, William, was born a year later, and during the summer of that year Mary and Percy went to stay with Byron and his doctor William Polidori on the banks of Lake Geneva.

Shelley later presented the genesis of her story as a dream that came to her after a late-night discussion with Byron, Percy and Polidori in Switzerland. They had talked of 'various philosophical doctrines . . . and among others the nature of the principle of life, and whether there was any probability of its ever being discovered and communicated'. While it would be uncharitable to doubt this account entirely, inspiration derived from dream-visions was a popular trope in the nineteenth century. Several scientists claimed to have made discoveries while in a dreamlike state, including August Kekulé (who deduced the structure of the benzene ring) and Dmitri Mendeleev (credited with the periodic table of elements). Besides, Polidori's account of the chronology of the Swiss excursion casts doubt on Mary Shelley's version of events, including the plausibility of the 'dream'.

Mary worked on the book until the following year, and when the proofs arrived she gave her husband (she and Percy were married at the end of 1816) 'carte blanche to make what alterations you please', an invitation of which he availed himself freely. The novel had been turned down by several publishers and ended up with Lackington, Hughes, Harding, Mavor and Jones, a somewhat disreputable firm specializing in occult and sensationalist literature. The book was published, in three slender but expensive volumes, on the day the Shelleys and their children, along with Mary's stepsister Claire, set sail for Italy; Percy Shelley never returned from that trip, for he was drowned in a storm while sailing to Livorno in 1822. Forced to support herself and her family, Mary became a professional writer, but her four other novels are more or less forgotten today.

A creation of science?

As the alleged prototype of the mad scientist, Victor Frankenstein is surprisingly sane. The question that motivates his anthropoetic experiments is no different from that which many scientists have pursued patiently and calmly for centuries, and continue to do so today: 'Whence did the principle of life proceed?' His downfall is not that he asks the question, but that (in absurdly short order) he answers it, finding himself burdened with perilous knowledge. But it is perilous only because Frankenstein decides to use this knowledge to make a man. The reason we have so much freedom to interpret Shelley's message is that she never explains this crucial decision – Frankenstein advances straight to this goal as though an understanding of the 'astonishing secret' of life permits no other. Although by no means lacking ego he seems heedless of fame, wealth and reputation, and indeed does not even consider, as a true scientist might, that by conducting the experiment he can put his (unexplained) theory to the test. He is excited only by the thought that 'a new species would bless me as its creator and source' – an apparently egotistical motive that is, however, immediately moderated by altruism: 'many happy and excellent natures would owe their being to me'. Even if this is hubris, it feels thrust upon the hapless young Frankenstein at the insistence of his own creator, for nothing previously in his character has adequately prepared us for it. Frankenstein's ambition might seem unwise, even reprehensible, and yet it lacks the pride and arrogance tradition imputes to Faust. Frankenstein, like Job, ends up a pawn of his maker.

The lack of any real hypothesis-testing, and in fact of any real theory, in Victor's studies furnishes just one of the objections to the idea that he is a 'scientist' at all, at least in the modern sense.* And if he is not, can Shelley really be deemed to be putting science on trial? There are disparate opinions on Frankenstein's credentials: science-fiction writer Brian Aldiss, for example, says that Victor 'turns away from alchemy and the past towards science and the future', while James Rieger asserts that Shelley 'skips the science'. Who is right?

The tension between old and new understanding of natural phenomena is made explicit in the book. Victor develops a youthful interest in alchemy, having chanced upon a book by the sixteenth-century exponent of natural magic Agrippa at the impressionable age of thirteen. For this he is ridiculed by his father, who tells him not to 'waste your time upon this; it is sad trash'. In the manner of all parental prohibitions, it merely spurs Victor to continue his studies with greater passion:

When I returned home my first care was to procure the whole works of this author [Agrippa], and afterwards of Paracelsus and Albertus Magnus. I read and studied the wild fancies of these writers with delight; they appeared to me treasures known to few besides myself. I have described myself as always having been imbued with a fervent longing to penetrate the secrets of nature.

Shelley knows that Victor's enthusiasm is anachronistic, and takes care to have him admit as much:

It may appear strange that such should arise in the eighteenth century; but while I followed the routine of education in the schools of Geneva, I was, to a great degree, self-taught with regard to my favourite studies. My father was not scientific, and I was left to struggle with a child's blindness, added to a student's thirst for knowledge.

---

* The word itself was not coined until 1833, by Michael Faraday's colleague William Whewell, but the profession was already becoming recognized by the start of the century.

It is not until he goes to study at the University of Ingolstadt that Victor's archaic convictions in natural philosophy are truly challenged. He meets first a professor named Krempe, 'an uncouth man, but deeply imbued in the secrets of his science', who pours scorn on Victor's interest in the alchemists:

The professor stared. 'Have you,' he said, 'really spent your time in studying such nonsense?' I replied in the affirmative. 'Every minute', continued M. Krempe with warmth, 'every instant that you have wasted on those books is utterly and entirely lost. You have burdened your memory with exploded systems and useless names . . . I little expected, in this enlightened and scientific age, to find a disciple of Albertus Magnus and Paracelsus. My dear sir, you must begin your studies entirely anew.'

It is inconvenient for a reading of *Frankenstein* as a simple critique of scientific hubris that what nonetheless leaves Victor disenchanted about modern science is its *lack* of ambition in comparison to the magical philosophers:

I had a contempt for the uses of modern natural philosophy. It was very different when the masters of the science sought immortality and power; such views, although futile, were grand; but now the scene was changed. The ambition of the enquirer seemed to limit itself to the annihilation of those visions on which my interest in science was chiefly founded. I was required to exchange chimeras of boundless grandeur for realities of little worth.

What finally changes Victor's mind is not Krempe's ridicule but the measured, sympathetic lectures of the chemist Professor Waldman. In comparison to the alchemists, Waldman says,

The modern masters promise very little . . . But these philosophers . . . have indeed performed miracles. They penetrate into the recesses of nature and show how she works in her hiding-places. They ascend into the heavens; they have discovered how the blood circulates, and the nature of the air we breathe. They have acquired new and almost unlimited powers; they can command the thunders of heaven, mimic the earthquake, and even mock the invisible world with its own shadows.

Waldman is here alluding to several recent developments in science and technology. Ballooning, using both hot air and the newly discovered hydrogen gas, dazzled spectators in the 1780s, and the first balloon crossing of the English Channel had been achieved in 1785. Oxygen, the breathable component of air, had been discovered by the English chemist Joseph Priestley in the 1770s,* and was named by Antoine Lavoisier in 1783. And Priestley had also written of the taming of electricity:

What would the ancient philosophers, what would Newton himself have said, to see the present race of electricians imitating in miniature all the known effects of that tremendous power, nay, disarming the thunder of its power of doing mischief, and, without any apprehension of danger to themselves, drawing lightning from the clouds into an [*sic*] private room and amusing themselves at their leisure by performing with it all the experiments that are exhibited by electrical machines.

Yet Shelley has Waldman express these things using the imagery of the early Enlightenment, popularized by Francis Bacon, in which nature is a coy woman who 'veils' her 'secrets' and has to be 'penetrated'.

Waldman is more sympathetic to the efforts of the alchemists, although adamant that they wasted their energies on a quixotic project: 'these were men to whose indefatigable zeal modern philosophers were indebted for most of the foundations of their knowledge'. Under Waldman's guidance, Victor's interests turn towards chemistry, 'that branch of natural philosophy in which the greatest improvements have been and may be made'. His fateful anthropoetic experiment, Shelley hints, happens at the juncture of chemistry, physiology and the study of electricity and galvanism.

Shelley's description of this discovery has elicited some understandable mockery, for it is plain that she has not the slightest notion of how to make it plausible. Victor is a young student who has until that point known barely any modern science. Simply from spending 'days and nights in vaults and charnel-houses' looking at decaying corpses, he finds that:

---

* Priestley was a radical dissenter, and a friend of William Godwin.

from the midst of this darkness a sudden light broke in upon me – a light so brilliant and wondrous, yet so simple, that while I became dizzy with the immensity of the prospect which it illustrated, I was surprised that among so many men of genius who had directed their enquires towards the same science, that I alone should be reserved to discover so astonishing a secret.

Well might he be 'surprised', and it is hard not to judge this a lame account of such a stupendous discovery. Victor's insight owes more to theological than to scientific tradition: the image of receiving special knowledge in a flash of divine illumination goes back to the Gnostic and Platonic beliefs of early Christianity, in which context it was invoked by St Augustine. This tradition was transferred in the Renaissance from the quest to know God to the quest to understand nature; Paracelsus and others spoke of becoming literally enlightened by the Light of Nature. This sort of revelation is won by grace alone, bypassing the need for endless prayer or, in the secular realm, endless study. And indeed, the enlightened Victor sounds very much like the religious convert, at once in a state of both bliss and convenient ignorance about how he obtained his insight:

The astonishment which I had at first experienced on this discovery soon gave place to delight and rapture . . . But this discovery was so great and overwhelming that all the steps by which I had been progressively led to it were obliterated, and I beheld only the result. What had been the study and desire of the wisest men since the creation of the world was now within my grasp . . . Life and death appeared to me ideal bounds, which I should first break through, and pour a torrent of light into our dark world.

And so, he says, 'I began the creation of a human being.' Because of the difficulty of joining the 'intricacies of fibres, muscles and veins' in the body parts he scavenges, Victor is forced to work at a super-human scale, making a creature eight feet tall. (It is not explained how this is possible using limbs and bones taken from corpses of lesser stature.) The animation itself is achieved in a single paragraph, with barely a word of explanation:

With an anxiety that almost amounted to agony, I collected the instruments of life around me, that I might infuse a spark of being into the lifeless thing that lay at my feet. It was already one in the morning; the rain pattered dismally against the panes, and my candle was nearly burnt out, when, by the glimmer of the half-extinguished light, I saw the dull yellow eye of the creature open; it breathed hard, and a convulsive motion agitated its limbs.

There is no thunderstorm, there are no crackling switches, no great laboratory filled with machines and flasks, no exultant cry of 'It's alive!'* The cynic is tempted to say: well, that was easy. But in retrospect this lack of detail plays in Shelley's favour, for there was no explanation she might have offered that would not now seem hopelessly crude and obsolete. Subsequent interpretations of the process have latched on to those suggestive words, 'a spark of being'. This alone intimates the involvement of electricity, and it is fairly clear that the hint was intentional. But did Shelley have anything more particular in mind for her anthropoetic procedure?

The Gothic science of reanimation

At the end of the eighteenth century, scientists found new ways to think about life. Previously regarded as a vague animating force described but not explained by the concept of vitalism, it now seemed that it might depend on the freshly discovered element oxygen, the gas dubbed 'fire air' when first isolated by the Swedish chemist Carl Wilhelm Scheele. Joseph Priestley discovered that breathing was easier when one inhaled pure oxygen, and in Lavoisier's view this substance was the principle of combustion and respiration, connecting it implicitly to the ancient notion that life was instilled either by fire or by breath.

Mary Shelley was familiar with the chemistry of the day, having read Humphry Davy's *Elements of Chemical Philosophy* in 1816. Davy was, however, suspicious of the 'chemical physiologists', who tried to reduce life to mere chemistry. He said of them, repeating the old, sexually charged metaphor of 'hidden nature':

---

* For that, we need to return to Goethe's homunculus, made in a laboratory 'after the style of the Middle Ages', full of 'extensive, unwieldy apparatus, for fantastical purposes', and attended by the exclamation of Wagner, Faust's Igor: 'It works!'

Instead of slowly endeavouring to lift up the veil concealing the wonderful phenomena of living nature; full of ardent imaginations, they have vainly and presumptuously attempted to tear it asunder.

Notice that here Davy seems to be objecting not so much that such ideas are wrong, but that they are indecorous. Life, in his view, seems still to be a blushing maiden who needs protecting so that her mysteries might remain intact. Davy's protégé Michael Faraday, later the foremost expert on electrical phenomena, felt, perhaps for religious reasons, that the innermost mysteries of nature lay outside the realm of science.

The emerging discipline of chemistry was sometimes accused of harbouring ambitions beyond its station. By simultaneously investigating the composition of organic, naturally occurring compounds and developing its skills in chemical synthesis, it was complicating distinctions between nature and art – much as alchemy had done previously, but with greater demonstrative force. At the same time, the suspicion of Faustian hubris that once attached to alchemy was now transferred to chemistry. This is apparent in Honoré de Balzac's novel *La recherche de l'absolu* (1834), which portrays an early version of the 'mad scientist' in the person of Balthazar Claes, allegedly a chemist but very obviously cut from the cloth of the legendary obsessive alchemist. Claes neglects his wife and family in his quest for the 'Absolute', the fundamental element of all matter which the ancient Greeks called *hyle* and which here stands obvious proxy for the philosopher's stone:

'I shall make metals', he cried; 'I shall make diamonds, I shall be a co-worker with Nature!'

'Will you be the happier?' she [his wife] asked in despair. 'Accursed science! Accursed demon! You forget, Claes, that you commit the sin of pride, the sin of which Satan was guilty; you assume the attributes of God . . . Analyse fruits, flowers, Malaga wine; you will discover, undoubtedly, that their substances come, like those of your water-cress, from a medium that seems foreign to them. You can, if need be, find them in nature; but when you have them, can you combine them? Can you make the flowers, the fruits, the Malaga wine? Will you have grasped the inscrutable effects of the sun, of the atmosphere of Spain? Ah! Decomposing is not creating.'

It is not, then, for the attempted creation of life, let alone specifically human life, that Claes's wife accuses him of impiety; rather, her implication is that *any* effort to replicate natural substances and objects is sinful. Alchemy had earlier been condemned for the supposed inferiority of its productions and for its suspected reliance on demonic assistance, but not because the fabrication in itself was improper. In Balzac's dialogue, in contrast, we find an early intimation of an inviolate, reified nature.

Balzac's anti-scientific agenda is clear enough. He abhorred chemistry, which he considered impious not because it could rival God's creation but because it destroyed it. In *La peau de chagrin* (1831) he calls chemistry 'that fiendish employment of decomposing all things', and deplores what he considers to be its totally materialistic view of life. Balzac says that, whereas the mechanistic conception of life advanced by René Descartes (the topic of the next chapter) at least requires a divine 'operator', chemists assert that life sprang from matter of its own accord – as Balzac puts it, that 'the world is a gas endowed with the power of movement'. Chemistry, then, could be interpreted as an atheistic science – even as a nihilist one, as personified in the fictional chemists Dr Sturler in Alexandre Dumas' *Le comte Hermann* (1849) and Bazarov in Ivan Turgenev's *Fathers and Sons* (1862).

Other writers criticized chemistry as an embodiment of narrow-minded, bourgeois acquisitiveness. In *Jezebel's Daughter* (1880), Wilkie Collins does not stray far from the medieval narrative of the fanatical, deluded alchemist in his portrait of a German chemist named Dr Fontaine, who ruins himself in pursuit of gold, diamonds and the philosopher's stone. A more contemporary image is presented in the person of Flaubert's petty, scheming pharmacist Homais in *Madame Bovary* (1857). In one way or another, then, chemistry was regarded with suspicion in nineteenth-century literary circles.

There was another new 'substance' besides oxygen that offered itself as a candidate for the 'astonishing secret' of life. Even before the monster is created, Shelley implies that electricity plays a part in Victor's discovery. While on holiday near the Jura mountains, he sees a thunderstorm during which an old oak tree is struck and shattered by lightning, and he is so impressed by this demonstration of electricity's primal power that it leaves him momentarily doubting his alchemical mentors:

I have never beheld anything so utterly destroyed. Before this I was not unacquainted with the more obvious laws of electricity. On this occasion a man of great research in natural philosophy was with us, and excited by this catastrophe, he entered on the explanation of a theory which he had formed on the subject of electricity and galvanism, which was at once new and astonishing to me. All that he said threw greatly into the shade Cornelius Agrippa, Albertus Magnus, and Paracelsus, the lords of my imagination.

When Victor encounters again his escaped creature in Geneva, it is during another thunderstorm: 'a flash of lightning illuminated the object and discovered its shape plainly to me'. Shelley is even more explicit about the electrical nature of her anthropoetic vision in the introduction to the 1831 edition of the novel: she describes here how, in the late-night conversation in Switzerland that preceded her dream, the idea was mooted that 'Perhaps a corpse would be reanimated; galvanism had given token of such things.'

The perilous power of electricity, particularly as manifested in lightning, was in the late eighteenth century compared to the fire of Prometheus stolen from Zeus' thunderbolts. After a Swedish scientist named Georg Wilhelm Richmann was killed by lightning in 1753 while conducting an experiment like that made famous (if not actually performed) by Benjamin Franklin, the *Gentleman's Magazine* said that 'we are come at last to touch the celestial fire, which if . . . we make too free with, as it is fabled Prometheus did of old, like him we may be brought too late to repent our temerity'. And the Frenchman Guillaume Mazéas wrote of his own studies of electricity that 'The fable of Prometheus is verify'd.'

An Italian physiologist named Luigi Galvani offered good reason to suppose that electricity might hold the key to life itself. Using static electricity stored in a 'Leyden jar', a kind of capacitor which provided the power source for most early experiments in electrical phenomena, Galvani showed that frog's legs and other animal limbs could be made to twitch by the flow of current. In 1791 he published his *Commentary on the Effects of Electricity on Muscular Motion*, in which he argued that a vital force called 'animal electricity' is what impels living things into motion – a theory subsequently known as galvanism. Galvani believed that there is an 'electric fluid' that is made in the brain and which passes along nerves to produce stimulation in the muscle fibres – an idea not so far from the truth (for nerves do carry electrical currents which activate muscle contraction), although this electrical activity is merely an aspect of life in higher organisms and not its cause. Galvani discovered that he could generate an electrical current from two different metals placed in contact with one another, constituting a primitive kind of battery. In 1800 this arrangement was developed by Galvani's compatriot Alessandro Volta into the so-called voltaic pile, a stack of copper and zinc plates connected by brine-soaked cloth or card. But whereas Galvani believed that the electricity in his experiments issued from the animal tissue and flowed into the metals, Volta believed (rightly) that it was produced at the junction of the metals themselves. He took a more materialistic view of electricity, seeing it not as some mysterious vital principle but as an ordinary property of matter. Volta's cell was studied by Davy, who deduced the fundamentally electrical nature of chemical reactions and used the pile to isolate several new elements by electrolysis.

Galvanism won public attention through the energetic and rather theatrical promotional efforts of Galvani's nephew and disciple Luigi Aldini, who demonstrated the electrical 'reanimation' of a severed ox's head in front of the Prince of Wales and other British nobles in 1802. Still more compelling and suggestive was Aldini's experiment in 1803, in which he instilled a ghastly semblance of life in the corpse of a criminal who had been recently hanged at Newgate Prison in London and brought swiftly to the College of Surgeons. As Aldini wrote:

The jaw began to quiver, the adjoining muscles were horribly contorted, and the left eye actually opened . . . The action even of those muscles furthest distant from the points of contact with the arc was so much increased almost to give an appearance of re-animation . . . Vitality might, perhaps, have been restored, if many circumstances had not rendered it impossible.

On another occasion, Aldini reported that he had made a corpse rise up as if about to walk.

It was commonly thought at this time that one could travel reversibly across the boundary of life and death; experiments with paralysing drugs had apparently demonstrated as much. In 1814 an explorer named Charles Waterton brought back from an expedition to the Amazon samples of the deadly poison curare, which he proceeded to test on animals. A donkey was 'killed' by a dose and then wholly resuscitated:

A she-ass received the wourali poison [curare] in the shoulder, and died apparently in ten minutes. An incision was then made in its windpipe, and through it the lungs were regularly inflated for two hours with a pair of bellows. Suspended animation returned. The ass held up her head, and looked around; but the inflation being discontinued, she sunk once more in apparent death. The artificial breathing was immediately recommenced, and continued without intermission for two hours more. This saved the ass from final dissolution; she rose up and walked about.

The implication is that 'death' may not be final and that surgical intervention can restore life.

Moreover, the association of electricity with life was widely suspected.* The popular science lecturer Adam Walker, a friend of Joseph Priestley, wrote of electricity that:

Its power of exciting muscular motion in apparently dead animals, as well as of increasing the growth, invigorating the stamina, and reviving diseased vegetation, prove its relationship or affinity to the *living principle*. Though, Proteus-like, it eludes our grasp; plays with our curiosity; tempts enquiry by fallacious appearances and attacks our weakness under so many perplexing subtilties; yet it is impossible not to believe it is the soul of the material world, and the paragon of elements!

Electricity was touted as a wonder cure: the Scottish doctor James Graham created a 'Temple of Health' in London at which people took electric baths (these are still popular in Japan) or sat in chairs that delivered mild shocks. Graham sold an 'aethereal balsam' which contained a gum allegedly mixed with 'ether, electricity, air, or

---

* Not everyone made this connection, however. The English anatomist William Lawrence was dismissive, writing in 1819 that 'the contrast between the animal functions and electric operations is so obvious and forcible that attempts to assimilate them do not demand further notice.' Such opinions were not rare.

magnetism'. He is said to have cured the Duchess of Devonshire's infertility with his 'electrotherapy' in 1779.

One of the most startling claims of 'electrically induced life' appeared in the 1830s, shortly before the second edition of *Frankenstein* was published. A wealthy English gentleman scientist in Somerset named Andrew Crosse reported that he had created insects by passing an electrical current through solutions of mineral salts. Crosse had previously been hailed by the eminent British geologist William Buckland for his studies of 'electrical crystallization' of minerals, and it was in the course of this innocuous work that he made his remarkable discovery. He dripped a solution of potassium silicate and hydrochloric acid over a lump of mineral iron oxide connected to the poles of a voltaic pile, and observed whitish protuberances sprout on the surface of his 'electrified stone'. These developed in a bizarre and quite unforeseen manner. 'On the 26th day', he later wrote, 'each figure assumed the form of a perfect insect, standing erect on a few bristles which formed its tail.' He went on:

It was not until the 28th day, when I plainly perceived these little creatures move their legs, that I felt any surprise, and I must own that when this took place, I was not a little astonished . . . In the course of a few weeks, about a hundred of them made their appearance on the stone.

The smaller of these insects had six legs, the larger ones eight, and they appeared to be of the genus *Acarus* (a kind of spider), 'but of a species not hitherto observed'.

Crosse's experiments were reported in the *Somerset County Gazette* in 1836, prompting a subsequent description in *The Times*. This brought them wide attention, and they became a topic of vigorous debate among British scientists. The findings, still at that stage unpublished, were discussed at the 1836 meeting of the British Association, and in February of the following year it was rumoured that none other than Michael Faraday had reproduced them. That was quite untrue – Faraday hadn't even tried, and he issued rapid denials. He later mentioned Crosse's claims dismissively in a letter to his friend Christian Friedrich Schönbein at Basle, saying, 'With regard to Mr Crosse's insects etc. I do not think anybody believes in them here except perhaps himself and the mass of wonder-lovers.' But the mere hint of Faraday's involvement was enough to sustain credulity, and several genuine attempts were made to reproduce the results. A surgeon named William Henry Weekes in Sandwich, Kent, claimed to have done so, and the relation between electricity and life was grist for the mill at the newly formed London Electrical Society, formed in 1837, of which Crosse became a member. A chemist and surgeon named Andrew Smee tried to establish a new discipline, electro-biology, in which electricity was considered the animating principle of all organic tissues. And almost as though it were required by some peculiar narrative symmetry, an explicit connection to *Frankenstein* was forged when Crosse was visited by Ada Lovelace, Byron's daughter and a friend of Michael Faraday, who portrayed him in an account of her visit as the archetypal dishevelled mad scientist.

Some commentators disapproved of Crosse's work because they considered it atheistic. One clergyman called it 'a very dark business, and such as no Christian man ought to engage in', and both Crosse and Weekes received threats of violence. Crosse vigorously denied these accusations, asserting that his experiments were simply realizing hitherto unknown facets of God's laws. He sought, with little success, to avoid being cast in the role of a modern Faust or Prometheus:

I have met with so much virulence and abuse, so much calumny and mis-representation, in consequence of the experiments which I am about to detail, and which it seems in this 19th century a crime to have made, that I must state . . . for the sake of truth and the science which I follow, that I am neither an 'atheist', or a 'Materialist', nor a 'self imagined creator', but a humble and lowly reverencer of that Great Being, whose laws my accusers seem wholly to have lost sight of.

Crosse's claims were never substantiated. But whatever else one makes of this strange episode, it reveals both how electricity was seen as a vital force and how a suggestion of impiety now clung to the alleged synthesis of any living thing.

The nature of the beast

It is possible that Mary Shelley meant to suggest links between her tale and the alchemical tradition of anthropoeia, although we can't be sure how much of the alchemical imagery in *Frankenstein* is intentional, unconscious, or just coincidental. There is, for example, the wedding motif, recalling the 'chymical wedding' of ingredients that creates the philosopher's stone: first in the monster's desire for a 'bride' and then in the terrible wedding night of Victor and Elizabeth, when the monster kills Elizabeth in retribution for Victor's reneging on his promise to make the creature a mate (a scenario proposed by Percy Shelley). And the dismemberment of this female creature by Victor parallels the mutilation of an allegorical figure in alchemical texts reaching back to Zosimos. There is probably more conscious use of alchemy in Victor's suggestion that 'To examine the causes of life, we must first have recourse to death.' He says that 'I became acquainted with the science of anatomy: but this was not sufficient; I must also observe the natural decay and corruption of the human body.' As we saw earlier, it was only through putrefaction and rebirth that the homunculus could arise.

The monster is, however, no homunculus, no shining little being of light. In some ways the vision Shelley offers, in considerable detail, of the artificial man is quite original, although aspects of it were foreshadowed. Importantly, the creature, although of disturbing size, is not obviously hideous – and yet everyone reacts to it as if it were. Or rather, the true source of its ugliness is never disclosed. Its appearance is all the more terrible because it was supposed to be beautiful: it is as if this aspiration makes the shortfall especially unbearable:

His limbs were in proportion, and I had selected his features as beautiful. Beautiful! Great god! His yellow skin scarcely covered the work of muscles and arteries beneath; his hair was of a lustrous black, and flowing; his teeth of pearly whiteness; but these luxuriances only formed a more horrid contrast with his watery eyes, that seemed almost of the same colour as the dun-white sockets in which they were set, his shrivelled complexion and straight black lips.

In what follows, it is not clear whether Victor's 'breathless horror and disgust' expresses his revulsion at the creature's appearance – in other words, the failure of his effort to make a thing of beauty – or comes from a dawning realization of the enormity of what he has done.

Shelley invites our sympathy for the monster, whose murderous acts are the consequence of Victor's rejection. This is no brute, but a creature who speaks with eloquent anguish of his fate. The justice of his claim to Frankenstein's love is beyond doubt:

How dare you sport thus with life? Do your duty towards me, and I will do mine towards you and the rest of mankind . . . I am thy creature, and I will be even mild and docile to my natural lord and king if thou wilt also perform thy part, the which thou owest me. Oh, Frankenstein, be not equitable to every other and trample upon me alone, to whom thy justice, and even thy clemency and affection, is most due.

As literary critic Chris Baldick says, he is perhaps more human than his creator – a theme revisited many times in subsequent stories of artificial beings, particularly in the robots and androids of science fiction (think, for example, of the mercy shown by the android Roy Batty to his hunter Rick Deckard at the climax of Ridley Scott's 1982 movie *Blade Runner*).

Shelley felt that the perversion of the creature's character brought on by Victor's refusal to 'parent' lay at the heart of the novel. Percy Shelley can be reasonably assumed to speak for his wife too when he wrote, in one of the 'reviews' that close associates of an author felt licensed to produce in the nineteenth century, that the most important moral of the story was this:

Treat a person ill, and he becomes wicked . . . It is thus that, too often in society, those who are best qualified to be its benefactors and its ornaments, are branded by some accident with scorn, and changed, by neglect and solitude of heart, into a scourge and curse.

Frankenstein indeed gives a poor showing of himself in comparison to his creation, and there can be no doubt that Mary Shelley intended that we should condemn his behaviour along Faustian lines: why else would she have subtitled the novel 'The Modern Prometheus'? As Victor warns Captain Walton when he dictates his tale in the Arctic:

Learn from me, if not by my precepts, at least by my example, how dangerous is the acquirement of knowledge, and how much happier that man is who believes his native town to be the world, than he who aspires to become greater than his nature will allow . . . Seek happiness in tranquillity, and avoid ambition, even if it be only the apparently innocent one of distinguishing yourself in science and discoveries.

Anne Mellor sees this as an admirable, 'feminist' critique of scientific over-ambition, asserting that *Frankenstein* is 'our culture's most penetrating literary analysis of the psychology of modern "scientific" man, of the dangers inherent in scientific research, and of the exploitation of nature and of the female implicit in a technological society'. There would certainly be value in a critique of this kind, both then and now; but the analysis on display in Frankenstein's comment is gauchely simplistic, betraying a censorious conservatism reminiscent of reactionary medieval clerics like St Bernard of Clairvaux inveighing against the evils of curiosity. Shelley, via Victor, seems to be saying not only that we must beware the temptation of hubris, but that science, and the acquisition of any knowledge, is inevitably corrupting. Philosopher of science Joachim Schummer argues that the novel 'suggests both psychological and historical determinism, according to which the "seeds of evil" necessarily develop in the course of the scientific endeavour'.

Besides, if Mellor is right to say that Shelley 'contrasted what she considered to be "good" science – the detailed and reverent description of the workings of nature – to what she considered "bad" science, the hubristic manipulation of the elemental forces of nature to serve man's private ends', she is hardly deserving of praise on that account. To Mellor, Shelley celebrates 'that scientific research which attempts to describe accurately the functions of the physical universe' while warning of the dangers of 'that which attempts to *control* or *change* the universe through human intervention'. Mellor thus presents *Frankenstein* as a harbinger of every subsequent technological nightmare, culminating in the Hiroshima bomb. And this, of course, merely reiterates the old prejudice against 'applied' science as opposed to 'pure' – the ancient suspicion of *techne*. Quite aside from the fact that these distinctions of pure and applied are imaginary in the first place (the scientific enterprise is not so neatly segmented), are we really being asked to believe that Erasmus Darwin's natural history did more to benefit humankind than Humphry Davy's miner's lamp?

Mellor's accusation that Victor's experiments 'violated the rhythms of nature' and 'transgressed against nature' – that they are, in a word, unnatural – must be read as another expression of the reified and moralistic vision of an idealized 'nature'. She argues that the novel anticipates the danger inherent in our new-found ability 'to manipulate life-forms in ways previously reserved only to nature and chance'. It is now only a matter of time and social will, Mellor says, before we see 'the replacement of natural childbirth by the mechanical eugenic control systems and baby-breeders envisioned in Aldous Huxley's *Brave New World* or Marge Piercy's *Woman On The Edge of Time*'. As we will see, there is no reason to suppose this at all, and one can't help being struck by the unreflecting conservatism that *Frankenstein* has unleashed in Mellor's otherwise thoughtful analysis. It has, you might say, touched a raw nerve.

Surprisingly, of all the sources for Mary Shelley's novel, the one Chris Baldick discounts is the Faust myth. He does so partly on the basis that there are no actual demons invoked in Frankenstein, and partly because it seems Shelley was unaware of Goethe's retelling until after she had written her book. That is itself surprising, given that she had clearly read *The Sorrows of Young Werther*, on which her monster discourses. In any case, neither point particularly favours Baldick's contention. If, as seems very likely, Shelley wished to make some comment on modern science, it stands to reason that she should eliminate supernatural elements from the story. And Goethe's Faust is of course not the only one – William Godwin himself wrote of the more traditional quasi-historical figure.

To the extent that *Frankenstein* is a Faustian tale, it can be positioned within a body of such works from the late eighteenth and nineteenth centuries that portray obsessive, secretive and over-ambitious chemists or skilled craftsmen – that is, Faust figures who are *artists* in the original sense. Balzac's *La recherche de l'absolu* is one such, and so is Herman Melville's 'The Bell-Tower' (p. 152), a tale of an ancient engineer who makes a mechanical man. In Nathaniel Hawthorne's 'Ethan Brand', the eponymous lime-maker returns to his kiln after having travelled the world in search of the Unpardonable Sin, a quest in which he is rumoured to have received diabolical assistance. Brand has found this sin in his own heart: it is 'the sin of an intellect that triumphed over the sense of brotherhood with man and reverence for God, and sacrificed everything to its own mighty claims'. He has become educated beyond all university professors, but in doing so has lost his humanity:

The Idea that possessed his life had operated as a means of education; it had gone on cultivating his powers to the highest point of which they were susceptible; it had raised him from the level of an unlettered laborer to stand on a star-lit eminence, whither the philosophers of the earth, laden with the lore of universities, might vainly strive to clamber after him. So much for the intellect! But where was the heart? That, indeed, had withered, – had contracted, – had hardened, – had perished! It had ceased to partake of the universal throb. He had lost his hold of the magnetic chain of humanity. He was no longer a brother-man, opening the chambers or the dungeons of our common nature by the key of holy sympathy, which gave him a right to share in all its secrets; he was now a cold observer, looking on mankind as the subject of his experiment, and, at length, converting man and woman to be his puppets, and pulling the wires that moved them to such degrees of crime as were demanded for his study.

Thus Ethan Brand became a fiend. He began to be so from the moment that his moral nature had ceased to keep the pace of improvement with his intellect.

In the end, the morally shattered Brand casts himself into his own lime-burning kiln, like the semi-legendary Empedocles, the cataloguer of the four classical elements, who is said to have leapt into Mount Etna.

The personal and the political

If Mary Shelley's novel is a parable about the dangers of a Promethean desire to make life, it is not that alone; there are many other ways to read it and its influence.

Mary's family history is inflected throughout the story. Most obviously, the murder of Victor's young brother William by the monster lends itself to several, not necessarily exclusive, interpretations. William was the name not only of Mary's father but also of her half-brother by his second marriage (who remained ever in her father's favour), and furthermore of her own son, born several months before the

Swiss trip. One can thus make arguments for fantasies of patricide, fratricide or infanticide – or, it must be said, simply admit that William was a common name, as for that matter was Elizabeth, the name of Victor's ill-fated fiancée but also of Percy Shelley's sister and his mother.

Some have claimed that Victor Frankenstein is a portrait of Percy, although it seems much more likely that Mary's husband is represented by Victor's faithful friend Henry Clerval. Victor seems more obviously interpreted as a representation of Mary's father, who at the time that the book was conceived had abandoned Mary much as Victor disowns his progeny. The ambivalence she must have felt towards William Godwin (to whom the book was dedicated) might, in this view, explain why we are not led wholly to condemn Victor, particularly in the revised 1831 text. The creature, Victor's 'child', is left desperately seeking an affective family bond, which at one point he almost finds in the surrogate family of the De Laceys. Like Mary, the monster has a father but no mother, and the questions he asks could easily have come from Mary's lips: 'Why was I? What was I? Whence did I come?'

Besides the personal, there are literary and religious resonances. For her novel's epigraph Shelley quotes from Adam in Milton's epic of creation and transgression, *Paradise Lost*:

> Did I request thee, Maker, from my clay
> To mould me man? Did I solicit thee
> From darkness to promote me?

This challenge to God's authority harks back to the defiance of the rebel angel Satan, which was regarded by the Romantics as a heroic act of rebellion. Godwin himself took that view: in his *Political Justice* he called Satan's insubordination a 'principled opposition to tyranny'. And Frankenstein's monster is intoxicated by *Paradise Lost*, saying that:

It moved every feeling of wonder and awe that the picture of an omnipotent God warring with his creatures was capable of exciting. I often referred the several situations, as their similarity struck me, to my own. Like Adam, I was apparently united by no link to any other being in existence; but his state was far different from mine in every other respect. He had come forth from the hands of God a perfect creature, happy and prosperous, guarded by the especial care of his Creator; he was allowed to converse with and acquire knowledge from beings of a superior nature, but I was wretched, helpless, and alone. Many times I considered Satan as the fitter emblem of my condition, for often, like him, when I viewed the bliss of my protectors, the bitter gall of envy rose within me.

Despite the monster's qualifying comments, this monologue and the novel's epigraph give ample cause to identify in the creature some spiritual kinship with Adam. But spiritual only; unlike all other humans, the creature is no biological son or daughter of Adam, for he is 'united by no link to any other being in existence'. For devout readers this again suggested the worrisome possibility that the monster might be free from original sin. He *does* commit grave sins, of course, yet these arrive not from his own impulse but because of the injustice meted out by his creator. Could Shelley be implying that Adam's sin too was ultimately his Creator's doing? Or that Satan's fall was spurred by God's indifference and tyranny? There is plenty here to discomfit the believer, although none of it was really new to the anthropoetic tradition.

In the Romantic circles of the day there was something of a cult of Prometheus, who was celebrated as a noble iconoclast. In the year that *Frankenstein* was published, Percy Shelley was working on his own version of the Prometheus myth, *Prometheus Unbound* (1820),* which in its very title promises to overturn Aeschylus' account by celebrating the Titan's defiance of Zeus (here Jupiter):

> Fiend, I defy thee! with a calm, fixed mind,
> All that thou canst inflict I bid thee do;
> Foul Tyrant both of Gods and Human-kind,
> One only being shalt thou not subdue . . .

Byron published his own Gothic Faustian tale *Manfred* in 1817, as well as a poetic telling of the Prometheus myth in which the ancient god exemplifies mankind's sublime ambition:

---

* Brian Aldiss's 1973 science-fiction novel was cheekily called *Frankenstein Unbound*, and it has a scientist from the twenty-first century transported back in time to 1816, where in postmodern fashion he meets Victor Frankenstein, Mary Shelley, Percy Shelley and Byron.

In the endurance, and repulse
Of thine impenetrable Spirit,
Which Earth and Heaven could not convulse,
A mighty lesson we inherit.

Anne Mellor argues, however, that Mary Shelley's novel does not so much draw on this Romantic tradition as challenge it, revealing it (and by extension her husband, Byron and her father) to be in thrall to lofty ambition while neglecting human relationships and failing to face up to responsibilities. It would be a fair criticism – both Byron and Percy Shelley showed beastly selfishness on occasion – but to find this message in *Frankenstein* seems to require not a little hindsight. If it were there, it seemed wholly to escape the notice of Mary's husband, who devoted much effort to promoting his wife's book.

That *Frankenstein* was written by a woman is central to its themes, but some feminist readings speak more to our own times than to Shelley's. The novel provokes a torrent of gender-based paranoia from Mellor.* 'By stealing the female's control over reproduction', she says,

Frankenstein has eliminated the female's primary biological function and source of cultural power. Indeed, for the simple purpose of human survival, Frankenstein has eliminated the necessity to have females at all . . . One of the deepest horrors of this novel is Frankenstein's implicit goal of creating a society for men only . . . there is no reason that the race of immortal beings he hoped to propagate should not be exclusively male.

Even if we set aside the implication here that women have no cultural power unless they reproduce (we will come back to that idea later), this is a strange interpretation of the plot. If by 'human survival' one means perpetuation of creatures made from human parts (a very odd definition, I think you'll agree), Victor Frankenstein seems to have

---

* A simple charge of misogyny and fear of women carries more weight. It's striking, for example, that Victor fears a female creature will, *pace* Eve and Pandora, be even more evil than a male: 'She might become ten thousand times more malignant than her mate and delight, for its own sake, in murder and wretchedness.' Or she might be capricious, deserting the creature for 'the superior beauty of man'.

done away with the necessity of 'ordinary' human males too, so long as the monsters learn how to make themselves, and their organs are endlessly recyclable. (There seems to be no suggestion that his creature is immortal.) More significantly, what the creature craves most is a female mate, and in Frankenstein's refusal to grant that wish it seems more fruitful to discern a metaphor for the father's refusal to allow his child to develop a sexual relationship, rather than the creator's literal refusal to make females. If Frankenstein's 'implicit' goal is to make a male-only society, it is so deeply implicit that he never once hints at it. And while it is certainly a recurrent theme of modern speculations on artificial procreation that they instil visions of single-sex utopias, these are, as Mellor points out, as often all-female as they are all-male (she doesn't specify whether both are equally 'horrific').

Besides, it is immensely important, from the perspective of how *Frankenstein* develops the legend of anthropoeia, that the creature *does* seem in principle to possess the ability to procreate. This, after all, is partly what drives Victor to refuse his monster's demand:

One of the first results of those sympathies for which the daemon thirsted would be children, and a race of devils would be propagated upon the earth who might make the very existence of the species of man a condition precarious and full of terror . . . I shuddered to think that future ages might curse me as their pest, whose selfishness had not hesitated to buy its own peace at the price, perhaps, of the existence of the whole human race.

This is today a very familiar narrative of anthropoetic science fiction: our artificial creations take over the world and cast us aside. Conquest by the synthetic organism is the fear behind a great deal of antipathy to biotechnology: what if genetically modified crops, or engineered microorganisms, or nanotechnological replicators proliferate out of control? This is not to say that such fears are always groundless; but we must recognize that they go back much further than the modern era of biological manipulation. And as far as artificial humans are concerned, their roots surely lie in the fear of the outsider: the terror that another culture or race will overwhelm and supplant our own, be it the Jews, the Muslim infidels or the immigrants. The xenophobia evident in Mary Shelley's letters and diaries may well have played its part in creating this vision: constantly depreciating the locals during her sojourns in continental Europe, she wrote that the French peasants are 'squalid with dirt, their countenances expressing every thing that is disgusting and brutal', while the Germans are 'exceedingly disgusting'.

On the other hand, Mellor's suggestion that '*Frankenstein* is a book about what happens when a man tries to have a baby without a woman' is, so to speak, pregnant with implications – not least because, as she goes on to say, it means 'the novel is profoundly concerned with *natural as opposed to unnatural modes of production and reproduction*' (my italics). It is not entirely obvious that this is indeed what the book was about for Mary Shelley, but there can be little doubt that it became a major theme in the story's later incarnations. Mellor's claim that *Frankenstein* articulates fears of pregnancy is especially pertinent. Will my child kill me, the book asks (as Shelley may at some level have felt she killed her mother)? Might it be monstrous and repulsive? Will that happen even if, or especially if, I try to make it perfect? It is precisely *because Frankenstein* emerged from older myths that it could serve these discourses.

There is also a broader social interpretation of the themes of monstrosity and unnaturalness. The idea that there is a natural order to the world and, by implication, something unnatural and undesirable about its contravention, played out not only in science and theology but in politics. To many people in England in the wake of the English Civil War, the usurpation and execution of the sovereign had the character of a disturbingly unnatural event, no matter what they felt about the iniquities of Charles I's reign. The same feeling surfaced during the turmoil of the French Revolution, which was condemned by conservatives in England for violating the natural order of monarchy. They considered that political systems deviating from traditional hierarchy were not just unnatural but, as a result, monstrous: not only grotesque, misshapen and liable to end badly, but also, following St Augustine's view of monsters, moral aberrations.

The association of political systems with the human body is ancient, but was popularized by Thomas Hobbes in *Leviathan* (1651), published just after and partly motivated by the Civil War. Here he referred explicitly to the state in anthropomorphic terms, and moreover stressed that this composite being was human-made:

For by Art is created that great Leviathan, called a Common-wealth, or State, (in Latine, Civitas) which is but an Artificiall Man; though of greater stature and strength than the Naturall.

To Friedrich Schiller, who along with Goethe laid the foundations of Romanticism, society is 'an ingenious mechanism' made from 'the piecing together of innumerable but lifeless parts'. And as a product of art, this composite state stood at risk of degenerating into monstrosity, as it did in times of revolution. To conservatives such as Edmund Burke, the republic in France was doomed to fail because it cobbled together an unnatural body politic as if by some grotesque anthropoetic magic:

[We] should approach to the faults of the state as to the wounds of a father, with pious awe and trembling solicitude. By this wise prejudice we are taught to look with horror on those children of their country who are prompt rashly to hack that aged parent in pieces, and put him in the kettle of magicians, in hopes that by their poisonous weeds, and wild incanta-tions, they may regenerate the paternal constitution, and renovate their father's life.

Burke even accused the Revolutionaries of being alchemists, sorcerers and 'fanatical chemists'. 'Out of the tomb of the murdered monarchy in France', he wrote, 'has arisen a vast, tremendous unformed spectre.' These arguments were vehemently opposed by those sympathetic to the Revolution, including Mary Wollstonecraft, and they were probably discussed in the Godwin household.

It's not surprising, therefore, that after *Frankenstein* was published, the monster and its presumed mode of creation were made a metaphor for the French republic. To the Calvinist historian and writer Thomas Carlyle:*

France is as a monstrous Galvanic Mass, wherein all sorts of far stranger than chemical galvanic or electric forces and substances are at work; electrifying

---

* The conservative Carlyle had dismissed *Frankenstein* in 1818 (while assuming it was penned by Percy Shelley) after having read only a review of it, saying that it seemed to be just 'another unnatural disgusting fiction'.

one another, positive and negative, filling with electricity your Leyden-jars, – Twenty-five millions in number! As the jars get full, there will, from time to time, be, on slight hint, an explosion.

Some stage adaptations of *Frankenstein* indeed implied parallels between the violence of the monster and the mob. Given Mary Shelley's distaste for the 'squalid' and 'brutal' masses, it is quite conceivable that she intended this. The association was made explicit by Elizabeth Gaskell in *Mary Barton* (1848), where she initiates the long tradition of confusing the monster with its creator:

The actions of the uneducated seem to me typified in those of Frankenstein, that monster of many human qualities, ungifted with a soul, a knowledge of the difference between good and evil.

The people rise up to life; they irritate us, they terrify us, and we become their enemies. Then, in the sorrowful moment of our triumphant power, their eyes gaze on us with a mute reproach. Why have we made them what they are; a powerful monster, yet without the inner means for peace and happiness?

As Baldick points out, this reactionary position rather undermines *Mary Barton*'s supposedly egalitarian message by expressing horror of the working classes if they are free to rebel against their masters. The workers, says Baldick, are here portrayed as 'an unfortunate but morally irresponsible creature which lashes out blindly and mutely at its begetter in the deluded belief that the employers are in some way to blame for its misery'. At the same time, Gaskell reiterates the belief that such an artificial creature can have no soul, and endorses the notion, by then well established, that the monster is a brute with a childlike intelligence.

Moulding the monster

Frankenstein was published anonymously; it was not until the second edition in 1823 that the author's identity was revealed. Unsurprisingly, many were scandalized by it. The *Quarterly Review* called it 'a tissue of horrible and disgusting absurdity', while the *Edinburgh Magazine* said (more significantly) that it was 'bordering too closely on impiety'. 'These volumes', thundered the *British Critic*, 'have neither principle, object, nor moral.'

But it's easy to overplay this kind of negative reaction. The simple fact is that the novel was not particularly widely read (the first edition ran to only 500 copies, and they did not sell quickly), and remained little known until, just five years after its first publication, the story was brought to the theatre. In 1823, the English Opera House on the Strand in London staged *Presumption: or the Fate of Frankenstein*, an adaptation by Richard Brinsley Peake. There was fresh outrage; one leaflet about the play implored:

Do not go to the Opera House to see the Monstrous Drama, founded on the improper work called FRANKENSTEIN!!! Do not take your wives, do not take your daughters, do not take your families!!! – The novel itself is of a decidedly immoral tendency; it treats of a subject which in nature cannot occur. This subject is PREGNANT with mischief; and to prevent the ill-consequences which may result from the promulgation of such dangerous Doctrines, a few zealous friends of morality, and promoters of the Posting-bill (and who are ready to meet the consequences thereof) are using their strongest endeavours.

Needless to say, this is the kind of publicity theatre managers dream of.

Indeed, even Peake recognized that, deliciously parodying his own staging in a piece of frippery written that same year for the London Adelphi Theatre called *Another Piece of Presumption*. The play features a tailor called Frankinstitch, and presents a discussion of the controversy between one Mr Devildum and the Adelphi's stage manager Mr Lee:

Lee: But Mr Devildum – have not I heard that there is something of an immoral tendency in this story?
Devildum: So much the better – every body will come and see it – The moment I told my wife of its being improper she went and laid out her last 2 shillings in the gallery.

The fact that Peake could write and stage such a self-referential frivolity testifies to the impact and notoriety of his *Presumption*. And this was not the only burlesquing of the story to follow in the wake of the play's success.

As the title implies, *Presumption* insisted on telling a simplistic, morally instructive fable: 'The striking moral exhibited in this story', the theatre's publicity stated, 'is the fatal consequence of that presumption which attempts to penetrate, beyond prescribed depths, into the mysteries of nature.' Seek not to know high things, as St Paul put it. Everything in the play is exaggerated to serve this message. Victor Frankenstein becomes the crazed lunatic of later tradition, the mad scientist working in a laboratory filled with flasks of bubbling liquid. Here he is given the wild exclamation when his experiment succeeds: 'It lives! It lives!' He even gets his grotesque (and significantly foreign) assistant, not yet Igor but called Fritz, who drives the point home when he announces that 'like Dr Faustus, my master is raising the Devil'. Frankenstein compares himself to Prometheus, while Clerval confesses to Fritz that he sometimes suspects Frankenstein of being more alchemist than chemist. Alchemy was the animating principle of all the nineteenth-century stage versions.

Despite all this, Mary Shelley seemed to enjoy the show when, notified of its existence by her father, she went along to a London performance. 'The story is not well managed', she said with considerable understatement, but she was impressed with the portrayal of the creature by the actor T. P. Cooke, and admitted that 'I was much amused, & it appeared to excite a breathless eagerness in the audience.' One wonders if her tolerance towards the liberties taken with her story had more than a little to do with the fame that the play brought her.

One of the most effective gimmicks of *Presumption* was to give the role of the monster in the programme no name or title: Cooke was indicated only as playing '——'. The monster was thus an almost ghostly presence, not quite a part of this world. More significantly, it was dumb and shambling, with the mind of an infant. This was no pristine invention, but drew on theatrical archetypes of the time: the Wild Man of preliterate cultures ('savages', as they would have been called, like those encountered by Darwin on the voyage of the *Beagle*), the bumbling and white-faced Clown, the 'fairground freak'. Like *The*

Tempest's Caliban, these characters have an ambiguous ontogeny: they are part human, part alien, powerful yet inarticulate. In fact, Steven Earl Forry, who has examined the stage adaptations in detail, suggests that Caliban was 'perhaps the most formidable influence on this role'. Like Caliban, he says, 'the Creature's deformed body mirrors an evil nature'. If this is so, then it turns Frankenstein by association into an old-fashioned magician, more Prospero than Humphry Davy gone bad.

The stage monster is typified in the frontispiece to Henry Milner's lurid 1826 stage adaptation, *Frankenstein: or, The Man and the Monster!*: ugly, with unkempt hair, and dark-skinned to signify his savagery (although in the play Frankenstein says of his creature 'Instead of the fresh colour of humanity, he wears the livid hue of the damp grave'). Yet the creature is also given the capacity for gentleness that was so notably used in James Whale's myth-defining 1931 film. In Milner's play he rescues Frankenstein's mistress Emmeline and her child when they become lost in a storm, and attempts to befriend the child.

Milner also cemented the tradition of giving Frankenstein a comic assistant, here an absurd, boasting clown called Strutt, recalling the

The frontispiece to Henry Milner's *Frankenstein: or, The Man and the Monster!*

peacock-like Face, the servant of the title character Subtle in Ben Jonson's *The Alchemist*. In a conversation between Strutt and an Italian peasant Lisetta, Milner introduces a crowd-pleasing bawdiness, at the same time suggesting that Frankenstein's asexual creation of a being is unnatural and perverted, even onanistic:

Strutt: I really do think, at least it seems so to me, that my master is making a man.
Lisetta: Making a man! – What is not he alone?
Strutt: Yes, quite alone.

It was said that, as is so often the case, the waggish or buffoonish side-kicks Fritz and Strutt often stole the show.

So Mary Shelley lost control of her creation just as quickly and surely as Victor Frankenstein did. And she capitulated to the consensus, rewriting the novel to make it more explicitly a morality tale about the tragic consequences of hubris. In the introduction to the 1831 edition, despite her insistence that her revisions had introduced no new ideas or circumstances, she offered an old-fashioned, quasi-theological and Faustian reading somewhat at odds with the original text. Her dream vision, she insisted here, was of a 'pale student of unhallowed arts' whose creation 'mock[s] the stupendous mechanism of the Creator of the world'.

But some of Shelley's revisions ended up blurring rather than simplifying the moral of the tale. More was attributed to fate – to predestination, the 'silent workings of immutable laws' that Victor senses in the groaning glaciers of the Alps – than to his free choice. Victor calls his decision to study chemistry a matter of 'chance – or rather the evil influence, the Angel of Destruction'. And his benign instructor in that subject, Waldman, now becomes a Mephisphelean figure, bewitching him with 'words of fate, enounced to destroy me'. As he listens, Victor feels 'as if my soul were grappling with a palpable enemy'. The narrative therefore now tugs in different directions: it is on the one hand more baldly Faustian, warning of the dangers of 'presumption' (Shelley even appropriates this word to describe Frankenstein's actions), while on the other absolving Victor by making him a victim of greater forces, as if in a Greek tragedy. To the extent that he is culpable, it is not because of his personal failure to parent his creation but because he has from the outset transgressed some unwritten 'natural law'.

Many of the changes introduced for stage versions of *Frankenstein* were retained when the story was brought to the cinema screen. The first movie adaptation was directed by J. Searle Dawley in 1910; it was simply called *Frankenstein* and featured the popular actor Charles Ogle as the monster. Ogle's character appears in clownish whiteface, with a misshapen body and twisted grimace and the Wild Man's shock of unruly hair. The sledgehammer morality continues: 'Instead of a perfect human being', reads the text frame, 'the evil in Frankenstein's mind creates a monster.'

The most influential retelling in modern times is James Whale's 1931 movie, which was itself based not directly on the novel but on another stage adaptation called *Frankenstein: An Adventure in the Macabre*, written by Peggy Webling. The genesis was complicated: Webling's play was adapted for Broadway by John Balderston, but Universal Studios bought the movie rights before the play opened. It was Balderston who introduced electricity to Frankenstein's laboratory ('a large intricate machine – like a galvanic battery'), and the crucial reanimation scene takes place in a splendidly Gothic laboratory (in the movie, Universal reused the set from their 1931 movie *Dracula*), witnessed by the hunchbacked servant Fritz as well as by Waldman and Victor's fiancée Elizabeth: Frankenstein's secrecy has given way to grandiose theatricality.

Balderston and Whale also insisted on the Faustian and religious themes. A prologue to the film states that 'We are about to unfold the story of Frankenstein, a man of science, who sought to create a man after his own image, without reckoning upon God.' To emphasize the point, Victor cries out as his creature stirs into life, 'Now I know what it feels like to be God!' – a line that proved too blasphemous for the censor when the movie was first released. Balderston turns Waldman into a moderating moral influence, in whom science and religion are in ideal balance. During the animation sequence, Waldman cries out: 'In the name of Religion, I forbid your experimenting' – to which 'Henry' Frankenstein* replies, 'In the name of Science – remain and verify it!'

---

* Why Frankenstein and Clerval (here given the surname Moritz) have their first names exchanged is not clear. Psychological interpretations beckon, but the names had been garbled in any case by earlier stage adaptations.

The adaptations of Webling and Balderston brought to the foreground a theme that can be discerned in Shelley's novel: that Frankenstein's creature is his monstrous doppelgänger. Henry formalizes the pre-existing conflation of creature and creator when he says, 'I call him by my own name – He *is* Frankenstein.' Webling even stipulated that the creature should appear onstage dressed just like Frankenstein: a parodic Mr Hyde. Forry sees this as a symbol of the warped, self-directed (a)sexuality that attached to contemporaneous research on artificial parthenogenesis and later to cloning: it suggests, he says, that 'the self can only engender the self in a parthenogenetic, even homoerotic form of creation'. Unlike Jekyll, however, Whale's Frankenstein is not destroyed by his shadow self. After plunging from the windmill in the final climax, Frankenstein is unconvincingly seen recovering in bed – an ending apparently imposed by Universal after preview audiences reacted badly to the scientist's demise in an early cut. And after all, this gave scope for the sequels *Bride of Frankenstein* (1935)* and *Son of Frankenstein* (1939).

The hardest blow to the psychological core of Shelley's tale comes from the crude way in which the monster's pathological nature is rationalized. There is no longer the bitter tumult of the rejecting father and abandoned progeny; instead, Fritz simply bungles his assignment to steal the 'perfect brain' from Waldman's medical school. He drops the intended specimen and quickly grabs a replacement jar, which happens to be the brain of a criminal (helpfully labelled '*Disjunctio cerebri* – Abnormal brain'). Needless to say, this all rather undermines the moral purported at the outset: Frankenstein's attempt to make a human is abortive not because it defies God and nature, but because he couldn't get the staff. That, however, is doubtless to think too hard about the matter.

Whale's monster is the lumbering Boris Karloff, his fleshy seams still exposed and his head held on by a bolt through the neck. As in the stage versions, he is inarticulate and brutish. On the one hand this combination of immense strength and childlike intelligence appeals

---

* *Bride of Frankenstein* plays further with the doppelgänger theme. It begins with Mary Shelley telling Byron of a planned sequel to her novel, in which Frankenstein is forced by the evil Dr Pretorius to make the female monster after all. This memorably coiffeured 'bride' was played by Elsa Lanchester – who also played Mary Shelley in the prologue.

to our sympathy in the same manner as poor Lennie in John Steinbeck's *Of Mice and Men*. But on the other hand it removes the creature further from the human sphere. With his jerky movements and semi-mechanical appearance, he is informed by the new, industrial archetype of the artificial human: the prototype is no longer the savage Wild Man, but the robot.

It's tempting to see all this as a reflection of Hollywood's insistence only to paint with the most elementary colours, and that's not altogether unfair. Perhaps we shouldn't make too much of it: few books can claim to have had their subtlety enhanced by a transfer to the screen. But Whale's movie is a reminder that a narrative which pitches science against God still had currency in the early twentieth century. And even if Victor Frankenstein is here explicitly Faustian, he is not the unsympathetic Faust of Spiess's biography in the late sixteenth century. Even as he became more ruthless, as in Peter Cushing's portrayal in Hammer Horror's *Curse of Frankenstein* (1957), he never degenerates into a caricature of pure evil. While superficially we are told to deplore the hubris of anthropoeia, we are still being dared to admire it.

Some critics argue that the 'coarsening' of Shelley's fable, including that which she wrought herself, takes us away from its 'true' meaning. But precisely the opposite is true. There is much in the original *Frankenstein* that is highly personal, wracked by the emotional convulsions of Shelley's own predicament. But in the retellings, in the elisions and misattributions, we see how the story both fits and alters the evolving myth of anthropoeia, just as the different versions of Greek myths or fairy tales elaborate on their psychological meanings. As Chris Baldick says:

The truth of a myth . . . is not to be established by authorizing its earliest versions, but by considering all its versions. The vitality of myths lies precisely in their capacity for change, their adaptability and openness to new combinations of meaning. That series of adaptations, allusions, accretions, analogues, parodies and plain misreadings which follows up on Mary Shelley's novel is not just a supplementary component of the myth; it *is* the myth.

Society had its own views about how this tale was 'meant' to go, and soon enough these popular preconceptions asserted themselves. As of 1982, there had been 130 works of fiction based on *Frankenstein*, along with fifty fiction series, more than forty straight adaptations in film (and eighty that use *Frankenstein* in some way) and more than eighty stage productions. The anthropoetic scientist and his creature have provided a channel through which we may examine our love–hate relationships with the monster and the artificial man.

The creator and his creation

Baldick suggests that Victor Frankenstein provided a template for all subsequent Mad Scientists: 'After *Frankenstein*, the figure of the scientist in fiction has, almost as a rule, to be that of an aspiring young medical student who dabbles in galvanism, and whose long hours in the seclusion of the laboratory engender or reinforce a misanthropic, or at best insensitive, disregard for his social bonds and duties.' And Andrew Tudor, an expert on the cultural influences of horror movies, says that Frankenstein 'towers above all others in the classical development of horror-movie science'. Victor's descendants are not, however, evil megalomaniacs driven by lust for power, but share the ambivalent idealism of defiant Prometheus:

Devoted to the pursuit of knowledge at the expense of humane values, he and his successors (whether or not they bear his name) are permitted the equivocal comfort of defensible scientific motives. 'Where should we be if nobody tried to find out what lies beyond?' asks Henry Frankenstein in [Whale's] *Frankenstein*. 'Have you never wanted to look beyond the clouds and stars, to know what causes trees to bud and what changes darkness into light? But if you talk like that people call you crazy.'

And it is not the mere pursuit of knowledge that drives these deluded savants, but the quest for the ultimate mystery, for Balzac's *absolu*, now not the philosopher's stone but the 'secret of life'. As Dr Niemann puts it in the 1933 movie *The Vampire Bat* (which cannibalized both *Frankenstein* and *Dracula*, not just in subject matter but for its sets and actors):

Is one who has solved the secret of life to be considered mad? Life, created in the laboratory. No mere crystalline growth, but tissue, living, growing tissue that moves, pulsates, and demands food . . . Think of it. I have lifted the veil. I have created life. Wrested the secret of life from life.

The deficiency in these men (all of them men) is not hubris alone, but the fact that their wilful goals are pursued with a furtiveness that erodes behavioural norms. That is certainly the case in the other modern myth of the scientific transmutation of life, Robert Louis Stevenson's *Dr Jekyll and Mr Hyde*, in which Jekyll confesses that:

Had I approached my discovery in a more noble spirit, had I risked the experiment while under the empire of generous or pious aspirations, all must have been otherwise, and from these agonies of death and birth I had come forth an angel instead of a fiend.

The same failing is found in H. G. Wells's monster-maker Dr Moreau, who, even while he was still working in England and enjoying the respect of the scientific community, conducted his abhorrent research stealthily. The 1932 movie *Doctor X* takes the image to camp extremes, creating for the eponymous doctor a medical academy in an isolated old mansion on Long Island where he conducts unorthodox cross-examinations on people suspected of a series of cannibalistic murders. 'The human mind will only stand so much,' says his butler, 'we're all a bit strange up here.'* (This movie, one of many attempts to cash in on the success of Whale's *Frankenstein* film, has a plot involving the creation of 'synthetic flesh'.) These remote, ancient settings establish a subliminal link with the magical roots of anthropoeia. We will later see how early research on IVF, and recent claims of human cloning, were also tainted by the suspicion they were being conducted in seclusion.

The surreptitious, obsessive behaviour exhibited by Frankenstein and his successors is often associated with sexual dysfunction. This is a subtext in *Frankenstein* itself, for the monster literally prevents Victor from finding sexual union with his new wife Elizabeth on their wedding night by murdering her. The whole tenor of that encounter is charged, in Victor's imagination, with sexual energy. He speaks of his monster 'consummating' his crimes by (as Victor anticipates) murdering him,

---

* The campness, including the creepy butler, informs *The Rocky Horror Picture Show* (1975), which namechecks the movie in its opening song: 'Doctor X will build a creature'. The star of *Rocky Horror* is of course Dr Frank-N-Furter, who seeks to create life so that he might make a perfect sexual plaything – an instance of many a myth being spoken in jest.

and there is a perhaps unconscious ambiguity in the way he describes this fear on his wedding night: 'I reflected how fearful the combat which I momentarily expected would be to my wife.' The convoluted syntax seems to betray a horror in Victor's mind at the 'consummation' of his own marriage in sexual struggle. And as Anne Mellor points out, Victor embraces his wife 'with ardour' only after the monster has strangled her: he 'most ardently desires his bride when he knows she is dead' – and therefore when she can no longer pose a sexual threat.

It is arguably a repression of sexual desire that has spawned the monster in the first place, just as Paracelsus warned that misdirected sexual imagination could cause sperm to grow into monstrous shapes inside the body. Mellor observes that 'Frankenstein dedicates himself to his scientific experiment with a passion that can be described only as sexual . . . In place of a heterosexual attachment to Elizabeth, Victor Frankenstein has substituted a homosexual obsession with his creature.' Such libidinal displacement or distortion is a common attribute of mad scientists in general, and of 'people makers' in particular. The sociologist of science Evelyn Fox Keller thinks it is in fact part of the mindset of the (predominantly male) modern scientist: she found in a psychological survey of physicists at Harvard University that many 'feel uncomfortable with their emotions and sexuality'. Make no mistake, then: anthropoeia is considered to come with a dash of sexual perversion or dysfunction.

So much for the maker; what did *Frankenstein* tell us about the created being? Shelley introduces a loathsomeness that was not present in the homunculus tradition, and also latent savagery and acute loneliness. The artificial being stands apart from humankind, and suffers on that account; but he is resentful for this same reason, and the resentment may turn murderous. Shelley's novel represents a bifurcation of the anthropoetic tradition in which the path taken is subsequently only the dark one: human perfectibility by human agency is no longer an option. As D. H. Lawrence wrote:

The magicians knew, at least imaginatively, what it was to create a being out of the intense *will* of the soul. And Mary Shelley, in the midst of the idealists, gives the dark side to the ideal being, showing us Frankenstein's monster. The ideal being was man created by man. And so was the supreme monster.

It remains unclear if Frankenstein's monster has a soul, but there is a strong suspicion that he does not. Martin Willis, a specialist in nine-teenth-century science fiction, argues that the monster should be regarded as little more than a sophisticated, fleshy version of the mechanical automata popular around this time, which I consider in the next chapter. In this sense, says Willis, he is a creation of materi-alist science, which has 'given him life but stripped him of a soul'. Yet Shelley herself leaves the question open: she straddles the materialism of scientists such as Volta and a transcendental vitalism evident in those inclined towards Romanticism (including Galvani), who suspected that electricity might be a spiritual substance, the basis indeed of the human soul. In refusing to be explicit about the crea-ture's soul, Willis says, Shelley artfully brings together both points of view and allows them to coexist in the novel:

In leaving the creation of the monster equivocal, the vital turning point between inertia and animation can be appropriated by either materialist or Romantic science. Moreover, without either of these opposing philosophies able to defend their position from textual evidence, they both exist simulta-neously, caught in a moment of equivalence at the very center of the novel.

Equally significant for the evolving concept of the artificial being are the characteristics that were forced by cultural consensus upon Shelley's 'hideous progeny' (as she called her book). The most impor-tant changes to the monster on which society at large seemed to insist were: that he no longer display the nobility and acute intelli-gence that Shelley's creature shared with Goethe's homunculus, but instead be shambling and inarticulate, an object of pity and fear rather than awe; and that he be very visibly non-human, to the point that his synthetic nature is impossible to misconstrue (he is green/flat-headed/quasimechanical/zombie-like).

These characteristics serve to accentuate what may be the most significant general attribute of the creature: his *unnaturalness*. As Mellor puts it, 'Nature prevents Frankenstein from constructing a normal human being. His unnatural method of reproduction produces an unnatural being.' Mellor even describes this as an act of 'raping nature'. To her, Mary Shelley 'envisions nature as a sacred life-force' (a vision that we are invited to applaud). As a result of his unnatural origin, says Mellor, the creature is evil – although she remains unsure whether he is created that way or is corrupted by the rejection of his maker. Again, this all makes naturalness a moral issue. To find compassion for the stumbling Boris Karloff, we are being challenged to have sympathy for the devil.

5 Descartes' Daughter

This is no Workmanship of Humane Skill, here is no *automaton* made by Art, no *Daedalus's walking Venus*, no *Archytas's Dove*, no *Regiomontanus's Eagle and Fly*. Here is none of *Albertus magnus* or *Frier Bacon's speaking head* or *Paracelsus's Artificial Homuncle*. Here is nothing but what proceeds from a divine Principle and Art, and therefore cannot be reckoned among those mechanical Inventions which have an external Shew of Sensation and Life for a time, but are destitute of a vital Spring.

> John Edwards, *Demonstration of the Existence and Providence of God* (1696)

This has been the fallacy of our age – the assumption that we . . . can create the perfect being and the perfect age . . . But we can *create* nothing. And the thing we can make of our own natures, by our own will, is at the most a pure mechanism, an automaton.

> D. H. Lawrence, *The Symbolic Meaning* (1918)

When a visitor once asked René Descartes if he might see his library, the French philosopher is said to have led the guest to his dissecting room, filled with specimens under examination. 'There', he allegedly said, 'is my library.'

The conceit that nature is the most profound reference source of the natural philosopher is hardly original to Descartes, but it is seldom that he is considered an anatomist. Yet to Darwin's colleague Thomas Henry Huxley, Descartes was 'a physiologist of the first rank'. Huxley argued that Descartes' dualism, the sundering of body and mind or spirit, was informed by first-hand exploration of the workings of living organisms. His studies in anatomy led Descartes to a concept of the human body that was to transform not just medical science but our entire picture of what humanness entails. For the more he looked, the more it seemed to Descartes that the body is an ingeniously wrought mechanism, different in substance but not in principle from the machines that were an increasingly familiar aspect of everyday life in the seventeenth century.

Thus Descartes turned the human body into a collection of tiny machines. Our movements are controlled by muscles arranged like springs and pulleys, activated by 'animal spirits' that flow down the intricate tubes of the nerves. 'I do not recognize any difference', he wrote,

between the machines made by artisans and the various bodies which nature alone constructs, other than that machines that depend only on the effects of certain tubes, or springs, or other instruments, that, having necessarily some proportion to the hands that make them, are always large enough that their shapes and forms can be seen, instead of which the tubes and springs which cause the effects of natural bodies are ordinarily too small to be seen.

Aside from the scale of the mechanics, however, the difference between man and machine is that man possesses a soul or spirit that commands the operations. It sounds like a crucial distinction, and Descartes wanted to make it so. But his mechanistic philosophy inevitably raised the question: can we actually tell apart man and machine?

The implications of the Cartesian conception of human life were shocking. If we are mere machines, albeit wrought with infinite delicacy and skill, then might not a craftsman of sufficient ingenuity produce a mechanical device that, however clumsily in comparison to God's work, lays claim to some semblance of genuine life? The idea removes the creation of life from the sphere of esoteric art, from the incantations of sorcerers and the bizarre recipes of alchemists seeking to capture some hidden vitality immanent in organic matter. Neither need one be a physician or surgeon delving into the recondite secrets of galvanism and chemical physiology. A mere engineer might make life, and do so transparently, the mechanisms exposed for all to see and marvel at. You need not putrefied semen, not blood and guts, but the springs and levers of clockwork.

And these mechanisms were precisely what some remarkable inventors and entrepreneurs used in the seventeenth and eighteenth centuries to produce lifelike machines – automata – that seemed to embody the Cartesian dream. Here was a simulacrum of human life made from inert matter, capable of miracles: playing the flute or playing chess, writing and sketching, speaking and, it would seem, thinking. And if a mechanism could indeed think, what then? *Cogito ergo sum.*

The machine-man

It would be wrong to conclude that Descartes made humankind merely the clockwork toys of God. Indeed, his central tenet was precisely that people were more than that. We are, he said, the union of body and soul: the formula of Cartesian dualism. The body is the most wonderful mechanism, its parts operating just as their engineered analogues do. The heart is a pump, the lungs are bellows, the throat is an acoustic resonator, the limbs are articulated structures pulled by the cables of the muscles. But none of this, Descartes insisted, can move of its own accord. Without the soul, there is no animation.

Beasts, Descartes granted, are truly not much more than mechanical. But what distinguishes humankind is the possession of a rational soul: a feeling, sensitive gift of God. It is not sufficient, he warned, that this soul

be lodged in the human body exactly like a pilot in a ship, unless perhaps to move its members, but that it is necessary for it to be joined and united more closely to the body, in order to have sensations and appetites similar to ours, and thus constitute a real man.

Here Descartes was taking care to avoid any hint of heresy or atheism, for he had been unsettled by the fate of Galileo, confined to house arrest by order of the Church. He emphasized that the endowment of humankind with a rational soul revealed to us God's agency and wisdom:

After the error of those who deny the existence of God, an error which I think I have already sufficiently refuted, there is none that is more powerful in leading feeble minds astray from the straight path of virtue than the supposition that the soul of the brutes [animals] is of the same nature with our own; and consequently that after this life we have nothing to hope for or fear, more than flies and ants; in place of which, when we know how far they differ we much better comprehend the reasons which establish that the soul is of a nature wholly independent of the body, and that consequently it is not liable to die with the latter and, finally, because no other causes are observed capable of destroying it, we are naturally led thence to judge that it is immortal.

Even with such disclaimers, he did not dare publish his mechanistic view of the human body, and his book *Treatise on Man* appeared only posthumously. He confessed that 'I wouldn't want to publish a discourse which had a single word that the Church disapproved of; so I prefer to suppress it rather than publish it in mutilated form.'

Yet a suspicion that Descartes held a purely clockwork view of life clouded his reputation. A story that circulated after his death described how he travelled by sea to Sweden accompanied, so he said, by his daughter Francine. But the ship's crew never saw the girl, and eventually they became unnerved and forced open their passenger's luggage. There they found a lifelike mechanical doll which, in superstitious terror, they threw overboard. Descartes did have a daughter named Francine, born illegitimately to a servant in 1635, who died at the age of five. But the idea that the philosopher sought to recreate her in clockwork has all the hallmarks of apocryphal legend.* That, of course, makes it no less apt a signifier of the preconceptions that Descartes' theories generated at the time.

Besides, others were less reticent about taking these ideas to their logical conclusion. Julien Offray de La Mettrie was a French physician who studied in Holland in 1733 under the great doctor Herman Boerhaave before becoming a military surgeon in Paris. In 1745 he published a book called *The Natural History of the Soul*, which argued for a mechanistic, materialistic position considerably more extreme than that of Descartes. To La Mettrie, there was no need of a soul to animate matter: life was an innate property of matter, not something

---

* Descartes did speculate about making a human automaton operated by magnets, but there is no evidence that he ever tried to do so.

breathed into it. What is the soul, La Mettrie demanded, but 'an empty word to which no idea corresponds'? The human body is a 'self-winding machine, a living representation of perpetual motion', in which the motive force 'resides in . . . the very substance of the parts, not including the veins, arteries, and nerves, in short, the organization of the entire body'. Consequently, 'each part contains in itself springs whose forces are proportioned to its needs.' Everything works automatically, under its own agency. In support of this belief that life did not require the impulse of a soul, La Mettrie cited recent observations of the regeneration of polyps from fragments of their bodies.

The Natural History of the Soul was regarded as outrageous and blasphemous, even atheistic: as historian Aram Vartanian says, 'In the eighteenth century, to deny the soul's immateriality meant more or less to doubt God's existence.' La Mettrie was forced to flee to Leiden in the Netherlands, where in 1747 he published an even more materialistic and inflammatory work. The title said it all: *L'homme machine* (*Man a Machine*). It portrayed men as puppets, no different from beasts, or as La Mettrie put it, 'perpendicularly crawling machines'. The Church ordered the book to be burnt (Protestants in The Hague willingly complied), and La Mettrie had to move again, this time to Berlin, where he was welcomed by the Prussian king Frederick the Great.

Despite his disparagement of the word 'soul', La Mettrie did not quite deny outright its existence in humans. But he did not regard it as divorced from the body and its matter; rather, the soul *emerged* from living matter when it attained sufficient complexity of organization. All that was needed for this to happen was the inherent propensity of organic matter to exhibit regular movement. La Mettrie considered that the fibrous tissues of the body exhibited 'irritability', a kind of characteristic oscillation. The various physiological functions arise from these oscillations in different modes of organization: certain arrangements permit sensation, others perception, others judgement, reason, imagination. One might call it a 'string theory of consciousness'. All of these movements ultimately originate from the heart, the 'mainspring of the watch'. The only reason man has consciousness while animals do not is that we have a more sophisticated arrangement of our irritable fibres.

In this respect La Mettrie can be considered a prophet of the modern concept of emergence as a property of so-called complex systems. The behaviour of such systems arises from the interplay of their component parts, and yet cannot be predicted by considering the components in isolation. Moreover, in assigning a self-determining origin to the order of nature he comes close to the modern scientific position by asserting that this organization need be due neither to design nor to blind chance, but to the operation of natural laws. To La Mettrie, Descartes' dualism was merely a face-saving exercise, 'a trick of skill, a ruse of style, to make theologians swallow a poison'. That poison was precisely what all pious people regarded as special about humanity, which for La Mettrie was simply a denial of the evident fact 'that these proud and vain beings, more distinguished by their pride than by the name of men . . . are at bottom only animals and machines which, though upright, go on all fours'. It seems a far more shocking portrayal of humankind than our modern equivalent of machines controlled by genes.

To the French philosopher Denis Diderot, the mechanistic view of humankind implied strict historical determinism, for it was then nothing more than external forces that make people behave as they do. Even consciousness was not necessary, although we may experience an illusion of it. He imagined a mathematician who gets up, works on geometry, writes and posts letters, and dines with his friends – and yet all the while he

knows nothing, but nothing of what he has done, and I see this man – machine pure and simple – experiencing different motives which have impelled it without having freed it; it has not produced a single expressly voluntary act.

Moving parts

La Mettrie's ideas were more than theoretical speculation, for in the eighteenth century it was widely considered that the mechanical model of man was being explored experimentally. While on the one hand surgeons and anatomists were, like Descartes himself, opening up bodies and finding that their workings could be regarded as so many pumps and cables, at the same time engineers and inventors were approaching the issue from the other end: creating out of inanimate matter 'creatures' that presented an uncanny semblance of life. This was the golden age of automata.

The creation of machines that move and display human-like behaviour is a very old art. As we saw earlier, Daedalus is said in legend to have animated statues by mechanical means, and the citizens of Rhodes were allegedly entertained by moving automata of animals placed for their amusement in public spaces (even the legendary Colossus was rumoured to have been animated). The great Alexandrian engineers Ctesibius and Hero made moving metal creatures, probably powered by pneumatics or hydraulics, while it has been claimed that Mark Antony had a wax model made of the murdered Julius Caesar which, impelled by a hidden mechanism, rose from its funeral bier in front of the terror-stricken population.

The marvellous inventors of ancient China are said to have developed the art to a state of high refinement. In his encyclopaedic *Science and Civilization in China*, Joseph Needham tells of an 'artificer' of the third century BC named Yen Shih, who was asked by a visiting king, Mu of Chou, to show him what he was capable of. Yen Shih promised to bring to the king his latest project, and he duly turned up the following day with an assistant. 'Who is that?' asked Mu, to which Yen Shih replied, 'That, sir, is my own handiwork. He can sing and he can act.' Then the inventor put his machine through its paces, whereupon the automaton went so far as to make advances to the king's concubines. The enraged king would have executed Yen Shih on the spot, had the inventor not immediately dismantled the device to show that indeed it was made of nothing more than 'leather, wood, glue and lacquer'. The machine mimicked the human form in anatomical detail, having 'liver, gall, heart, lungs, spleen, kidneys, stomach and intestines; and over these again, muscles, bones and limbs with their joints, skin, teeth and hair, all of them artificial'. Each of the organs supplied a particular sensible function: without a heart, the automaton could not speak; without a liver, it could not see. The king exclaimed in wonder, 'Can it be that human skill is on a par with that of the great Author of Nature?'

This tradition persisted, if we can trust the notoriously marvel-prone accounts of medieval Europeans. When the fourteenth-century Italian traveller Odoric of Pordenone visited China, he wrote that the khan's palace in Peking contained dancing golden peacocks which, reiterating the demonic suspicions raised by ingenious artifice, he attributed either to 'the diabolical art' or, more prosaically, 'a device under the ground'.

It was commonly thought in the Middle Ages that the ancient world had thronged with mechanical marvels. A twelfth-century account of the Trojan War, *Roman de Troie*, by the French poet Benoît de Sainte-Maure describes how Hector was amused in his 'Chamber of Beauties' by a female acrobatic automaton that 'performed and entertained and danced and capered and gambolled and leapt all day long on top of the pillar, so high up that it is a wonder it did not fall'. And the thirteenth-century writer Gervase of Tilbury told of a bronze fly created by one Bishop Virgilius of Naples, a legendary Christianized amalgam of the Roman poet Virgil, Hero and Merlin. This device chased away all the other insects in Naples, so that food in the city's shops did not spoil for eight years.

Stories of artificial men that seem to be varieties of automata often grew up around famous philosophers and magicians in the Middle Ages. The eleventh-century Jewish scholar Solomon ibn Gabirol was said to possess a female servant made 'only of pieces of wood and hinges'. Albertus Magnus was reputed to have spent thirty years building one from metal, wax, glass and leather in the thirteenth century, 'which appeared so wonderful to the ignorant multitude, as to draw upon the inventor the dangerous imputation of being addicted to magic'. Many of these attributions were no doubt nothing more than the stock rumours attached to anyone who dabbled in experimental science, such as the thirteenth-century Franciscans Robert Grosseteste and Roger Bacon (see p. 147). But this period did delight in mechanical ingenuity of the sort that spawned automata. Nobles delighted in 'water toys' with hydraulic moving parts, such as that made in the thirteenth century for Duke Philippe, Count of Artois, which spouted water, flour and soot over the onlookers and incorporated moving models of apes covered in real hair. The renowned Strasbourg clock, constructed in 1350, sported a bronze cock which opened its beak, stuck out its tongue, and crowed to mark the hour. In the fifteenth century, the German mathematician and astronomer Regiomontanus is said to have presented the Holy Roman Emperor Maximilian I with an iron fly, while Leonardo da Vinci created an animated bronze lion to mark the victorious entry of the French king Louis XII into Milan.

As these examples illustrate, by the time of the Renaissance humanoid automata were no longer quasi-magical objects tainted with suspicions of demonic assistance, but a means for ingenious craftsmen to display their virtuosity and impress potential patrons. Those made by Hans Bullmann of Nuremberg played instruments, and a lute-playing lady fashioned by the clockmaker Gianello Torriano of Cremona still survives in the Kunsthistorisches Museum in Vienna. Torriano made automata to entertain Charles V after his abdication as emperor in 1556, and to judge from the claims made for them, rumour was apt to invest the devices with near-miraculous abilities: his mechanical soldiers fought battles, it was said, while his wooden birds flew around the room.

But if such stories were exaggerated, it was probably not just because that is what tends to happen to any marvellous accounts transmitted by word of mouth. For one thing, the skills of the best automata-makers in the early modern period really were such as to make almost anything seem possible. Furthermore, the new mechanical philosophy seemed to place no attribute of living things beyond the reach of mechanical mimicry, so that these virtuoso engineers seemed truly to be on the verge of breaching the boundary between inert and animate matter. In eighteenth-century Europe the undoubted master of this secretive art, a French engineer named Jacques Vaucanson, regarded himself (and was regarded by others) as a new Prometheus who 'dared to investigate the secrets of creation which . . . had been considered beyond the reach of mankind'.

Vaucanson spent his early life as a monk, but a most unconventional one: as a member of the order of the Minims in a monastery in Lyons, he occupied himself by making mechanical 'androïdes' that could perform astonishing feats. Some are said to have served dinner to a visiting Church dignitary and cleared away the plates afterwards, leading the visitor to declare subsequently that the inventions were profane. Vaucanson quit the religious life and set himself up in Paris, hiring a room in the Hôtel de Longueville where he charged people the considerable entrance fee of three livres to see his machines. The main attraction was a life-sized automaton, made of wood and painted white to resemble a marble statue, which could play several tunes on the flute. This was no easy matter – anyone who has tried to learn that instrument will aver that it demands careful breath control, as well as dexterity and accuracy in the fingering. But Vaucanson's Flute Player had bellows for lungs, and to achieve reliable stopping action on the fingerholes the wooden fingers had to be covered with skin. (It is not clear whether this was human skin – the French word *peau* can also mean leather.)

Vaucanson's salon was a hit in Parisian society. When they compiled the first edition of their famous *Encyclopedia*, Diderot and d'Alembert included the subject 'androïde' (defined as 'an automaton in human form'), under which they described the Flute Player's mechanism in great detail. Yet Vaucanson's follow-up to this popular musical automaton was a very curious invention: a duck that ate food and defecated.

Why would he stoop to making a mere duck, after having created an artificial human capable of playing music? The answer reveals much about Vaucanson's ultimate programme. There is no doubting a strong element of showmanship in his machines, but he was also determined that they should offer more than mere spectacle. If Descartes was right, wasn't he in some sense making, however primitively, an analogue of the real human body? True, the materials were different, and as Descartes had said, it wasn't possible to work at the minute scale of nature. But the basic mechanical principles were the same. What was an automaton, then, but a body without a soul?

If that were so, Vaucanson hoped it would be possible to mechanically simulate all the body's functions. Digestion was seen at that time as one of the most important principles of life: as the initial stage of metabolism, it was essential to all animals, and the way it transformed lifeless matter into living flesh was as mysterious as it was fundamental. A duck that could exhibit digestion would therefore be displaying a crucial aspect of life's processes. It is possible that a duck, as well as being smaller and more manageable than a human, was also far less risky a proposition – one can imagine the response Vaucanson would have received had he paraded before the Parisian gentry a shitting android.

The Duck was allegedly exquisite in its artificial anatomy, every bone of the wings being faithfully replicated. But the digestion turned out subsequently to be achieved by subterfuge: the 'food' given to the automaton passed into one chamber inside the machine, while the bird excreted a pulpy substance pre-prepared and dyed green, held within a different vessel.

(a)

(b)

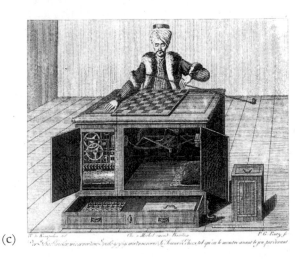

(c)

Eighteenth-century automata: Jacques Vaucanson's Flute Player (*a*), an early photograph of the nineteenth-century remnants of his Duck (*b*), and Kempelen's chessplaying Turk (*c*).

Unaware of this, audiences were astonished. Louis XV came to view the machine-bird, and afterwards he put to Vaucanson a remarkable proposal: could the engineer make a human automaton with circulating blood? Vaucanson was encouraged to think that a vascular automaton might be feasible using rubber, newly discovered in the Amazon and brought back to France in the 1730s. Here, it seemed, was the perfect material for making small, flexible tubes. But it was far from easy to do so from the raw rubber resin, and the plan came to nothing.*

These contrivances have been dismissed by some historians of science and technology as mere gimmicks and toys, irrelevant to serious explorations of how nature worked. But that underestimates the intentions of their makers. Vaucanson's Flute Player, for example, was meant to illustrate his theories of how humans play the instrument. The philosopher Jean Theophilus Desaguliers, who translated into English Vaucanson's description of the device, claimed that it 'gives a better and more intelligible Theory of Wind-Musick than can be met with in large Volumes'. More broadly, eighteenth-century automata offered direct experimental 'proof' that the workings of biology could be given a mechanistic explanation. According to historian Jessica Rifkin,

[they] made manifest, not a reduction of animals to machinery, but a convergence in people's understanding of animals and machines. Not only did they begin to understand animals as machine-like, but they also, at the same time, began to understand machines as animal-like: soft, malleable, sometimes warm, with fluid parts that acted, not only by constraint, but by inner purpose.

Vaucanson sold off his automata before he died, and they were passed around Europe like mummified relics. Gottfried Christoph Beireis, doctor to the Duke of Brunswick and a collector of curiosities, bought them in the late eighteenth century, and Goethe saw them in his

---

* It has been claimed that Vaucanson used rubber tubing for his duck's intestines, a notion supported by the drawing of the duck's mechanism in *Le Monde des Automates* (1928) by Alfred Chapuis and Edouard Gelis. Their reconstruction may, however, be rather fanciful, not least because it hints at a genuine 'simulated digestion' inside the creature.

possession in a dilapidated state. Napoleon offered to buy them, but Beireis would not sell.

Vaucanson had several rivals. The Swiss watchmaker Pierre Jaquet-Droz created a startlingly lifelike automaton in the 1770s that could write with a quill any message up to forty characters long – because of this processing versatility it is sometimes alleged to be the first computer. (It was also so marvellous as to arouse suspicions of its being animated by witchcraft – Jaquet-Droz was investigated on these grounds by the Spanish Inquisition.) Jaquet-Droz and his sons also made an automaton that drew four different sketches in charcoal and another that played the harpsichord. But the claims made by the Hungarian nobleman and engineer Wolfgang von Kempelen went further still. Whereas the movements of the Flute Player, however refined, were predetermined, Kempelen claimed to have constructed an automaton that could *think*. His Chess Player, unveiled in 1769, was, he said, 'for the mind . . . what the Flute Player of M. de Vaucanson is for the ear'. The automaton was about life-size (some say a little smaller, others a little bigger), and dressed like a Turk. It sat behind a cabinet on which a chess set was arrayed, and would play all comers – and most often, defeat them. It beat Napoleon, but lost to the French chess champion François André Danican Philidor.

How was it possible that the chess-playing Turk could make decisions? In the mechanistic era this did not seem as wholly improbable as it must today, although there were rumours that it was operated by a demon. Yet it was not long before Kempelen was accused of fraud; among the sceptics was Edgar Allan Poe, who encountered the device when it was brought to the United States in the nineteenth century after Kempelen's death. These doubts were warranted. In 1834, the French *Magazin pittoresque* published a confession from one of the Turk's controllers – the chess expert Jacques-François Mouret, one-time chess tutor to King Louis Philippe – who had fallen on hard times and was in desperate need of the magazine's fee. Mouret explained that the operators ('directors') would hide away inside the cabinet, moving from one compartment to another as Kempelen opened the box bit by bit apparently to reveal nothing but mechanism. The director would move the Turk's hand by levers while watching the game through a mirror hidden in the automaton's body. This forerunner of the chess-playing computer was, then, not after all a demonstration of the mechanical nature of mind. But it testifies to the spirit of that age that so many were prepared to believe it was.

What is man?

Had Vaucanson succeeded in making an automaton that bled, it would have blurred still further the boundary between artefact and organism. Indeed, it was supposed that it might be used for medical research, for example helping to resolve the argument then raging among French physicians about the value of the ancient practice of bloodletting. In other words, mechanisms that reproduced the workings of the body in sufficient detail could actually stand proxy for them.

Descartes had encouraged this two-way correspondence of man and machine by suggesting that not only was the human body the model for automata, but automata provided a kind of existence proof for the mechanical nature of life. Writing of the way in which severed heads, such as those of executed criminals, continue to move for a few moments, he said in his *Discourse on Method* (1637):

Nor will this appear at all strange to those who are acquainted with the variety of movements performed by the different automata, or moving machines fabricated by human industry, and that with help of but few pieces compared with the great multitude of bones, muscles, nerves, arteries, veins, and other parts that are found in the body of each animal. Such persons will look upon this body as a machine made by the hands of God, which is incomparably better arranged, and adequate to movements more admirable than is any machine of human invention.

The Flute Player and the fraudulent Turk were thus considered to be more than mere clever machines. There was a widespread perception that such automata stood at the brink of life itself, and the more so the more closely they mirrored the human form inside and out. An 'artificial man' was also a goal of the eminent physician Claude-Nicolas Le Cat, who shared a Cartesian, mechanistic view of physiology, and it was claimed that his machine 'will have respiration, circulation, quasi-digestion, secretions and chyle, heart, lungs, liver and bladder, and God forgive us, all that follows from it'. There is a note of anxiety in this remark about the propriety of such an exercise, which was doubt-less exacerbated by the way automata-makers surrounded their creations with an aura of mystery and magic that played to old notions of artificial life conjured by occult powers.

Yet to La Mettrie, the progression from machine to man was simply a matter of introducing greater mechanical complexity:

Man is . . . to the ape . . . what Huyghen's [*sic*] planetary clock is to a repeater watch made by Julien le Roi.* If more gadgets, more cogwheels, more springs were needed to register the movements of the planets than to register the hours, or repeat them: if Vaucanson needed more ingenuity to construct his *flute-player* than for his *duck*, he would have needed still more to construct a *talker* – a machine which can no longer be dismissed as impossible, particu-larly in the hands of a new Prometheus.

Quite what La Mettrie had in mind as a 'talker' is not clear, though some automata-makers did try to give their creations a capacity to make human-like sounds. Kempelen attempted around 1780 to make a speaking head, and in the nineteenth century his Turk was given a primitive 'vocal passage' that allowed him to enunciate, probably indistinctly, the word *echec*, 'check'. Others sought for ways to produce the basic elements of speech, such as vowel sounds, and Friedrich von Krauss of Vienna is said to have made four 'talking heads' in the 1770s. La Mettrie, however, seems to have been thinking not just of a crude mechanical analogue of a voice box, but the ability to frame things to say: to think.

We saw that Descartes was determined to maintain a distinction between automata and humans: only the latter have a rational soul. But how can we tell whether or not an autonomously motile, vocal humanoid houses a soul? Descartes proposed that there are several tests by which true humanity might be revealed. Automata cannot use language, he says, nor can they adapt behaviour to circumstances without prior experience:

---

* The Dutch physicist and mathematician Christiaan Huygens is credited with inventing the pendulum clock, and also devised a balance spring clock, priority for which was furiously disputed by the English scientist Robert Hooke. Huygens owned a marvellous astronomical clock for predicting the motions of the heavens. Julien Le Roy was a famous French watchmaker of the time.

were there such machines exactly resembling in organs and outward form an ape or any other irrational animal, we could have no means of knowing that they were in any respect of a different nature from these animals; but if there were machines bearing the image of our bodies, and capable of imitating our actions as far as it is morally possible, there would still remain two most certain tests whereby to know that they were not therefore really men. Of these the first is that they could never use words or other signs arranged in such a manner as is competent to us in order to declare our thoughts to others: for we may easily conceive a machine to be so constructed that it emits vocables, and even that it emits some correspondent to the action upon it of external objects which cause a change in its organs; for example, if touched in a particular place it may demand what we wish to say to it; if in another it may cry out that it is hurt, and suchlike; but not that it should arrange them variously so as appositely to reply to what is said in its presence, as men of the lowest grade of intellect can do. The second test is, that although such machines might execute many things with equal or perhaps greater perfection than any of us, they would, without doubt, fail in certain others from which it could be discovered that they did not act from knowledge, but solely from the disposition of their organs: for while reason is an universal instrument that is alike available on every occasion, these organs, on the contrary, need a particular arrangement for each particular action; whence it must be morally impossible that there should exist in any machine a diversity of organs sufficient to enable it to act in all the occurrences of life, in the way in which our reason enables us to act.

This is nothing less than a primitive version of the famous Turing test, devised by the British mathematician and pioneer of computing Alan Turing to probe a computer's ability genuinely to think. In a paper published in 1950 called 'Computing Machinery and Intelligence', Turing proposed an 'imitation game' in which a computer attempts to fool an interrogator into thinking it is human by supplying artfully constructed answers to questions. If a computer could pass such a test, Turing said, on what grounds could we insist that it cannot 'think'? The test provided the conceptual basis for the Voight-Kampff machine in Philip K. Dick's 1968 novel *Do Androids Dream of Electric Sheep*, a demonstration of which opens the seminal movie adaptation *Blade Runner*. Here the machine is used to spot the absence, in the test subject, of involuntary human physiological responses when asked emotive questions. The test is used to identify renegade 'replicants', artificial humans who have escaped into the human population. The Voight-Kampff test is thus predicated on the notion that these 'artificial humans' will display an inhuman response to emotional situations, even though the distinction between replicants and humans is never precisely defined: the replicants are flesh and blood, although their means of fabrication is never made clear. Here their artificiality is not so much betrayed by as *defined by* a lack of human empathy.

Huxley's automaton

The Cartesian picture of human biology was adopted in the nineteenth century both by mechanists who ascribed every behaviour to proximate mechanical causes, and by Romantics who, informed by Friedrich Schelling's *Naturphilosophie* that depicted the world as an interplay of cosmic forces in opposition, considered the body to be impelled by some hidden, insubstantial life force. The body was in either event a machine; the only question was what made it move and act.

The curious thing about Descartes' insistence that the soul ultimately pulls the strings of the body is that he imagined it doing so at one remove. For in the Cartesian view it is not the soul itself that causes motion, but rather, a fluid running through the nerves. The soul merely directs the flow. 'It appears to me', he wrote,

to be a very remarkable circumstance that no movement can take place, either in the bodies of beasts, or even in our own, if these bodies have not in themselves all the organs and instruments by means of which the very same movements would be accomplished in a machine. So that, even in us, the spirit, or the soul, does not directly move the limbs, but only determines the course of that very subtle liquid which is called the animal spirits, which, running continually from the heart by the brain into the muscles, is the cause of all the movements of our limbs, and often may cause many motions, one as easily as the other.

This seems to leave, at best, only a very weak role for the soul as the prime mover. It becomes little more than a spectator – and perhaps, by validating a person's humanity, a kind of watermark of God's handiwork. Maybe it has indeed no more of a role than this; in the eighteenth century the Swiss naturalist Charles Bonnet cautiously averred that the hypothesis that 'the soul is a mere spectator of the movements of its body, though perhaps of an excessive boldness, nevertheless deserves some consideration'. Thus, he wrote,

Give the automaton a soul which contemplates its movements, which believes itself to be the author of them, which has different volitions on the occasion of the different movements, and you will on this hypothesis construct a man.

In fact, Bonnet implied that, with sufficient skill, an automaton on its own might mimic every appearance of a human – that it might, to speak anachronistically, pass the Turing test. 'It is not to be denied', he said, 'that Supreme Power could create an automaton which should exactly imitate all the external actions of man.' This was a chilling idea, for it implied that soulless automata could walk among us and we would not know it. The argument works the other way too: if the soul is, as Descartes insisted, separate from the body, and if it is superfluous to human action, perhaps it might be removed from the individual altogether without causing death. That is, of course, precisely the way a zombie is made.

Thomas Huxley sang the praises of Descartes' skill as an anatomist in an essay of 1874 provocatively titled 'On the hypothesis that animals are automata, and its history'. Of course, by Huxley's time – and certainly in Huxley's Darwinian view – 'animals' could be considered to include humans. Using the contemporary habit of calling non-human animals 'brutes', he wrote:

What proof is there that brutes are other than a superior race of marionettes, which eat without pleasure, cry without pain, desire nothing, know nothing, and only simulate intelligence as a bee [in making a geometric comb] simulates a mathematician?

But humankind does know pleasure and pain – or at least, says Huxley, each one of us knows it for ourselves, and is 'justified in assuming its existence in other men'. These things are the products of consciousness, which for Huxley was all that separated man from machine. Yet about consciousness he could say little other than that it appeared in nature by degrees, being evident in some small way in various higher animals too. It was *not* this consciousness, or spirit or soul – call it what you will – that induced a volition to move, however. Rather, said Huxley, these movements are compelled by outside forces acting on the brain, and our sense of free will is simply the mental representation that the brain develops in response to our actions. In other words, volition is little more than an illusion: it is, in fact, a 'mental symbol' of the particular state of the brain that causes us to act a particular way. If that makes me a fatalist and materialist, Huxley wrote, so be it – but then I am no more so than predestinarian theologians such as St Augustine and Jean Calvin, who already implied that man is a 'conscious automaton'. There is, Huxley insisted (citing Gottfried Leibniz in support) no antagonism between this view and Christian orthodoxy.

Huxley considered his supposition that the soul might be equated with consciousness to be not only plausible but demonstrable. He described a soldier mentioned by the French physician Ernest Mesnet, who had suffered temporary partial paralysis after a gunshot wound to the head and subsequently experienced trance-like states in which he would go about his business with no apparent sensation or consciousness. This unfortunate soldier would drink vinegar without a grimace, be insensible of pinpricks, and, if he were writing a letter, would doggedly continue with the text in a mechanical way even if the page he was working on were substituted with a blank one halfway through. Although he would later emerge from this state and resume normal interactions and behaviour, Huxley said unnervingly that the soldier would during that time be a mere machine: 'in the condition of one of Vaucanson's automata – a senseless mechanism worked by molecular changes in his nervous system'.

A person in such a state, said Huxley, is really living two lives: one in which he may be 'eminently virtuous and respectable', the other in which he is capable of anything, of acting without conscience or restraint, of being 'guilty of the most criminal acts'. It isn't clear whether Robert Louis Stevenson knew of Huxley's comments, but this is very evidently the blueprint for his *Dr Jekyll and Mr Hyde* (1886). Mesnet's soldier was also, when entranced, displaying the kind of behaviour that in the twentieth century became associated with the zombie. For philosopher Simon Blackburn, a zombie is a hypothetical conundrum for the philosophy of mind, in that it represents precisely the situation that Huxley depicts: it looks and behaves like you or me but has no true consciousness.* The stereotypical zombie betrays itself with jerky movements and low intelligence (the trope that became attached to Frankenstein's monster); but there is no reason why its 'machine body' should not show all the complex outward behaviour of a fully functioning human. Artificial automata of sufficient ingenuity are thus zombies in all respects save the fact that, in the terms of the Afro-Caribbean culture in which they were originally conceived, they had never once been human.

Living puppets

The carefully stage-managed blend of magic and technical brilliance in the exploits of the automata-makers forged a link between the alchemical and magical claims of Albertus Magnus and Paracelsus and the new mechanistic philosophy. That heady concoction was stirred by the tales of the German writer E. T. A. Hoffmann, which are often considered to be early examples of science fiction.

Hoffmann read about the automata of Vaucanson and Kempelen in a book called *Instruction in Natural Magic* by the German apothecary Johann Christian Wiegleb. A book with such a title would, two centuries earlier, have undoubtedly been a manual of the occult arts, but now it signified nothing more than a how-to account of stage trickery. Hoffmann first deployed the figure of an automaton in a short story simply entitled 'The Automata', in which the narrator encounters such a being in the form of a girl named Adelgunda. The story goes on to describe an automaton called the Talking Turk, invented by a mysterious Professor X, that is clearly based on Kempelen's Chess Player.

Hoffmann's most renowned and influential exploration of the

---

* Blackburn proposes another vision of difference that serves also as an archetype for artificial beings: the Mutant. The Mutant looks and behaves like us, and is conscious, but its consciousness is wholly different from ours: it may experience sensations, colours, emotions, all in ways utterly different from us but which are nonetheless compatible with an outward appearance of normality. Such beings might be deemed to represent the vampire, the faerie changeling, the demon. And again the question arises: how shall we recognize them?

automaton-as-human occurs in his story 'The Sandman'. Here the protagonist, a student named Nathaniel, falls obsessively in love with the beautiful but spiritually blank Olympia, daughter of one Dr Spalanzani, and, inevitably, in fact another automaton. Both Adelgunda and Olympia are more than machines: indistinguishable from humans apart from a disturbingly wooden quality of action and speech, they are semi-magical creatures vitalized by apparently supernatural means (perhaps mesmerism – see below).* Olympia is not so much 'artificial' as 'incomplete': there is no ghost in the machine. Her eyes are 'peculiarly fixed and lifeless', her posture stiff and rigid, like a 'beautiful statue'.

In the nineteenth century, writers of automata narratives saw no need to render their devices scrupulously mechanical, and often spiced these tales with a whiff of the supernatural. Several of Edgar Allan Poe's short stories explore the concept of the body-machine animated by spirit. Poe was fascinated by the idea of zombie-like reanimation of the dead, as in the ghastly apparition of Madeline, buried alive in a tomb, in 'The Fall of the House of Usher'. Elsewhere he examined the machine-man in sophisticated, even satirical ways. 'The Man Who Was Used Up' tells of an old soldier, General Smith, whose body is gradually replaced by mechanical parts over the course of his career. The more machine-like he becomes, the more 'artificial' he sounds. Here he is singing the praises of the machine age:

There is really no end to the march of invention. The most wonderful – the most ingenious – and let me add, Mr – Mr – Thompson, I believe, is your name – let me add, I say the most *useful* – the most truly *useful* – mechanical contrivances are daily springing up like mushrooms, if I may so express myself, or, more figuratively, like – ah – grasshoppers – like grasshoppers, Mr Thompson – about us and ah – ah – ah – around us!

---

* Hoffmann's implausibly lifelike automata are pastiched in Allen Kurzweil's 1992 novel *A Case of Curiosities*, in which a mechanically talented young man named Claude Page sees his mentor, the spiritually wayward Abbé Jean-Baptiste-Pierre-Robert Auget (whose background is reminiscent of the young Jacques Vaucanson), apparently murder a woman in frustration at her ability to play the harpsichord. Claude runs away to become a maker of automata himself, only to discover that the woman is in fact an automaton made by the abbé. Claude goes on to make a 'talking head', a familiar emblem of nineteenth-century automata, that is bizarrely sentenced to the guillotine during the French Revolution. Although Kurzweil has no serious intention to write a Faustian morality tale, nevertheless this tradition is clearly at work here.

This stilted and repetitive articulation anticipates the way, in later decades, we would imagine robots might speak.

When the narrator eventually sees the general without his prosthetics, there is little that is recognizably human about him at all: he is an 'odd looking bundle' who proceeds to construct himself out of artificial hair, eyes, teeth and limbs. He has become metaphorically Cartesian, a mechanical body animated by a shapeless human spirit. But Poe's story prompts the question: which is now in control, the mind or the machine?

Poe was captivated by a further ingredient in the debate over whether man is a conscious machine. The eighteenth-century German physician Franz Anton Mesmer asserted that the spiritual life force that animated the man-machine was 'animal magnetism', a concept that drew on Luigi Galvani's demonstration of the movement of dead limbs induced by 'electrical fluid'. Animal magnetism, although deemed in some ways to be as 'mechanical' as electricity – which could be stored in the Leiden jar like so much water – was an altogether more occult force in the literal sense. Its flow could, according to Mesmer, be controlled by the mind of another individual, so that people could be placed in a trance-like state in which they lost control of their consciousness and volition – 'mesmerized', as this condition was later known. The mesmerized individual was like a zombie impelled by someone else's will, a human puppet jerked this way and that. Here was another way of reducing the human to a mechanism – and by implication, perhaps animating mechanics into a kind of humanity.

In 'The Case of M. Valdemar', Poe, writing at a time when enthusiasm for mesmerism was at its height in the United States, implies that the mesmeric trance can bridge the gulf between life and death. For Valdemar is dead in body, but by his own dying request kept alive in consciousness by a mesmerist. This dreadful state of limbo, which arrests physical decay, is again the condition of the zombie. The mesmeric force described here is an incorporeal influence capable of sustaining the flesh, which disintegrates in an instant once the force is relinquished.

Poe returned to mesmerism in 'A Tale of the Ragged Mountains', in which a man named Augustus Bedloe is mesmerized remotely by a Dr Templeton and forced to do his bidding until Templeton's power is broken and Bedloe regains volition with 'a shock as of a galvanic battery'. There were stories circulating at that time of remote mind control: the British mesmerist Chauncey Hare Townshend, a friend of Charles Dickens, was alleged to possess this power.

Industry mesmerized

Towards the end of the nineteenth century, the quasi-scientific animation of a mechanical body was given an outlandish exposition in the 1886 symbolist novel *L'Eve future* (*Tomorrow's Eve*), written by the hitherto unknown Frenchman Auguste Villiers de l'Isle-Adam. Villiers astutely understood that the dawning age of industrial technology was not about to banish the older dreams of mesmerism and magical automata but would merge with them to yield new incarnations of the artificial being. Villiers had these myths cohere in the person of a real-life engineer and inventor who combined the technical skill of Vaucanson with the mystique of Albertus Magnus: a bona fide 'electrified Paracelsus', the American entrepreneur Thomas Alva Edison.

In *Tomorrow's Eve*, Edison creates an artificial woman, a 'magneto-electric entity' called Hadaly who is part robot (made from unearthly new materials) and part spiritual avatar. But, as Villiers carefully explained, this virtuoso is not the real Thomas Edison but a fictional portrayal of the mythical persona that the famous inventor had deliberately constructed. By the 1880s, Edison was widely renowned as a wizard who could do anything: he was dubbed the 'Sorcerer of Menlo Park', and was compared to Faust. The inventor of the electric light bulb was in many ways a practical man, interested in solving problems so as to make modern life more convenient and entertaining. But he actively cultivated his image as a wonder-worker from whose laboratories came marvels that defied the comprehension of the ordinary man. This, said Villiers in defence of his novel's gambit, made Edison fair game for reinvention in fiction: 'If Doctor Johann Faust had been living in the age of Goethe and had given rise to his symbolic legend at that time', he said in the novel's preface, 'wouldn't the writing of *Faust*, even then, have been a perfectly legitimate undertaking?'

As an example of his almost supernatural power, the real Edison could conjure up the ghostly presence of particular individuals by recording their voice with his phonograph, a device that imprinted sound as grooves on a tinfoil sheet wrapped around a cylinder. From the outset Edison was keen to create the impression of a guiding intelligence behind the disembodied voice: he demonstrated the machine in 1877 at the offices of *Scientific American* by having an employee carry it in, place it on the desk, and turn the crank to relay the message: 'Good morning. How do you do? How do you like the phonograph?'

There was a suspicion that the phonograph had the power to challenge death itself. A reporter from the *New York Post* said to Edison:

What a pity you hadn't invented it before. There is many a mother mourning her dead boy or girl who would give the world could she hear their living voices again – a miracle your phonograph makes possible.

The talk of 'miracles' makes it plain that this invention was indeed regarded as something quasi-magical: the implication here is that the phonograph could work retrospectively as though summoning a spirit in the manner of a Victorian medium (for would a mother really have taken the precaution of pre-recording her child as an insurance against death?).

Having separated the human presence (or at least, the illusion of it) from the body, Edison's obvious next step was to combine this artificial voice with an artificial body. But whereas the likes of Vaucanson might have attempted to create a unique, bespoke automaton with a synthetic voice box, for Edison that artefact had to be mass-produced and marketed to the world: a commodity anyone could own. In 1878 he announced plans for the first talking doll, its body stamped out of metal sheet and a miniature phonograph inserted in its guts, wound with a rather obtrusive key jutting from its back. The Edison dolls not only looked hideous but were also too expensive and too heavy to be playthings for children. They rolled off the purpose-built production line, but it is not clear how many were ever sold, or what they were used for, or indeed why they have all now disappeared.

Edison's phonograph mechanism rendered obsolete the plans of automata-makers to create human speech by reproducing the human vocal apparatus: no longer was it deemed necessary for the automaton to be a copy of the human being inside and out, to copy the physiological mechanics as well as their effects. Nonetheless, there was something almost more uncanny about the way Edison made his talking dolls. Every phonograph cylinder had to be recorded individually, and so each was a trace of a unique human vocalization. In

Edison's doll factory, cohorts of women sat in booths reciting nursery rhymes into a machine, a rite enacted for the purposes of assembly-line infantile gratification.

In *Tomorrow's Eve*, the fictionalized Edison is approached by the English aristocrat Lord Ewald, to whom he is indebted for some past favour, with a request for help. Ewald is in love with a beautiful woman, Alicia Clary – or rather, Ewald adores her physically but knows that the woman is naive, vulgar, stupid and devoid of emotion. The despairing Ewald is driven to contemplate suicide; but Edison says he can help. He takes Ewald to see his android Hadaly, who exists in a nebulous state somewhere between the mechanical and the spiritual: a 'possibility', a 'magneto-electric entity' waiting to be given physical form. Like Edison's dolls, her voice is recorded on a phonograph (albeit etched in gold) which contains twenty hours of 'celestial chatter' – apparently as much as any man might require from a woman. Her joints are made of metal, her body powered by wires ('exact imitations of our nerves, arteries and veins') that carry electromagnetic currents. She is called the 'legacy of Prometheus' – by which we are to understand that she is a scientific mechanism, partaking of nothing divine. To Ewald's worries about the impiety of this creation, Edison replies that the age of science has banished such fears:

Since our gods and our hopes are no longer anything but *scientific*, why shouldn't our loves be so too? In place of that Eve of the forgotten legend, the legend despised and discredited by science, I offer you a scientific Eve.

Edison proposes that Hadaly be incarnated in the form of Alicia: that the human be merged with the 'machine', although this is presented simply as a process in which Alicia acts as a model so that Hadaly's body might perfectly replicate hers. Ewald frets over whether this is proper, and when one evening he meets Alicia and talks with her, he finds that after all she is more alluring than he had remembered. He warns her of Edison's plan, whereupon she replies: 'Dear friend, don't you recognize me? I am Hadaly.'

Yet this is after all no mere machine, for Hadaly has been somehow imbued with the spirit of an entranced woman called Anny Anderson (who calls her 'spirit' self Sowana), by means of the 'very ancient and very new' science of Human Magnetism. In this way,

Hadaly-Alicia possesses something like a soul. She and Ewald leave
Menlo Park and the United States to return to Britain, with Hadaly
packed, like Descartes' daughter, in the luggage as though in a coffin.
But their ship catches fire and Ewald is restrained by the crew from
rescuing his 'bags'. So Hadaly-Alicia is destroyed.

Villiers displays an acute awareness of the pedigree of his story.
Ewald refers to Faust, and says when he first enters Edison's laboratory at Menlo Park, 'It seems to me that I've come into the world of
Flamel, Paracelsus, or Raymond Lull, the magicians and alchemists
of the Middle Ages'. When Edison explains his scheme, Ewald protests,
'But my name is not Prometheus' – to which Edison replies, 'Every
man bears the name of Prometheus without knowing it.' That is to
say, making people is no longer the work of the gods, but may be
done by anyone who is prepared to roll up his sleeves.

Vaucanson is mentioned too, only to be dismissed by Edison as the
creator of crude clockworks. And the epigraph to the book's second
chapter comes from Hoffmann's 'The Sandman': ''Tis he! . . . Ah! said
I, opening my eyes wide in the dark, it is the Sand-Man!' Meanwhile,
Villiers positions his artificial being not only in the context of modern
science but of the popular belief, shared by several eminent Victorian
scientists, in a spiritual world. At one stage he invokes the English
chemist William Crookes, a pioneer of the cathode-ray tube that led
to the discovery of X-rays (a phenomenon itself seen as quasi-mystical),
and a supporter of psychic research. Hadaly is, then, a confluence of
alchemy, galvanism, Frankenstein's monster, Victorian spiritualism,
automata and industrialization, all gathered into an image that prefigures the ambiguously sexual female robot of Fritz Lang's *Metropolis*.

And that too is a significant innovation in Villiers' tale: the artificial being is not a monster, nor a creepy machine, but an object of
erotic desire. Hoffmann's hapless Nathaniel felt passion for Olympia,
but not until Villiers' time (and even then controversially) was it possible
to allow naked lust to enter the picture. Hadaly is an entity made to
fulfil sexual longing and, more specifically, to indulge the misogynistic
dream of a 'perfect woman'. Villiers introduces the notion that one
reason to make artificial beings is for sexual recreation, an idea that
has become standard fare in science fiction and grist for the contemporary anthropoetic mill. Hadaly represents the belief that through art
the human form can be perfected, idealized, made to measure – in

itself an ancient image, recorded in the myth of Pygmalion. It is no coincidence that at the same time as Villiers' book was published and Edison's talking dolls began to reach the shops, a market was emerging for sex dolls and sex toys, now made not from cold, uninviting wood and steel but from a new commodity material: soft, yielding rubber. Some appreciated at once the connection to the old art of automata-construction. According to the German sexologist Iwan Bloch:

There exist true Vaucansons in this province of pornographic technology, clever mechanics who, from rubber and other plastic materials, prepare entire male or female bodies . . . Such artificial human beings are actually offered for sale in the catalogue of certain manufacturers of 'Parisian rubber articles'.

And Bloch understood that the allure of these 'artificial human beings', as he so strikingly characterized them, was their mindless compliance. He describes an 'erotic romance' called *La Femme Endormie* (which could be translated either as the 'sleeping' or the 'deadened' woman) by one Madame B., published in Paris in 1899, 'the love heroine of which is such an artificial doll, which, as the author in the introduction tells us, can be employed for all possible sexual artificialities, without, like a living woman, resisting them in any way.'

The uncanny

These sexual fantasies have a decidedly transgressive aspect. In part, they confront the taboo of coupling with non-human beings; in part they elicit the frisson of necrophilia. But perhaps the spiciest ingredient in this illicit mix is something that only artificial humans can offer. For while the synthetic being made by Victor Frankenstein provokes revulsion and fear, the automaton that verges on a lifelike state evokes a quite different response: a sense of the uncanny. We recoil not out of physical fear but because there is something about these creatures that is 'not right', that makes our flesh creep and, in the end, leaves us uncertain about the nature of our own humanity.

Hoffmann describes this feeling in 'The Automata':

Picture to yourselves the most exquisite figure, and the most marvellously beautiful face; but her cheeks and lips wear a deathly pallor, and she moves gently, softly, slowly, with measured steps; and then, when you hear a low-toned word from her scarcely opened lips you feel a sort of shudder of spectral awe.

Spalanzani's Olympia is said to be disliked, on account of her vacant gaze and 'unpleasantly perfect' appearance, by everyone but the besotted Nathaniel. And even the horror that Frankenstein's monster awakens in all who see him also goes beyond what might be expected from his physical appearance alone, seeming instead to stem from the sense of unnaturalness that is always held to emanate from artificial people. Villiers' Edison states explicitly that the early mechanical automata had this effect (he claims that his electromagnetic creation does not, but the reader might disagree):

Vaucanson . . . and all that crowd were barely competent makers of scarecrows. Their automata deserve to be exhibited in the most hideous of wax museums . . . Just call to mind that succession of jerky extravagant movements, reminiscent of Nuremberg dolls! The absurdity of their shapes and colours! Their animation, as of wigmakers' dummies! That noise of the key in the mechanism! The sensation of vacancy! In a word, everything in these abominable masquerades produces in us a sense of horror and shame.

The same response is documented by today's robot designers, who talk of an 'uncanny valley' that is encountered when humanoid machines resemble real people too closely. This effect was first noted in 1970 by Japanese robotics engineer Masahiro Mori, who claimed that, while making industrial robots more human-like at first increases our sense of familiarity and ease, these feelings turn to intense distaste as the likeness approaches totality. It is in the depths of the uncanny valley, Mori said, that we meet the zombie, the person without a soul. If our robots and automata ape the human form too well, they threaten our view of ourselves as distinct from them: the Cartesian conceit vanishes and we see, as if in a mirror darkly, La Mettrie's perpendicularly crawling man-machine. The uncanny-valley effect was illustrated in the 1970s when the United States government devised an early teleconferencing system for their emergency procedures. To represent the person speaking at any moment, a plastic head was added to the video display. The result, according to computer scientist

Nicholas Negroponte, was that 'video recordings generated this way gave a so realistic reproduction of the reality that an admiral told me that those talking heads gave him nightmares'.

In an essay called 'Das Unheimliche' ('The Uncanny'), Sigmund Freud considered the origin of this unsettling feeling. It comes, he said, not from a fear of the unknown (which is what we might experience in the presence of monsters), but from a recognition of that which is known very well, yet somehow mutated. The uncanny, he wrote, 'is that class of the frightening which leads back to what is known of old and long familiar'. The German word acknowledges this, for *heimlich* means homely; and as I indicated earlier, the prefix 'un-' is not a mere negation – a way of saying 'not familiar' – but expresses condemnation and a sense of horror.

Freud's essay, published in 1919, was indebted to a 1906 article by the psychiatrist Ernst Jentsch, 'On the psychology of the uncanny'. Jentsch cited as examples of the uncanny 'doubts whether an apparently animate being is really alive; or conversely, whether a lifeless object might not in fact be animate' – for example, waxwork figures, ingenious dolls and automata. To these, Freud added the uncanny effect of epileptic fits and manifestations of insanity, because, he says they 'excite in the spectator the impression of automatic, mechanical processes at work behind the ordinary appearance of mental activity' – they give us, as Mesnet's soldier gave Huxley, a glimpse of humankind bereft of will but not of animation, a mere mindless machine.

What we sense as absent from these beings is, once more, a soul, whether that be regarded as (*pace* Huxley) consciousness or (*pace* Mesmer) an autonomous flow of 'animal magnetism' or (*pace* Descartes) something extracorporeal and spiritual. And this does not merely discomfort but in fact outrages us. We will turn against these creatures, tear them apart as Frankenstein did the bride of his monster, determined that we will either find what we seek in the fragments or wreak just retribution if it is not there. Baudelaire saw this tendency in children playing with their toys:

The child twists and turns his toy, he scratches it, shakes it, bangs it against the wall, throws it on the ground. From time to time, he forces it to resume its mechanical motions, sometimes backwards. Its marvellous life comes to a stop. The child . . . finally prises it open. But *where is its soul?*

6 Protoplasmic Beings

We must not read into [living organisms] either a chemical retort or a soul: we must read into them what is there.
    Claude Bernard

'You would seek me with less fervour', life might say to the biologist, 'if you were sure of finding me.'
    Jean Rostand, *Can Man Be Modified?* (1959)

Percy Shelley wrote in his introduction to the first edition of *Frankenstein* that 'the event on which this fiction is founded has been supposed by Dr Darwin and some of the physiological writers of Germany as not of impossible occurrence'. Dr Darwin here is, needless to say, not Charles, but his grandfather Erasmus, one of the leading zoologists of the late eighteenth century, who had hinted that life might be created by artificial means. Those speculations appeared in Erasmus Darwin's seminal work *Zoonomia, or the Laws of Organic Life*, which also hints at a process of evolution linking all living organisms. In what seems a nod towards old alchemical theory, Darwin argued that decomposing organic matter may be the matrix of new life. Victor Frankenstein can be considered to have, in a sense, merely attempted to speed up this process of rebirth – to quicken dead flesh.

There was clearly nothing new in the idea of making life by art. But since the Renaissance the character of that task looked quite different. The ancient model of the cosmos was biological: it was one vast super-organism, waiting only to channel its vitality into suitable vessels. So for Plato, and also for Paracelsus, life was immanent in all of nature, and it was hardly surprising that they felt it could be awakened in inanimate matter. But this picture was reversed by Enlightenment mechanists such as René Descartes and Thomas Hobbes: man was modelled after the inanimate universe, not the universe after living man. The ancient Neoplatonic dictum 'As above, so below' took on a new meaning: the mechanical, clockwork universe was reflected in a mechanical model of the body.

In the eighteenth and early nineteenth centuries, however, there emerged yet another way to regard the nature of life: not as a spiritual, transcendental endowment, nor as the operation of mechanical processes, but as a corollary of chemical composition. While the Cartesian picture spawned hopes that something resembling human life might be produced in a sufficiently complicated machine, chemical physiology implied that making life was more akin to cookery. It needed the right ingredients. How the most primitive manifestations of life, such as fungi, were related to humankind – let alone whether one could progress from the former to the latter – was as problematic theologically as it was scientifically. But if Charles (or indeed Erasmus) Darwin was to be believed, this progression was merely a matter of degree: the first step towards making humans was to make life from inanimate matter.

Vital thoughts

The ability of living matter to move, grow and replicate itself was what made it special, distinguishing it from stone and steel. While Newtonian mechanics explained the movements of bodies by the action of external forces, the impulse and motion of living organisms was autonomous, coming from within. It was literally an occult force, not only invisible to the eye but refusing also to reveal itself beneath the scalpel. Take an organism apart, and there is nothing there to betray the origins of its animation. Faced with this conundrum, natural philosophers could do no better than Aristotle in simply giving a name to that which they could neither observe nor understand: Aristotle's *psyche* became a mysterious 'vital force'.

There was never one single doctrine of vitalism, but rather many diverse ways in which the substance of living organisms was deemed to be 'alive'. Were the vital forces natural or supernatural? Do they involve sentience, or just movement? Or are sentience and intelligence in fact emergent properties of matter that has the self-determining ability to move?

To the French naturalist George-Louis Leclerc, the Comte de Buffon, living bodies are distinct from inanimate matter in being composed of *matière vive*, whose particles have a propensity to move from the centre to the edge of the bodies they constitute. Inanimate matter (*matière brut*), he said, is but *matière vive* that has died, expressing a belief in the continuity of all matter. To Buffon, the familiar manifestations of life are caused by the accumulated behaviour of all the living particles:

The life of the whole (animal or vegetable) would seem to be only the result of all the actions, all the separate little lives, if I may be permitted so to express myself, of each of those active molecules whose life is primitive and apparently indestructible.

The French philosopher Pierre Louis Moreau de Maupertuis went further, ascribing to these 'corpuscles' not only life but intelligence – for how else could the beings made from them be intelligent? He argued that this intelligence appears by degrees in matter, from stones to plants to animals. And Maupertuis maintained that the degree of intelligence attained by the whole organism was not merely a matter of *amount*, but depends also on the arrangement of the particles: it was a consequence of organization.

This belief in the immanent sensibility of the ingredients of organic matter was shared by Denis Diderot, and it led him to regard each individual organ of the human body as a kind of 'animal'. He explores this notion in his curious drama *D'Alembert's Dream*, in which he depicts his former friend and collaborator on the famous *Encyclopedia* musing on the nature of life while lying delirious in bed. In this story d'Alembert's musings are reported by Julie de Lespinasse, a socialite who ran a fashionable salon attended by d'Alembert and with whom he became infatuated, eventually taking up residence in her house. D'Alembert is seemingly trying to reason how an animal can be constructed from the union of 'living molecules':

Nothing at first, then a living point . . . Another living point attaches itself to this one, and then another – and from these successive conjoinings a single living unity results, for I am certainly a unity . . . It's certain that contact between two living molecules is something different from the contiguity of two inert masses . . . and this difference – what could it be? . . . a customary action and reaction . . . That way everything comes together to produce a sort of unity which exists only in an animal . . . My goodness, if this isn't the truth, it's really close to it.

He compares the coherent motion of this aggregate of 'living points' to that of a swarm of bees. If this is how organisms are constructed, then it should be possible to take apart the aggregate and put it back together in new configurations. This leads Diderot's d'Alembert to muse on the possibility of breeding people from fragments of tissue – thoughts that bear an uncanny presentiment of later visions of tissue culture and cloning (see Chapter 11):

Man splitting himself up into an infinity of atomized men which we could keep between sheets of paper like eggs from insects which spin their cocoons, remain for a certain period in the chrysalis state, pierce through their cocoons, and escape as butterflies – a human society formed and an entire region populated by the fragments of a single individual – all that is very pleasant to imagine . . . Who knows what new race could result some day from such a huge heap of sensitive and living points?

Indeed, d'Alembert ends up imagining a factory that makes people to order – a kind of cosy Enlightenment version of Aldous Huxley's chilling Hatchery (p. 165): 'A warm room, lined with small container cups, and on each of these cups a label: warriors, magistrates, philosophers, poets, cup of courtiers, cup of prostitutes, cup of kings.'

These speculations bring d'Alembert to a remarkable conclusion:

If there's a place where the human being divides itself into an infinity of human animalcules, people there should be less reluctant to die. It's so easy to make up for the loss of a person that death should cause little regret.

In other words, the ability to remake ourselves offers a consolation for mortality. Indeed, mortality itself becomes an illusion, for life never departs from the vital corpuscles, but resumes once they adopt a new configuration.

To the extent that vitalism 'atomized' the view of life, locating it in the agitation of its constituent particles, it removed the need for a 'spirit' or 'soul' to be added to base matter, as Descartes had supposed. Rather, the life force was something already present in organic matter, provided that it had the right composition and organization. In this sense, vitalism promised to reduce the mystery of life to a scientific question: it did not necessarily exclude the idea of a human soul, but that was no longer demanded. The German embryologist Caspar Friedrich Wolff supposed that the basic life-matter in cells is organized and arranged during embryonic development by a vital force called *vis essentialis*, while the Swedish chemist Jöns Jakob Berzelius suggested in 1812 that 'This *power to live* belongs not to the constituent parts of our bodies, nor does it belong to them as an instrument, neither is it a simple power; but the result of the mutual operation of the instruments and rudiments on one another.' These words, prescient in retrospect, were struggling to find a foothold in meaning at the time, since the phenomenon of life seemed so remarkable and imponderable that Berzelius spoke for many in proclaiming that its cause 'certainly will never be found'.

But doctrines of vitalism had a glaring flaw: in the end they amounted to little more than a tautology. Organic matter, they insisted, was alive because its constituents were alive. The difficult questions were simply displaced, and 'life' was still ultimately a mysterious, numinous, unknowable quality. As the German physiologist Johann Christian Reil, a sceptic of vitalism, put it in 1796, 'We look for the cause of animal phenomena in a transcendental substrate, in a soul, in a general world spirit, in a vital force, which we imagine as something incorporeal.' And from there, he implied, we can go no further: vitalism is an admission of defeat.

Belief in vital forces was gradually eroded in the nineteenth century by the demonstrations of chemists that there was no substantial difference in the chemical nature of organic and inorganic matter. Its demise is generally traced back to the synthesis of urea, hitherto a product of living organisms, from inorganic ingredients by the German chemist Friedrich Wöhler in 1828. But in fact it took the greater part of the century – and experiments such as Louis Pasteur's disproof of spontaneous generation in the late 1850s – before most scientists abandoned the principle of vitalism. It was still evident in some scientific writings in the early twentieth century.

Stuff of life

Several vitalistic theories in the eighteenth century dwelt on the notion that life exists in a single, fundamental 'living material', like Buffon's *matière vive* – the eminent English physician John Hunter merely translated this idea into Latin by postulating a *materia vitae*. Such life-matter was in itself a remnant of the medieval identification of life with a fluid substance, sometimes called *liquor vitae*. In 1835 the French anatomist Felix Dujardin claimed to have made the primordial life-matter by crushing microscopic animals to obtain a sort of living jelly: a 'pulpy, homogeneous, gelatinous substance', insoluble, sticky and slimy, which he called *sarcode*. And in 1839 the Czech physiologist Jan Evangelista Purkinje introduced a term for the fundamental vital substance that, in various forms, persisted until the early twentieth century: protoplasm. For Purkinje this was the jelly-like substance held within ova and embryonic cells, granules of which represented the building blocks of all organisms.* This protoplasm eventually became equated with Dujardin's *sarcode*. Austrian botanist Franz Unger claimed that it was a kind of protein – then known only as an organic material with a particular chemical composition, rich in nitrogen – and called it a 'self-moving wheel', able to propel itself.

In the 1850s Thomas Henry Huxley identified protoplasm as an organic slime in which tiny marine organisms were apparently embedded in sediments on the sea floor. This, he said, was 'the physical basis of life', the 'one kind of matter which is common to all living things'. In a phrase unconsciously redolent of the old debates about nature and art and the myths of men made from animated clay, he said that protoplasm 'is the clay of the potter: which, bake it and paint it as he will, remains clay, separated by artifice, and not by nature, from the commonest brick or sun-dried clod'.

Only plants, said Huxley, can make protoplasm from raw, lifeless ingredients, even though it contains nothing much more than the four elements carbon, hydrogen, oxygen and nitrogen. His view was profoundly materialistic: 'I can find no intelligible ground for refusing to say that the properties of protoplasm result from the nature and

---

* Purkinje's cellular view of living tissue was similar to the more famous cell theory of Theodore Schwann (see next page), except that Purkinje believed the term 'cell' should be used only for the walled compartments of plants.

disposition of its molecules.' Others objected to such a reductive view of life's material component, pointing out how varied living tissues are in form and composition. The Scottish philosopher James Hutchinson Stirling complained of Huxley's protoplasm that 'if the cornea of the eye and the enamel of the teeth are alike but modified protoplasm, we must be pardoned for thinking more of the adjective [modified] than the substantive [protoplasm]'. The key property of protoplasm, he jeered, seems to be 'infinite non-identity'.

In the event, Huxley's marine protoplasm ended ignominiously, for it turned out to be nothing more than a jelly produced by chemical reaction between seawater and the alcohol used to preserve his specimens. But the general idea, and Huxley's claim, were staunchly championed by Ernst Haeckel, an influential German zoologist working at Jena. Haeckel spread a curious sort of Darwinism in his native country, in which he argued that all matter has a quasi-Hegelian compulsion to organize itself into ever more complex life forms. This was in truth another theory of immanent vitalism. Haeckel attributed 'souls' to all entities known to science: not only to organisms but to their cells, tissues and nerves, and still further down the scale of complexity to crystals, atoms and the light-bearing ether. When liquid crystals were discovered in the 1880s, in which the rod-shaped molecules become spontaneously organized into aligned arrays like logs in a river-borne logjam, Haeckel considered this evidence of the primal lifelike destiny of matter, a position he argued in the strange, semi-mystical book *Crystal Souls: Studies on Inorganic Life* in 1917. For Haeckel, protoplasm itself was a kind of liquid crystal.

That was a decidedly unorthodox view, but the basic principle remained. According to science historian Gerry Geison:

Probably the most popular theory about the nature of life throughout the rest of the nineteenth century was that it resided in a material substance, as Huxley proclaimed, but that this substance when alive was of a composition so uniquely complex that it was unlike ordinary matter.

This idea that all life springs from primordial protoplasm sat somewhat uncomfortably alongside the thesis of German physiologist Theodore Schwann, advanced in the mid-nineteenth century, that all life is composed of cells. The cell theory offered the view that what was special about living matter was not so much a question of chemical composition as of *organization*: cells could be seen under the microscope to be minutely structured, having an encompassing membrane and various indistinct internal structures. They were not, it seemed, mere blobs of matter.

Organization implies a static picture, whereas natural philosophers since Aristotle had tended to make movement one of the defining characteristics of life. Perhaps the movements of 'living particles' could generate vital attributes when harnessed in organized arrangements, rather as the flow of traffic along organized highways permits the business of cities to progress? For the great German organic chemist Justus von Liebig, the 'life force', while differing from other forces in being self-generating, probably had a mechanical origin that resulted from some extraordinary configuration of ordinary matter. And experiments by the Scottish botanist Robert Brown in the 1820s seemed to show that all microscopically particulate matter is spontaneously motile. Brown discovered that pollen grains suspended in water incessantly jump around in random agitation. He thought at first that he was seeing 'vital forces' at work, acting on the 'elementary molecules of organic bodies, first so considered by Buffon' – until he discovered that inanimate particles, such as mineral dust, behave this way too. Movement, it seemed, was ubiquitous in the microscopic world.

Yet could science ever hope to penetrate life's mysteries, or to make life from scratch? The French physiologist Claude Bernard thought not. He pointed out that when chemists recreated products of living systems such as oils and fats, they used processes different from those of nature. 'It can be said', he wrote,

that a chemist in his laboratory and the living organism with its own apparatus both work in the same way, but each with its own tools. The chemist may be able to fabricate the products of the living organism, but will never be able to make its tools, because these are the result of organic morphology, which is ... outside chemistry proper; and under this aspect the chemist cannot make the simplest ferment any more than he can manufacture a whole living being.

Bernard was neither a vitalist nor a materialist: he did not believe that life could be explained solely with reference to immediate, proximate causes. Rather, he felt that life today depended on life yesterday, and likewise back through time to an origin that must forever remain obscure. This was a deterministic process which, once set in train, will keep on running. 'Determinism is the only scientific philosophy possible', he said:

It does in fact forbid us the search for the 'why', as this 'why' is illusory. In return it exempts us from doing as did Faust, who from affirmation threw himself into negation. Like those religious orders where men mortify their flesh by privations, we too, in order to perfect our mind, must mortify it by excluding certain questions and admitting our inability to answer them.

To other scientists, such defeatism was anathema. They would not be told to leave the veil untouched, nor discouraged by comparisons with Faust.

For at least half a century the protoplasm model – life as a fundamental vital substance – remained pre-eminent. But it became more sophisticated than the early hypothesis that protoplasm had only to achieve the right blend of elements in order to become animate. There was in fact no reason to regard cell theory as fundamentally incompatible with the idea of protoplasm: perhaps this substance itself had some special sub-microscopic organization? When the American cell biologist Edmund Beecher Wilson developed Thomas Huxley's theme of 'the physical basis of life' in a book of that title in 1923, he included many images of the delicate substructures of the cell. The term 'protoplasm' was still used, but as Wilson explained, this was now seen as 'a collective term to designate the substances that constitute the active or living material of which cells are composed', adding that 'the more critically we study the question, the more evident does it become that we cannot single out any one particular component of the cell as the living stuff, par excellence'. Life, Wilson said, is a question of 'organization', about which researchers could then do little more than watch the consequences and wonder: 'we are forever conjuring with the word "organization" as a name for that which constitutes the integrating and unifying principles in vital processes'.

Although the microscopic techniques of the Victorian age were incapable of revealing much about the structural details of the cell, chemists were already becoming aware of the role of molecules too small to see. They deduced that chemical processes such as fermentation which seemed to be unique to living systems are effected by reagents called enzymes. The German chemist Eduard Buchner found in 1897 that fermentation enzymes could be extracted from cells and still do their job: their function did not require the integrity of the cell.

Enzymes were found to be members of the class of biological substances known as proteins, which appeared to be highly complicated in chemical terms. The first proteins to be identified – collectively called albumins, and found in substances such as egg white, blood and wheat – were recognized in the late eighteenth century, and skilful analyses of the elemental composition of albumin by the Dutch scientist Gerrit Mulder revealed a formidable chemical formula: $C_{400}H_{620}N_{100}O_{120}P_1S_1$, where the subscripts denote relative proportions of carbon, hydrogen, nitrogen, oxygen, phosphorus and sulphur. Mulder believed that this was the fundamental composition of all albumins, and his colleague Berzelius proposed that they be known by the collective name 'proteins', meaning primary or foremost.

By the start of the twentieth century it was understood that enzymes and other proteins in fact have rather diverse compositions (there is nothing unique about Mulder's formula) and are built up from small-molecule building blocks, called amino acids. The German chemist Emil Fischer showed in the late nineteenth century how amino acids could be joined together by artificial means; but even the tour de force of synthetic chemistry that allowed Fischer by 1906 to make 'pseudo-proteins' composed of amino-acid chains eighteen units long fell far short of the scale of true proteins. Besides, chemists knew next to nothing about how the sequence of amino acids related to a protein's biochemical function.

The German chemist Franz Hofmeister commented in 1901 that studies of the chemistry of enzymes, sugars, proteins and the other classes of biomolecules were converging with microscopic investigations of the structure of cells and their 'protoplasmic' contents:

If on the one hand the morphologist strives to elucidate the structure of the protoplasm up to its finest details, on the other hand the biochemist endeavours, with his apparently grosser and yet more penetrating aids, to ascertain the chemical performance of the same protoplasm, on the whole this still concerns only two different sides of the same coin.

Gradually, attention began to focus on the organization of the enzymes themselves. As the British biochemist Frederick Gowland Hopkins put it in 1913:

It is clear that a special feature of the living cell is the organization of chemical events within it. So long as we are content to conceive of all happenings as occurring within a biogen* or living molecule all directive power can be attributed in some vague sense to its quite special properties. But the last fifteen years have seen grow up a doctrine of a quite different sort . . . I mean the conception that each chemical reaction within the cell is directed and controlled by a specific catalyst [an enzyme] . . . We make a real step forward when we escape from the vagueness which attaches to the 'bioplasmic molecule' considered as the seat of all change.

Compelling evidence emerged that the amino acids in proteins are linked together by strong chemical bonds, rather than, as some thought, being more loosely aggregated. This made proteins 'macromolecules', of enormous size by molecular standards. By the 1930s, X-ray crystallography had started to reveal that these giant molecules have complex three-dimensional structures, on which their precise and delicate chemical behaviour depends. The fundamental ingredients of life were starting to look like tiny machines, every bit as intricately wrought as a watch or a motor car.

This issue of microscopic structure and organization eventually become the dominant framework for understanding what life is. Today the molecules and molecular aggregates of living matter are revealed with atomic precision by X-ray beams and electron microscopes, and the structure of one particular biomolecule – DNA – has been awarded the status of an 'instruction manual' for making a living being. We will see in Chapter 12 why this is, however, a potentially misleading image.

---

* The 'biogen' hypothesis was formulated in 1903 by the German physiologist Max Verworn. He proposed that biogen is a kind of living molecule which is constantly modified by chemical processes within the protoplasm.

Cooking up life

If vitalism made life an intrinsic (though ineffable) property of certain kinds of matter rather than something that had to be mysteriously 'breathed into it', the dawning realization that there was not even a privileged class of 'living matter' at all made it conceivable that life might be created from scratch by chemistry. Indeed, if Darwinian evolution was to be believed then life *must* once have arisen from lifeless material. Some scientists began to speculate about how this quickening commenced in primordial ichor. In an echo of the ancient idea that life on earth began in mud warmed by the sun, Charles Darwin suggested that it might have arisen in some 'warm little pond' containing a fortuitous blend of chemical compounds.

And so scientists set about a task that will have seemed at the same time more and less ambitious than that of making an artificial being: to produce life from inert matter. All of the 'recipes' for human-like beings in the past had relied on the addition of human seed to set the process underway: they were, one might say, a primitive analogue of IVF that began with material already charged with potential life. But Darwin's theory had made humans just a special case of a more general phenomenon: the origin and successive elaboration of life.

An early hint of a chemical synthesis of life was reported in 1890, when Otto Bütschli of the University of Heidelberg claimed to have created a substance resembling protoplasm by the action of potassium carbonate on various oils. This stuff was said not only to look like the protoplasm of amoebae but to move like it too. Drops of the substance moved about in water – an effect most probably due to the way the ingredients affect the liquid's surface tension, and not at all connected to 'vital forces'. Bütschli was circumspect but nonetheless convinced he was on to something, saying:

I am the last person to defend the view that these drops, exhibiting proto-plasma-like movements, are directly comparable to protoplasm. Composed as they are of oil, their substance is entirely different to protoplasm. They may be, however, compared with the latter, in my opinion, firstly with regard to their structure, and secondly with regard to their movements . . . I hope, therefore, that my discovery will be a first step towards approaching the problem of life from the chemico-physical side.

The issue of a chemical origin of life on earth was sharpened by the speculations of the German physiologist Eduard Pflüger, who made pioneering contributions to the understanding of respiration and nerve action in the second half of the nineteenth century. Pflüger believed that life depended on a certain mode of molecular organization in protein, in contrast to the disorganized 'dead' protein in cooked food (such as boiled egg white). He believed that the difference was due to the presence of reactive chemical compounds called cyanogen, composed of carbon and nitrogen, in the 'living' protein. This reduced the origin of life to a relatively simple chemical question: how did cyanogen form? Pflüger wondered whether that might have happened through the combination of primitive substances such as carbon monoxide and ammonia 'when the Earth was still wholly or partially in an igneous or heated state' – one of the first intimations that life's inception might be considered in *geological* terms.

Such a chemical genesis must have happened under very crude conditions. In 1906 the German chemist Walther Löb showed that a mixture of carbon dioxide, carbon monoxide, ammonia and water would react to make simple organic molecules, including the amino acid glycine, when energized by electrical discharges. Several later experiments using similar 'primitive' mixtures and energy sources such as electricity or ultraviolet radiation produced a range of organic substances including amino acids and sugars. An article in *Harper's Magazine* in 1900 captured the mood of these forays into what later became known as prebiotic chemistry: 'It seems well within the range of scientific explanation that the laboratory worker of the future will learn how so to duplicate telluric [geological] conditions that the play of universal forces will build living matter out of the inorganic in the laboratory.'

To make a convincing story, however, one needs to know what those conditions really were: what substances and energy sources were available on the prebiotic earth? J. B. S. Haldane brought his inventive acumen to bear on that question in an article published in the *Rationalist Annual* in 1929. He pointed out that the atmosphere of the early Earth would have contained little or no oxygen, but was mostly (he presumed) carbon dioxide, ammonia and water vapour. Without oxygen, there would have been no ozone layer to block most of the sun's ultraviolet rays, as there is today. 'Now, when ultra-violet light acts on a mixture of water, carbon dioxide, and ammonia,' he wrote, a vast variety of organic substances are made, including sugars and apparently some of the materials from which proteins are built up* . . . The first living or half-living things were probably large molecules synthesized under the influence of the sun's radiation, and only capable of reproduction in the particularly favourable medium in which they originated.

Haldane was already known by that time as a gifted popularizer of science; but his article, published in a relatively obscure magazine, was seen by very few and made little impact. He wasn't alone, however; quite independently a young Russian chemist named Aleksandr Ivanovich Oparin in Moscow was thinking along the same lines. In 1924 Oparin published a booklet called *The Origin of Life*, which again considered the nature of the geological environment on the early Earth and how this might have contained the ingredients for life. He argued that compounds of nitrogen and carbon, as well as hydrocarbons and related organic molecules containing oxygen, could have been formed in the atmosphere and been flushed out when the planet became cool enough for water to fall and fill the oceans. These molecules, Oparin wrote, would have gradually coagulated into complex aggregates called colloids: 'material which had formerly been structureless first acquired a structure and the transformation of organic compounds into an organic body took place . . . With certain reservations we can even consider that first piece of organic slime which came into being on the Earth as being the first organism.' Echoes of the Greek creation myth are clearly audible.

To turn the haphazard cookery of crude prebiotic ingredients into a plausible story about our chemical genesis, one needed to know what our planet looked like before it was colonized and transformed by living organisms. The American chemist Harold Urey proposed in the 1950s that the environment on the early Earth had been 'reducing' – dominated by compounds of hydrogen rather than those of oxygen. In 1952 he and his graduate student Stanley Miller sent electrical discharges through a mixture of reducing gases – hydrogen, nitrogen, methane and water – and found that the sticky red residue that accumulated in their apparatus within a matter of days contained a rich

---

* This was demonstrated in the late 1920s by Edward Charles Baly, a chemist at Liverpool University.

variety of amino acids. The associations with the electrified anthro-poeia of Frankenstein no doubt helped to foster the idea, evident in some contemporary news reports, that Urey and Miller were on the verge of making life itself.

Lab life

If all this seems nonetheless a long remove from making people, remember that – as scientists and indeed most lay people were by then perfectly well aware – humans begin their lives as single cells. Louis Pasteur made the connection explicit in a lecture in 1864 at the Sorbonne in Paris, which contained a clear allusion to the protoplasmic view of life:

Take a drop of seawater . . . that contains some nitrogenous material, some sea mucus, some 'fertile jelly' as it is called, and in the midst of this inani-mate matter, the first beings of creation take birth spontaneously, then little by little are transformed and climb from rung to rung – for example, to insects in 10,000 years and no doubt to monkeys and man at the end of 100,000 years.

In his 1929 article Haldane went so far as to ponder the possibility of synthesizing living creatures 'in the biochemical laboratory'. He admitted that this goal was a 'long way' off, saying, 'I do not think I shall behold the synthesis of anything so nearly alive as a bacterio-phage [a bacterial virus] or a virus, and I do not suppose that a self-contained organism will be made for centuries.' But events were already suggesting otherwise. For in 1899 the *Boston Herald* carried the headline:

Creation of Life. Startling Discovery of Prof. Loeb. Lower Animals Produced by Chemical Means. Process May Apply to Human Species. Immaculate Conception Explained.

The report described the work of Jacques Loeb, a German working at the research centre for marine biology in Woods Hole, Massachusetts. Now, Loeb had most certainly not created life. Rather, he had caused an unfertilized sea-urchin egg to divide by treating it with a mixture of simple salts: this was, in other words, artificially induced parthenogenesis. On that basis the newspaper report could sound like hysterical over-extrapolation, the kind of thing that makes scientists roll their eyes in despair. But here such an accusation would be unfair, for this is what Loeb himself had written in his account of the discovery:

The development of the unfertilized egg, that is an assured fact. I believe an immaculate conception may be a natural result of unusual but natural causes. The less a scientist says about that now the better. It is a wonderful subject, and in many ways an awful one. That the human species may be made artificially to reproduce itself by the withdrawal of chemical restraint by other than natural means is a matter we do not like to contemplate. But we have drawn a great step nearer to the chemical theory of life and may already see ahead of us the day when a scientist, experimenting with chemicals in a test tube, may see them unite and form a substance which shall live and move and reproduce itself.

What Loeb had done was to replace the function of sperm in initiating development of the egg with the chemical action of salts such as sodium chloride and magnesium chloride – in essence, with a kind of reformulated seawater. In organisms that reproduce sexually, the development of an egg into a new organism usually proceeds only when triggered by the entry of a spermatozoon. But Loeb's discovery revealed that this stimulus is – in the sea urchin, at any rate – strictly optional. The sperm's role in starting growth of an embryo is quite distinct from its role in providing genetic material for the developing organism, and the former function can be mimicked rather crudely. The implication is very profound: it means that a lone egg, a single unfertilized cell, is already a potential new being. And when it is provoked to develop into that being through parthenogenesis, the resulting organism is a *clone* of the female that supplied the egg, with identical genetic constitution. Loeb had therefore not so much created life as invented cloning. As historian Philip Pauly puts it, 'The invention of artificial parthenogenesis represented an attack on the privileged status of natural modes of reproduction.' This was unsettling enough to provoke defensive mockery from other scientists: the French zoologist Camille Viguier called the progeny of Loeb's sea urchins 'chemical citizens, the son of Madame Sea-Urchin and Monsieur Magnesium Chloride'.

Loeb's discovery was no chance affair. He had been experimenting for several years on the control and manipulation of sea-urchin development using salts, at first in association with the American biologist Thomas Hunt Morgan (whose supporters later accused Loeb of stealing his ideas) at Bryn Mawr College in Pennsylvania and then at the University of Chicago. But his breakthrough precipitated Loeb into the limelight, a position that he seemed at first rather to relish. Despite some early scepticism, his discovery was widely lauded, and in 1901 he narrowly missed out on being awarded a Nobel Prize.

The work was quickly followed up elsewhere. In 1910 the French scientist Eugène Bataillon in Dijon discovered that frog eggs will begin developing into embryos if pricked with a needle, implying that the initiating role of sperm might be a purely mechanical process.* The embryologist Frank Rattray Lillie, also at the University of Chicago and later founder of the Woods Hole Oceanographic Institution, was particularly interested in whether the same tricks would work for humans, and hinted that this should be possible. (It is in fact not so easy at all – human embryogenesis differs in some important ways from that of sea urchins and frogs.)

Loeb and other biologists viewed the prospect of human parthenogenesis triggered by salt with somewhat uneasy humour, joking that 'maiden ladies' might feel compelled to stop bathing in the sea. This casual extrapolation to humans led to the perception that Loeb had revealed males to be redundant in reproduction. He received letters from women asking him to induce parthenogenesis in their own ova, while the French embryologist Yves Delage, who worked on the same problem, was sent correspondence congratulating him for freeing women from 'the shameful bondage of needing a man to become a mother'. These are prescient themes, as we shall see: artificial means of conceiving a baby, however hypothetical, are now seen both as removing women's control of their own reproductive destiny (and

---

* Loeb made a frog by chemical parthenogenesis too, but it met a sad fate, as the *Chicago Daily Tribune* explained in 1912: 'alas, the professor, so learned in biology and chemistry did not know that a frog could not live in water and he let the poor thing drown.' But the newspaper reported that Loeb had deduced the flaw in his experiment: 'The next one will live', he said, 'because I will bring him up on dry land.'

placing it under the management of male scientists) and as liberating them to take unilateral control. Equally telling as a taste of what was to come was a report speculating about the possibility of raising 'domestic animals and children born without help of a male through an operation which would be regulated scientifically and almost commercially, similar to raising the fry of trout'.

Loeb himself was a committed materialist, convinced that life could be understood in mechanistic terms and that, as a result, it could be woven together synthetically. 'We must either succeed in producing living matter artificially', he wrote in 1912, 'or we must find the reasons why this is impossible.' But this creation of life would be only the first step in a general programme of manipulating and shaping living matter to our own designs – a Baconian vision that Loeb knew would be perceived as a hubristic form of *techne*, and vulnerable to the charge of being *contra naturam* even among other scientists:

The idea is now hovering before me that man himself can act as a creator, even in living nature, forming it eventually according to his will . . . Man can at least succeed in a technology of living substance. Biologists label that the production of monstrosities; railroads, telegraphs, and the rest of the technology of inanimate nature are accordingly monstrosities.

At this early stage in the evolving understanding of an inherited origin of the instinctive behaviour of organisms, Loeb posited a conceptual bridge between the Cartesian vision of human automata and the modern picture of genetic behaviourism: 'We eat, drink and reproduce', he wrote,

not because mankind has reached an agreement that this is desirable, but because, machine-like, we are compelled to do so . . . We struggle for justice and truth since we are instinctively compelled to see our fellow beings happy.

Pre-empting modern game-theoretic views about altruism and co-operation, Loeb believed that the basis of ethics lay within these hard-wired instincts: when they are inhibited by economic, social and political circumstances, he said, the result is 'a civilization with a faulty or low development of ethics'.

The mythic resonance of Loeb's research was captured in an

article by Carl Snyder (a journalist who later became an influential economist), 'Bordering the mysteries of life and mind', published in *McClure's Magazine* in 1902. Snyder presented Loeb's artificial parthenogenesis as almost akin to 'the manufacture of life in the laboratory.' He had a sharp sense of the pedigree of Loeb's ambitions, making comparisons with Prometheus and Faust. If his interview with Loeb is accurately reported, the biologist gave him ample reason to make those allusions: he attributes to Loeb the comment that

I wanted to take life in my hands and play with it. I wanted to handle it in my laboratory as I would any other chemical reaction – to start it, stop it, vary it, study it under every condition, to direct it at my will!

One can almost hear the deranged cackle. Comparisons with Frankenstein followed soon enough; even *Scientific American* subtitled a 1909 article on Loeb 'The Achievements of the Scientific Frankenstein'. The *San Francisco Examiner*, meanwhile, argued (somewhat hazily) that Frankenstein's creation was a monster because Mary Shelley had the science wrong: she 'premised the necessity of a soul where she should have predicated the fermentation of the enzyme'. A description of Loeb's laboratory in *Cosmopolitan* evidently aimed to suggest a link to the occult tradition, saying that 'the tanks full of prisoners kept in running water, and the great microscopes standing in stiff, staring attitudes . . . suggest the spirit of investigation and of prying into Nature's secrets'.

Before long Loeb was being credited with what he had claimed to desire: almost complete mastery of life, an ability to rearrange the bodies of animals and even to make them immortal. He became a well-known figure in American science, and was the inspiration for Max Gottlieb, the mentor of the eponymous hero of Sinclair Lewis's 1925 novel *Arrowsmith*, one of the first literary studies of modern scientific culture. But although the fame brought Loeb academic status, he eventually grew weary of newspaper hype and began to resent the way he was sometimes presented as a borderline crank.

Loeb's findings, however important for the field of reproductive biology, did not in any sense amount to a chemical synthesis of life. They suggested instead that one of the organic components involved in sexual reproduction could be replaced by inorganic materials. But the discovery became conflated with speculations about the genuinely *de novo* creation of life. In 1912 the physiologist Edward Albert Schäfer, president of the British Association for the Advancement of Science, described Loeb's results at the association's annual meeting in Dundee within the context of 'the possibility of the synthesis of living matter'. Schäfer claimed that living matter has a simple composition in principle: the elements carbon, hydrogen, oxygen, nitrogen and phosphorus, plus some inorganic salts. 'The combination of these elements into a colloidal compound', he said,

represents the chemical basis of life; and when the chemist succeeds in building up this compound it will without doubt be found to exhibit the phenomena which we are in the habit of associating with the term 'life'.

Reporting on this address, the *Lancet* began with an allusion to the homunculus of Paracelsus, but was subsequently careful to clarify that 'the most that we can look for is no more than to create the dim beginnings of vital activity, to mould the faint transitional forms between living and non-living matter'. Yet although the press recognized that Schäfer had been talking about something other than the creation of human life, it lacked any other cultural reference points within which to frame the notion. And so the *Spectator* magazine remarked that

it is some consolation to think that the life-producing scientist, whose advent Professor Schäfer confidently anticipates, will be no Frankenstein. The utmost he can produce is something many millions of years removed from the level of humanity.

Besides, the synthesis of even the most primitive of life forms was sometimes deemed to warrant comparison to legendary sources. According to the *Manchester Guardian*, if 'we should be able to turn out living germs of our own creation – wholescale from the test-tube in which they had been made', this would be 'like the homunculus in Part II of Goethe's *Faust*, nay, if they proved noxious, it would be like a new version of *Frankenstein*'. According to the London *Evening News*, most people deemed the creation of *any* human-made life to be distasteful and blasphemous, while the *Daily Mirror* was convinced that a synthesis of protoplasm was only the precursor to imminent anthropoeia – indeed, that the shapeless jelly might even form itself into a human being:

In time, this *thing* might, by industrious coaxing, be induced to become an *animal* . . . Then, at last, after much more coaxing, a person, a rudimentary savage person might emerge; and so the whole process of evolution be conveniently shortened, in lecture time, by an ingenious gentleman on a platform. 'Ladies and gentlemen, I propose to turn this mass of weltering matter into a man.'

As so often with the tabloid press, one can detect in these remarks not genuine ethical concern but a journalistic delight in concocting images that will titillate its readers into expressions of outrage. The *Daily Herald* launched a similar flight of fantasy, ascending freely beyond any bounds set by 'men of science':

This idea of artificially creating life . . . may lead to the manufacture of living men . . . Nice, well-intentioned scientists will successfully perform, in fact, what Frankenstein achieved only in fiction. Can't we hear them talking? 'Better give this one a little more bicep: looks as if coal-miners are running short.' Or 'You're allowing that man too much brain space; he'll be joining a trade union if you don't take care.'

While there is whimsical pleasure to be found in the image of protoplasm joining a trade union, it is plain enough that the *Herald* (at that time positioning itself as the popular voice of the left wing) is really just manoeuvring to score political points through a paranoid extrapolation that anticipates *Brave New World*:

The life creators will join with the eugenicists . . . The necessary population, calculated in Whitehall, will be supplied with all the vices and virtues – or shall we say their virtues, and leave it at that? – ready made.

It is within this context that we can appreciate how the *New York Times* was prepared to announce in 1910 that 'The Mexican consul in Trieste reports that Prof. Herrera, a Mexican scientist, has succeeded in forming a human embryo by chemical combination.' The claim was nonsense, but it was a sign of the times that such an august period-ical would take it seriously.

Such sensationalism apart, Loeb's speculations about 'the artificial production of living matter' let the genie out of the bottle. As late as 1959, the French biologist and philosopher Jean Rostand deemed it necessary to write that

The general public is prone to believe that to have caused the birth of animals without a father is the same as to have half created life, and looks forward to the time when science, by replacing both the generative cells, will have wholly created life. But in fact it has created nothing, for it has replaced the father merely as an agent of stimulation, not as father . . . We combine, we transpose, we interpose, we interlard, but at every stage we are using what exists, at every stage we are exploiting the *really creative* power of the vital, we are embroidering on the pre-existent frame which is the real masterpiece, we are ingeniously making use of the genius of the cells, and in doing so, we are rather like revue artists who win cheap applause by parodying a scene from *Le Cid* or a speech from *Cyrano de Bergerac* . . . Let us not give ourselves the airs of demigods, or even of demiurges, when we have merely been petty magicians . . . The longer, the more assiduously one has lived on familiar terms with living things, the more one feels out-distanced, bewildered and put to shame.

The inevitable inferiority of the 'artificial' product was also asserted (with no obvious rationale) by Eugène Bataillon, who said, 'We merely plagiarize nature, and our plagiarism has not the perfection of the original.' Rostand even asserted that biologists, by persisting sub-consciously in attributing 'a privileged value to the living thing', were still vitalists at heart. One can argue that they still are.

But humble wisdom was not now going to hold back the Promethean tide that Loeb had unleashed. In his 1927 book *Archimedes, or the Future of Physics*, physicist Lancelot Whyte argued that 'As the result of the alteration in physical conceptions biology will soon cease to draw a definite line between inanimate and living systems.' The question which then arises, he said, is this: 'Could an infinitely wise physicist order the necessary chemicals to-day, and to-morrow put together a synthetic man?' Whyte's answer acknowledged what had been learnt about evolutionary timescales, while again making clear the historical associations of the image: 'it is *shortage of time* that our ambitious scientist is up against in his haste to create a homunculus. Only the synthetic alchemy of time can build up organisms, each bearing within itself a long heredity.' It was, then, just a matter of degree – of *time* – that separated protoplasm from an artificial person. Or perhaps not even that, Whyte implied, but of money and manpower: 'Only an International Institute of Evolutionary Research under the most stable Leagues of Nations could hope to create an artificial man.' That goal was attainable in Utopia.

If so, the theological musings in a 1912 *Cosmopolitan* article titled 'Creating life in the laboratory' carried full force:

Life is a chemical reaction; death is the cessation of that reaction; living matter, from the microscopic yeast spore to humanity itself, is merely the result of accidental groupings of otherwise inert matter, and life can actually be created by repeating in the laboratory nature's own methods and processes! Think for a moment what this declaration signifies. If it be true, where is the theology? If you and I are merely physico-chemical compounds, slightly more complex than a potato, a little less durable than a boulder, what is the basis of our moral code? If man can lump together sand and salt and by pouring water on them create life, what becomes of the soul?

With a small change in terminology, this could have been written by medieval theologians condemning the propositions of Aristotle, for it is ultimately asking: What role is left for God?

The old questions were not going away.

7 Animating Clay

Living beings are always imbued with magic to a certain extent.
There is a sort of fetishism attached to them. They sum up all
the forces of nature. In them, matter possesses miraculous properties: it is activated, influenced, transformed.
  François Jacob, *The Logic of Life* (1989)

Making artificial people is an industrial secret.
  Harry Domin, in Karel Čapek's *R.U.R.* (1921)

The real Prometheus, it could be argued, does not just mimic the
gods' creations but commandeers divine methods to do so. Dust is all
he requires; that, and a means to summon the Breath of Life. And so
there is a tradition of making people that is entirely distinct from the
pseudo-gestational process by which homunculi are generated or the
galvanic anatomizing of Frankenstein, and which involves the fashioning of human-like beings by the animation of lifeless material: clay,
stone, metal.

Since it could not utilize the life force immanent in human sperm
or eggs, this feat required supernatural agency. The rumour that the
thirteenth-century Franciscan friar Roger Bacon, a renowned inventor
of mechanical marvels, made a talking head out of brass secured his
notoriety as a wizard.* Similar stories circulated about several other

---

* Bacon describes some of his mechanical marvels in his *Letter on the Secret Works of
Art and Nature* (c.1260), including self-propelled ships and carts, flying machines and
submarines. There is no evidence that he ever tried to make any of them; historians
Katharine Park and Lorraine Daston aptly call him an 'armchair inventor'.

medieval natural philosophers, including the tenth-century scholar Gerbert of Aurillac (later Pope Sylvester II), Bacon's mentor Robert Grosseteste, and Albertus Magnus. The 'heads' may have been brass vessels used for alchemy, which made noises as they discharged gases.*
Or the stories may have been garbled allegorical accounts of alchemical recipes – Bacon's brass head is said to have exploded (as certain chemical mixtures might) after uttering the phrase 'Time was'. But it is equally possible that such explanations are just attempts to rationalize a recurrent mythic form, in the same way that Aristotle 'explained' Daedalus' moving statue of Aphrodite as a purely mechanical contrivance ingeniously activated by a flow of mercury. For the animation of statues and sculptures was a well-established magical act, which, according to the semi-legendary progenitor of alchemical art Hermes Trismegistus, could be achieved using precious stones, plants, incense and ritual. The result, he said, was a 'living god'.

These accomplishments seem to lie outside a scientific or even proto-scientific approach to anthropoeia – there is no biology in them, whether it is Aristotelian, Paracelsian, Cartesian or Darwinian. But they share common roots insofar as they involve human *art*, and the myths of growing humans from seed and of animating them from clay have a long history of cross-fertilization. We cannot understand some of the themes and preconceptions in contemporary efforts to fabricate humankind without acknowledging this interaction.

The mark of life

The animation of raw earth features in the Jewish tradition of the golem, a clay figure brought to life by religious magic. *Golem* is Hebrew for 'unformed matter', and in the early Christian era the word was used for the first moments of human existence, before the body has acquired its form – many modern versions of the Bible translate this word, *golmi*, (improperly) as 'embryo'. Adam himself is said to have existed in a 'golem' state in the initial stage of his creation, when he was mere clay (*'adamah* is Hebrew for 'earth')

---

* The seventeenth-century English writer Thomas Browne suggested this prosaic explanation in his debunker's manual *Pseudodoxia Epidemica* (1646).

untouched by the breath of God. Until that point, he had no soul. One Jewish text says that God fashioned the golem-Adam early on, but then left him soulless until everything else was made so that he should not be said to have been accompanied by a living being during the Creation – or perhaps, so that Adam should not see how it was done.

A golem is thus a kind of imperfect or incomplete human body. In the patriarchal Hebrew tradition, an unmarried woman might be described as a golem, requiring a husband to attain completion. This state of being was also linked to infertility; one anonymous Jewish text says that

When the sages say, whoever has no sons is like a dead man, they mean that he is like a Golem, without form . . . it is possible that, using wondrous powers, someone will make a man who speaks, but he cannot [confer the faculty] of procreation nor that of intellect, for this is unattainable for any creature to accomplish except God.

Recipes for making (and unmaking) a golem vary, but most invoke the magical power of words – one of the core principles of the Jewish mystical tradition called the Kabbalah, according to which God made the entire cosmos from letters. Letters, says the fifteenth-century north Italian rabbi Yohanan Alemanno, 'are forms and seals [that] collect the supernal and spiritual emanation as the seals collect the emanations of the stars'. Some sources say that adepts could animate a statue of inchoate matter by pronouncing over it the name by which God identified himself to Moses. A seventeenth-century account by the German astronomer Christoph Arnold tells how it was done:

After saying certain prayers and holding certain fast days, they make the figure of a man from clay, and then when they have said the *shem hamephorash* over it, the image comes to life. And although the image itself cannot speak, it understands what is said to it and commanded; among the Polish Jews it does all kinds of housework, but is not allowed to leave the house. On the forehead of the image, they write: *emeth*, that is, truth.

The real purpose of making a golem was not to provide a domestic servant but to imitate God's creation of Adam: as the Jewish scholar Moshe Idel puts it, 'we may describe the Golem practices as an attempt of man to know God by the art He uses in order to create man'. The propriety of such an act was unclear, notwithstanding the fact that the main manufacturers of golems were rabbis. The Talmud tells how, in the third century BC, a rabbi known as Rava created a man out of earth and sent him to the great sage Rabbi Zeira, who, discerning that the being was mute and imperfect, decided that his conception was impious and commanded him to return to dust.

But there seems to be little explicit suggestion that the intention was ever to *usurp* God's power. The fabrication of a golem seems more an act of homage, like making a clumsy, humble copy of an Old Master – or perhaps it was done just to enjoy a taste of the same artisanal satisfaction. This impulse is described delightfully by the narrator's father in Bruno Schulz's haunting, fabulous account of life in the Jewish community in Poland, *The Street of Crocodiles*. This strange, crazy man is musing about bringing his tailor's dummies to life 'for one gesture, for one word alone':

'We have lived too long under the terror of the matchless perfection of the demiurge,' my father said. 'For too long perfection of his creation has paralysed our own creative instinct. We don't wish to compete with him. We have no ambition to emulate him. We wish to be creators in our own, lower sphere; we want to have the privilege of creation, we want creative delights; we want – in one word – Demiurgy.'

The most famous of these creatures, made by Rabbi Yehudah Löw of Prague in the late sixteenth century, was employed in his maker's household and protected the inhabitants of the Jewish ghetto. This tradition of the golem as the guardian of the Jews never vanished, as indicated by the story told by a Holocaust survivor from Prague to a Jewish soldier in Bologna in 1945:

The Golem did not disappear and even in the time of the war it went out of his hiding-place in order to safeguard its synagogue. When the Germans occupied Prague, they decided to destroy the Altneuschul [the oldest synagogue in Europe]. They came to do it; suddenly, in the silence of the synagogue, the steps of a giant walking on the roof, began to be heard. They saw a shadow of a giant hand falling from the window onto the floor . . . The Germans were terrified and they threw away their tools and fled [*sic*] away in panic.

But making a golem could be dangerous in itself. It grows larger and stronger each day, and 'ultimately becomes a threat to all those in the house'. To return the golem to lifeless clay, the first character of *emeth* was erased to leave *meth*, Hebrew for 'dead'. 'When this is done', says Christoph Arnold, 'the golem collapses and dissolves into the clay or mud that he was.'

His account goes on to illustrate the perils of making a golem. It tells how Rabbi Elias Baal Shem of the Polish city of Chelm made a creature that grew so large the rabbi could no longer reach the forehead to erase the *e*. He had to trick his creature by telling it to remove its boots, whereupon he dashed at the kneeling golem and obliterated the letter. But as the golem returned to mud, its weight fell on the rabbi and killed him. (This narrative was later transferred to Löw by Jakob Grimm.)

There are two key elements to this fable. First, the golem is stupid and brutish, demonstrating the gulf between the creative powers of mankind and those of God. Drawing on the Aristotelian notion of the soul, the sixteenth-century philosopher Joseph Ashkenazi claimed that the human creators of golems can only give their creatures an animal soul; 'to give him a real soul, *neshamah*, is not in the power of man, for it comes from the word of God'.

Second, the deed is apt to end badly. The artificial being threatens constantly to overwhelm its maker, and in the case of Elias (or Löw) eventually brings about his demise. This aspect of Faustian retribution is far more explicit than it is in the generation of homunculi, which are often examples of man's creative prowess rather than his inferiority, and which need not be ultimately destroyed. The Jewish scholar Gershom Scholem believes that the medieval tradition of golem-making, which originated among the German and French Hasidic Jews, began as a purely allegorical ritual whose purpose was to attain mystical experience – no genuine being was created at all. Nevertheless, this ritual seems to have been fraught with danger: if it was conducted improperly, the person who enacted it might lose their life.

Western intellectuals developed an interest in the Kabbalah during the revitalization of Hermeticism and Neoplatonism in the early Renaissance. The fifteenth-century German humanist Johann Reuchlin was fluent in Hebrew and an expert on Kabbalistic doctrines, and the dazzling young Italian scholar Pico della Mirandola was the first to popularize them for gentiles. These efforts brought the golem legend into contact with alchemical ideas about medicine and human generation. According to the fourteenth-century Jewish physician Nissim Gerondi of Barcelona, a golem, like a homunculus, can only be made in some kind of vessel. Yohanan Alemanno, who taught Pico, wrote that the Kabbalah reveals a way to combine the four classical elements into a mixture which acts like semen, allowing the experimenter-magician 'to give birth to a man without [using] the male semen and the blood of the female', and without sex. This is not exactly a homunculus, since it specifically does not use semen; rather, true to the golem tradition it makes a being out of only lifeless components. Historian William Newman perceptively characterizes the distinction in contemporary terms:

If it were not rash to draw a modern comparison, perhaps one could say that the golem belongs to the realm of 'hard' artificial life, the world of robotics, cybernetics, and artificial intelligence, where ordinary biological processes are obviated or simulated by nonbiological means. The homunculus proper is a child of the 'wet' world of *in vitro* fertilization, cloning, and genetic engineering, where biology is not circumvented but altered.

Iron golem

Since I am primarily concerned here with the second of these worlds, I propose to say very little about the 'hard' artificial life of computer intelligence and robotics. While robots are in one sense descendants of the marvellous mechanical automata of the Enlightenment, we have lost sight of their connection to the machine-man of Descartes and La Mettrie: no one now considers robots to be alive merely because they move autonomously or make voice-imitating sounds. While artificial computer intelligence continues to be evaluated in relation to human intelligence, and while the distinctions between the two, as explored in the Turing test, continue to be matters of philosophical debate, these themes generally lie outside the mythical tradition of anthropoeia.

Yet the magical animation of dead matter represented in legends of living statues and golems does link to this tradition. It is easy to see how the golem archetype informed *Frankenstein*, in which the monster is said to be hideous, eight feet tall, and possessed of tremendous strength. Mary Shelley became familiar with the 'animated statue' legend when, shortly before she wrote her novel, she read the dramatic sketch *Pygmalion et Galatée* by the French writer Madame de Genlis. In legend, Pygmalion is a sculptor who falls in love with the statue of a beautiful woman he carves from ivory, identified as Galatea in later versions. In answer to his prayer, Aphrodite brings the statue to life and the couple are married.

Quickly divested of its eloquence and capacity for compassion, and made into an inarticulate hulking savage, Frankenstein's monster drew closer to the conventional image of the golem as Shelley's tale was retold. Yet Boris Karloff's stiff-limbed, rectilinear portrait forges a link to a more contemporary breed of *faux* inorganic being – the robot.

Arguably one of the first literary robots appears in Herman Melville's short story 'The Bell-Tower'. The tower of the story's title was built in some unspecified but long-past era by a 'great mathematician' called Bannadonna. To strike the massive bell in the belfry, Bannadonna fashions a man of iron called Haman. Constructed by an engineer out of metal parts, this being sounds superficially like the automata of Vaucanson and Jaquet-Droz, but in contrast to the explicitly mechanical automata of E. T. A. Hoffmann's Spallanzani, the principle of Haman's movement is never explained. The iron man clearly has more in common with magically animated statues than it does with the mechanic's assemblies of machine parts. That might seem to be at odds with the assertion that Bannadonna was 'without sympathy for any of the vain-glorious irrationalities of his time', were it not for the fact that the same can be said of Albertus Magnus and Agrippa (to whom he is compared). Since he denies 'that between the finer mechanic forces and the ruder animal vitality some germ of correspondence might prove discoverable', he explicitly rejects the Cartesian idea that a sufficiently complex automaton displays rudimentary life. Nor was he any sort of 'chemical physiologist': 'As little did his scheme partake', Melville's narrator adds,

of the enthusiasm of some natural philosophers, who hoped, by physiological and chemical inductions, to arrive at a knowledge of the source of life, and so qualify themselves to manufacture and improve upon it.

Yet even while insisting that Bannadonna is nothing more than a skilled mechanic, a 'practical materialist', Melville implies that there are occult forces coursing through his productions: 'With him, common sense was theurgy; machinery, miracle; Prometheus, the heroic name for machinist; man, the true God.' For Bannadonna's miraculous man of iron not only has the ability to move but also 'the appearance, at least, of intelligence and will'. There can be no question, then, that in Haman the animated statue blends into the metallic robot: immensely strong but bovine, lacking a soul (as Bannadonna makes clear), a servant to humankind.

And so this creation is at root Promethean and hubristic after all; it is seen as

supplying nothing less than a supplement to the Six Days' Work; stocking the earth with a new serf, more useful than the ox, swifter than the dolphin, stronger than the lion, more cunning than the ape, for industry an ant, more fiery than serpents, and yet, in patience, another ass . . . Talus,* iron slave to Bannadonna, and, through him, to man.

As we might anticipate from this context, Bannadonna's ambition is his downfall. The narrator, speaking long after the bell tower has fallen into ruin, explains that Bannadonna was somehow killed by his creation, most probably because he set Haman to strike the bell at a certain time but then forgot about him and was caught unawares still working on the bell mechanism, his head coming between the bell and the heedless Haman's hammer. 'So the creator was killed by his creature . . . And so pride went before the fall.' Melville somewhat dilutes the Faustian message, however, with a story about Bannadonna

---

* The reference here is to the bronze man Talos created by Hephaestus (see p. 34).

having struck and killed a worker in anger during the casting of the bell: blood from the slain man had fallen into the foundry and been mixed with the molten metal, causing a flaw, yet Bannadonna's crime had been excused because of the almost superstitious awe with which his mechanical skill was regarded. So did he die as punishment for this crime, or for challenging God? Whatever the case, 'The Bell-Tower' warns us about the dangers of iron men.

Robot rising

The first robots to be so called, appearing in the 1921 play *R.U.R.* ('Rossum's Universal Robots') by the Czech writer Karel Čapek, were not metal machines but genuine artificial people of flesh and blood. Although Čapek is now remembered primarily for this play, he was an influential intellectual of his day who wrote several other plays and books expressing wariness of technology and cautioning about the dehumanizing effects of the modern industrial economy.

In the course of inventing the concept of the robot (derived from the Czech for 'labourer'),* Čapek's remarkable play establishes one of the dominant narratives of anthropoeia in the twentieth century – its industrialization and commercialization – as well as anticipating the themes of just about every robot story that has followed since. What is more, the play recapitulates most of the old narratives about artificial people: their imperfection and lack of a soul, the tension between the sympathy and revulsion we feel for them, the Promethean ambition of their creators, the Faustian quest to improve man's estate and its tragic consequences.

Čapek even makes a connection with the chemical physiology that had dominated the nineteenth-century science of life. The robots' inventor is a scientist named Rossum, his name stemming from the Czech *rozum*, generally meaning 'reason' or 'wisdom' although Čapek indicated that it was to be interpreted here more neutrally as 'Brain' or 'Intellect'. Wishing to prove that life could have arisen without divine intervention, Rossum tries to synthesize protoplasm by chemical methods. His nephew, hearing that it may take ten years to make a man this way (considerably less than Lancelot Whyte's estimate),

---

* Čapek attributed the invention of the word to his brother Josef, a painter and poet.

grows impatient and devises a simpler process that can be conducted on an assembly line. In the play's prologue, Harry Domin, the director general of the eponymous company Rossum's Universal Robots, explains the process to an idealistic young woman named Helena Glory, who is visiting the factory to plead for robots to be given rights:

Domin: Well, this is what happened. It was in 1920 when old Rossum, still a young man then but a great scientist, came to live on this isolated island in order to study marine biology. Alongside his studies, he made several attempts to synthesize the chemical structure of living tissues, known as protoplasm, and he eventually discovered a material that behaved just the same as living tissue despite being, chemically, quite different. That was in 1932, exactly four hundred and forty years after the discovery of America . . . And then, Miss Glory, this is what he wrote down in his chemical notes: 'Nature has found only one way of organizing living matter. There is however another way which is simpler, easier to mould, and quicker to produce than Nature ever stumbled across. This other path along which life might have developed is what I have just discovered.' Just think: he wrote these words about a blob of some kind of colloidal jelly that not even a dog would eat. Imagine him sitting with a test tube and thinking about how it could grow out into an entire tree of life made of all the animals starting with a tiny coil of life and ending with . . . ending with man himself. Man made of different material than we are. Miss Glory, this was one of the great moments of history . . . Next he had to get this life out of the test tube and speed up its development so that it would create some of the organs needed such as bone and nerves and all sorts of things and find materials such as catalysts and enzymes and hormones and so on and in short . . . are you understanding all of this?

Helena: I . . . I'm not sure. Perhaps not all of it.

Domin: I don't understand any of it. It's just that using this slime he could make whatever he wanted. He could have made a Medusa with the brain of Socrates or a worm fifty metres long. But old Rossum didn't have a trace of humour about him, so he got it into his head to make a normal vertebrate, such as a human being. And so that's what he started doing.

Helena: What exactly was it he tried to do?

Domin: Imitating Nature. First he tried to make an artificial dog. It took him years and years, and the result was something like a malformed deer which died after a few days. I can show you it in the museum. And then he set to work making a human being.

Evidently the ancestry of the robot is not, then, among the machines and automata of Vaucanson and his rivals, but in chemical speculations about protoplasmic life-matter. By the time in which the play is set, Rossum's process has been perfected, streamlined and reproduced on an industrial scale. The organs of the robots are made individually in vats from the appropriate ingredients, and then assembled like the components of Henry Ford's automobiles. This assembly of organic parts recalls the methods of Victor Frankenstein, but there is no scent of the charnel-house – instead it is the astringent odour of the chemical laboratory that reaches our noses. Domin explains to Helena about the 'mixers':

Helena: What mixers?
Domin: For mixing the dough. Each one of them can mix the material for a thousand robots at a time. Then there are the vats of liver and brain and so on. The bone factory. Then I'll show you the spinning-mill.
Helena: What spinning-mill?
Domin: Where we make the nerve fibres and the veins. And the intestine mill, where kilometres of tubing run through at a time. Then there's the assembly room where all these things are put together, it's just like making a car really. Each worker contributes just his own part of the production which automatically goes on to the next worker, then to the third and on and on. It's all fascinating to watch. After that they go to the drying room and into storage where the newly made robots work.
Helena: You mean you make them start work as soon as they're made?
Domin: Well really, it's more like working in the way a new piece of furniture works. They need to get used to the idea that they exist. There's something on the inside of them that needs to grow or something. And there are lots of new things on the inside that just aren't there until this time. You see, we need to leave a little space for natural development. And in the meantime the products go through their apprenticeship.

Significantly, while Rossum himself is portrayed as the hubristic mad scientist who ended up making a monster, he has long since left the stage. In effect, Domin implies that Frankenstein is a figure of the past; he, in contrast, presents himself as a practical man, a businessman – not even an Edisonian genius, but something more like a factory manager:

Domin: That old Rossum was completely mad. Seriously. But keep that to yourself. He was quite mad. He seriously wanted to make a human being.
Helena: Well that's what you do, isn't it?
Domin: Something like that, yes, but old Rossum meant it entirely literally. He wanted, in some scientific way, to take the place of God. He was a convinced materialist, and that's why he wanted to do everything simply to prove that there was no God needed. That's how he had had the idea of making a human being, just like you or me down to the smallest hair . . . In the museum I'll show you the monstrosity he created over the ten years he was working. It was supposed to be a man, but it lived for a total of three days. Old Rossum had no taste whatsoever. This thing is horrible, just horrible what he did. But on the inside it's got all the things that a man's supposed to have. Really! The detail of the work is quite amazing. And then Rossum's nephew came out here. Now this man, Miss Glory, he was a genius. As soon as he saw what the old man was doing he said, 'This is ridiculous, to spend ten years making a man; if you can't do it quicker than Nature then you might as well give up on it.' And then he began to study anatomy himself.

But in an explicit attempt to use art to improve nature and rival God, the younger Rossum also made monsters. In its time, Čapek's frank and even humorous suggestion that God knew too little about making men sailed close to the wind:

Domin: Young Rossum created something much more sophisticated than Nature ever did – technically at least!
Helena: They do say that man was created by God.
Domin: So much the worse for them. God had no idea about modern technology. Would you believe that young Rossum, when he was alive, was playing at God.

In the prologue, Helena enunciates both the fears and the empathy on which all subsequent robot narratives draw. As Domin makes it clear that the robots are deemed to have no feelings and are treated with indifference, it becomes evident that they represent the labourers of industrial complexes, the 'monstrous' masses both pitied and feared in Elizabeth Gaskell's *Mary Barton*:

The best sort of worker is the cheapest worker. The one that has the least needs. What young Rossum invented was a worker with the least needs possible. He had to make him simpler. He threw out everything that wasn't of direct use in his work, that's to say, he threw out the man and put in the robot. Miss Glory, robots are not people. They are mechanically much better than we are, they have an amazing ability to understand things, but they don't have a soul.

This satire on the dehumanization of mechanized industrial labour is emphasized when Helena later asks Fabry, one of the other managers, why they make robots. He replies:

So that they can work for us, Miss Glory. One robot can take the place of two and a half workers. The human body is very imperfect; one day it had to be replaced with a machine that would work better ... [People] were very unproductive. They weren't good enough for modern technology ... to give birth to a machine is wonderful progress. It's more convenient and it's quicker, and everything that's quicker means progress.

All this exposition is merely the background for the main story, which opens in Act I after the roguish Domin has seduced and married Helena. She has accepted R.U.R.'s business for some time, but has second thoughts about the wisdom of making robots surface after she hears the remarks of her maid Nana. 'It's against the will of God', Nana insists. 'It's blasphemy against the Creator, it's an offence against the Lord who made us in His own image, Helena. And you've dis-honoured the image of God, that's what you've done. You'll suffer a terrible punishment from God for that.'

She is right, of course, for the robots revolt and begin to massacre the human race. 'Robots of the world!' they cry. 'We, the first union at Rossum's Universal Robots, declare that man is our enemy and the blight of the universe.' They claim to be 'more developed than man, more intelligent and stronger ... man is a parasite on them.' And so the robot revolutionaries entreat the others 'to exterminate mankind'.

Eventually Domin, Helena and the other managers are besieged by robots at the factory on Rossum's Island. But Domin feels no remorse; rather, he reveals that he too had ambitions that were truly Promethean: 'I don't feel sorry for what I did . . . Not even now, on the last day of civilisation. It was a magnificent undertaking'. He rejects any suggestion that his motives were rooted in the profits of the R.U.R. shareholders:

Domin: I did it for myself, d'you hear? For my own satisfaction! I wanted mankind to become his own master! I wanted him not to have to live just for the next crust of bread! I wanted not a single soul to have to go stupid standing at somebody else's machines! . . . I wanted mankind to become an aristocracy of the world. Free, unconstrained, sovereign. Maybe even something higher than human.
Alquist: Superhumans, you mean.
Domin: Yes. If only we'd had another hundred years. Another hundred years for the new mankind . . . we can't always be thinking about the things we lost by changing the world as Adam knew it. Adam had to gain his bread by the sweat of his brow, he had to suffer hunger and thirst, tiredness and humiliation; now is the time when we can go back to the paradise where Adam was fed by the hand of God, when man was free and supreme; man will once more be free of labour and anguish, and his only task will once again be to make himself perfect, to become the lord of creation.

It is now no longer pure science, the pursuit of knowledge, that stands proxy for Eve's apple, but *techne*, technology. As Alquist, R.U.R.'s head of Construction, says, 'Technology I blame! . . . We thought we were doing something great, giving some benefit, making progress . . . And now all our greatness is bursting like a bubble!' *R.U.R.* returns to the notion that *art* sits at the nexus of the Faustian pact, and by locating it in the industrial enterprise the play offers a much more apt parable for modernity than does *Frankenstein*.

The humans have only one bargaining counter to save them from the wrath of the robots: Rossum's papers describing the secret process of how the beings are put together from the raw material. Without this, the robots know they cannot continue to replicate. The problem is that Helena, spooked by Nana's worries, has burnt the document. The canny Alquist, however, convinces the robots that he still knows the secret – and so when the others are killed, he alone is spared and held prisoner by the robots. The robot leader Damon (none of these names lacks symbolic cargo) decides they will devise a new machine for making themselves, but when Alquist dismisses the idea, Damon reveals the denouement:

Damon: We will make ourselves by machine. We will erect a thousand steam machines. We will start a gush of new life from our machines. Nothing but life! Nothing but robots! Millions of robots!
Alquist: Robots aren't life! Robots are machines.
Robot 3: We used to be machines, sir; but by means of pain and horror we have become . . .
Alquist: Become what?
Robot 2: We have obtained a soul.

Finally Alquist encounters two robots, called Primus and Helena, who have been altered by Dr Gall, R.U.R.'s head of Physiology and Research, at the original Helena's request. They have fallen in love, and he realizes that they are now capable of feeling and are as 'human' as he is. He tells them to go forth like a new Adam and Eve, and sees that through them life will go on. 'Life begins anew, it begins naked and small and comes from love; it takes root in the desert and all that we have done and built, all our cities and factories, all our great art, all our thoughts and all our philosophies, all this will not pass away.'

Given that Čapek's play sets the template for robot tales in the same way that *Frankenstein* does for 'mad scientist' stories, it is surprising that it is now known only as the source of the word. Perhaps this is partly because Čapek's robots were quickly divested of their flesh and blood. Were organic robots just too uncanny? Despite constant reference to them as 'machines', they are physically much closer to humans than the vast majority of robots that subsequently marched across page or screen, beginning with Fritz Lang's gleaming brass Maria in *Metropolis* only six years later. From *The Forbidden Planet*'s Robbie to Schwarzenegger's Terminator, it has been deemed important to remind us of the cold steel at the robot's core. The replicants of *Do Androids Dream of Electric Sheep?* and *Blade Runner* are a rare example of androids that appear to be human down to the bone (except, crucially, in their capacity for empathy) – and it is their proximity to our race that supplies the story's moral pinion, as Blade Runner Rick Deckard becomes increasingly uncertain about what manner of being he is 'retiring'. Curiously, R.U.R.'s managers and engineers show no such qualms despite the fact that their robots are, to all appearances, indistinguishable from humans. In the 1920s, marking the distinction by means of a soul still sounded plausible. Yet Čapek's concept of 'soul' appears to be decidedly secular, and his characters' talk of playing God seems already to be more rhetorical than a genuine accusation of blasphemy. For Čapek, in fact, the human soul seems to be associated with a willingness to act in inhuman ways: as Domin exclaims, 'Nobody could hate man as much as man!' It is as if the acquisition of a soul is a Fall rather than an endowment of divine grace. But then, was it not always suspected that artificial people, lacking souls, are free of original sin?

What also distinguishes *R.U.R.* as a perceptive contribution to the anthropoetic tradition is its focus on sex and fertility. Čapek was able to go further than earlier writers in exploring openly the question of what these things mean for artificial people. Robots, it appears, do not have functioning sex organs. Like God creating Adam with a spurious navel, Domin explains, Rossum 'got it into his head to make everything, every gland, every organ, just as they are in the human body. The appendix. The tonsils. The belly-button. Even the things with no function and even, er, even the sexual organs.' He admits that they do, of course, have a function, 'but if people are going to be made artificially then, er, then there's not really much need for them'.

Domin relishes this racy banter as a prelude to his seduction of Helena. But it also prefigures a significant later theme: what do artificial people imply for normal human sexuality? In the prologue, Helena is disgusted not so much that the robots *cannot* mate sexually but that they lack the desire to do so. Thus they violate the norms of human sexuality by being gendered yet sexless: 'It's just so . . . so unnatural!' she cries. Nana shares this distaste for the unnatural asexuality of robots, for she feels this places them emphatically beyond nature, more alien than the beasts: 'They don't have children, but even a dog has children, everyone has children.'

So, once again, the artificial being is sterile and unable to propagate.* Because they can't reproduce sexually, the robots rely on manufacture to propagate their species. But like Frankenstein's monster, they lack the knowledge to assemble themselves. When they try to use Rossum's factory equipment, one of them laments that 'The machines produce nothing but pieces of bloody meat.' Ordered to hand over the crucial manufacturing information, Alquist explains: 'I've told you time and again that you need to find some people. It's only people that can procreate, renew life, put things back to how they used to be.' Even when Alquist admits he does not know Rossum's secret, the robots persist in thinking he could save them. 'Do experiments', one of them implores Alquist. 'Search out the formula of life.' Alquist replies: 'There's nothing to search for. You'll never get the formula for life from a test tube.' In this regard, Čapek's tale truly belongs to the anthropoetic tradition rather than to the more recent mythology of the robot, for it is in their ability to self-replicate prodigiously that inorganic robots pose one of their gravest threats. In the modern myth of nanotechnology, this self-replication of microscopic machines literally devours the planet.

But it's not just the robots who cannot reproduce. In a startling theme that anticipates the way reproductive technologies provoke fears about the 'natural' future of the human species, the rise of R.U.R.'s robots is accompanied by a decline in human fertility. The reasons are never made clear; Helena's maid Nana suspects it is the retribution of God, and some of R.U.R.'s executives seem to share that suspicion. When Helena asks Alquist why so few children are being born, he replies in quasi-religious terms:

Because there's no need for them. Because we've entered into paradise . . . Because there's no need for anyone to work, no need for pain. No-one needs to do anything, anything at all except enjoy himself. This paradise, it's just

---

* It is implied in the play's final act that Alquist will be able to give them this ability by experimenting on robots to find 'the secret of life'. It is presumably by this means that Primus and the robot Helena will be capable of furthering the new species like the Edenic couple. But we are left to infer that their love, perhaps more than Alquist's experiments, bestows this fertility.

a curse! Helena, there's nothing more terrible than giving everyone Heaven on Earth! You want to know why women have stopped having children? Because the whole world has become Harry Domin's Sodom! . . . There's no need for men any more, Helena, women aren't going to give them any children!

When Helena asks Dr Gall, she gets a different answer, but equally mystical: it is because we have violated nature, because 'there are robots being made. Because there's an excess of manpower. Because mankind is actually no longer needed.' It is almost as if, Gall says reluctantly, 'making robots were an offence against Nature'.

I have considered *R.U.R.* at some length, not because it is a work of great literary distinction but because it offers such a thorough exegesis of modernist anthropoeia. When the play was premiered in London in 1923, its implications were publicly debated by George Bernard Shaw and G. K. Chesterton, prompting Čapek to explain his intentions in the *Saturday Review*. In doing so, he demonstrated profound understanding of how the myths of making people were changing:

I wished to write a comedy, partly of science, partly of truth. The old inventor, Mr Rossum, is no more or less than a typical representative of the scientific materialism of the last century. His desire to create an artificial man – in the chemical and biological, not the mechanical sense – is inspired by a foolish and obstinate wish to prove God unnecessary and meaningless. Young Rossum is the modern scientist, untroubled by metaphysical ideas; for him scientific experiment is the road to industrial production; he is not concerned about proving, but rather manufacturing. To create a homunculus is a mediaeval idea; to bring it in line with the present century, this creation must be undertaken on the principle of mass production . . . [In the robot], a product of the human brain has at last escaped from the control of human hands. This is the comedy of science.

Čapek wanted to show that in the modern age, truths can be mutually contradictory. Helena is right, he said, but so is Alquist, and so even are the robots:

I ask whether it is not possible to see in the present societal conflict an analogous struggle between two, three, five equally serious truths and equally noble ideals? I think it is possible, and this is the most dramatic element of modern civilization, that one human truth is opposed to another truth no less human, ideal against ideal, positive value against value no less positive, instead of the struggle being, as we are so often told it is, one between exalted truth and vile selfish wickedness.

In the anthropoetic controversies which were to follow in the late twentieth century, one might wish sometimes that Čapek's words had been remembered.

8 The Bottled Babies

> Developments in this direction are tending to bring mankind
> more and more together, to render life more and more complex,
> artificial, and rich in possibilities – to increase indefinitely man's
> powers for good and evil.
>
> J. B. S. Haldane, *Daedalus, or Science and the Future* (1924)

The opening scene of Aldous Huxley's *Brave New World* (1932) is the
one that makes the strongest impression on the reader:

A squat grey building of only thirty-four storeys. Over the main entrance
the words, CENTRAL LONDON HATCHERY AND CONDITIONING
CENTRE, and, in a shield, the World State's motto, COMMUNITY,
IDENTITY, STABILITY.

The enormous room on the ground floor faced towards the north. Cold
for all the summer beyond the panes, for all the tropical heat of the room
itself, a harsh thin light glared through the windows, hungrily seeking some
draped lay figure, some pallid shape of academic goose-flesh, but finding
only the glass and nickel and bleakly shining porcelain of a laboratory.
Wintriness responded to wintriness. The overalls of the workers were white,
their hands gloved with a pale corpse-coloured rubber. The light was frozen,
dead, a ghost . . .

'And this,' said the Director opening the door, 'is the Fertilizing Room.' . . .

'These,' he waved his hand, 'are the incubators.' And opening an insulated
door he showed them racks upon racks of numbered test-tubes. 'The week's
supply of ova. Kept,' he explained, 'at blood heat; whereas the male gametes,'
and here he opened another door, 'they have to be kept at thirty-five instead

of thirty-seven. Full blood heat sterilizes.' Rams wrapped in thermogene beget no lambs.

Gone are the thunderstorms and electric sparks of the Gothic novel (although the reek of the graveyard remains). Now the creation of artificial people takes place in funereal silence. And there is no need to cobble them together from scavenged body parts, because human reproductive biology has been mastered and divorced from the human body. It happens in test-tubes under scientifically controlled conditions, and it even has a technical name: Bokanovsky's process, in which many individuals can be created from a single fertilized egg. It is a factory process, which the director of Hatcheries and Conditioning describes explicitly as 'the principles of mass production at last applied to biology'.

Huxley's book is often described as a visionary masterpiece, a prescient glimpse into the age of IVF and cloning. But the fact is that Huxley simply put into novelistic form what his scientific friends, including his brother Julian, were forecasting in all seriousness. As Joseph Needham said when he reviewed *Brave New World*, the biology in it 'is perfectly right, and Mr Huxley has included nothing in this book but what might be regarded as legitimate extrapolation from the knowledge and power we already have . . . the production of numerous low-grade workers of precisely identical genetic constitution from one egg, is perfectly possible'.

One might almost infer from these comments that Needham also considered it desirable. Neither he nor his peers went quite that far, but what Aldous Huxley presented as dystopia was the flipside of a vision that many considered appealing. The 1920s and 30s saw an intense debate among scientists and other intellectuals about the possibilities of new reproductive technologies of the sort Huxley described, in which humans are conceived and grown entirely outside the human body. This would permit new forms of social engineering in which a population's gene pool could be managed, tailored and optimized. And many social prophets framed the prospect not in Huxley's terms of a totalitarian state that has all but extinguished the spark of human creativity, but as a route to a better society.

Gravid bottles

Bokanovsky's process is a form of cloning, producing identical embryos that are gestated outside the womb – or rather, in an artificial womb. That process had become known as ectogenesis, and Huxley had already alluded to the idea before; in his 1921 novel *Crome Yellow*, the character of Mr Scrogan forecasts that

An impersonal generation will take the place of Nature's hideous system. In vast state incubators, rows upon rows of gravid bottles will supply the world with the population it requires.

Brave New World took this fleeting suggestion and opened it up into a world – one set in the twenty-sixth century, when a World State grows its population in hatcheries, the embryos chemically regulated to determine their physique and intelligence according to a strict five-tier caste system. The population is kept docile by the drug soma, and is conditioned to accept complete sexual freedom, the notion of the family being considered obscene. In truth, the plot itself – the story of the 'Savage' named John who encounters and is gradually destroyed by this corrupted society – is secondary to the novel's impact. John's struggles and demise are hardly needed to illustrate the dystopian satire, and Huxley himself puts rather little effort into his characters: the novel derives its impetus almost entirely from its context, founded on the World State's mastery of biological social engineering and its realization through ectogenesis. A lazy reading presents all this as a prescient feat of brilliant imagination. But while there is little question that Huxley's vision of a dystopia based on mindless, state-sanctioned apathy is as relevant today as Orwell's vision of totalitarian coercion (if not even more so), his political setting for ectogenetic anthropoeia has scant application to the twenty-first century. As we will see, the social structures that support technological people-making today are of a quite different nature.

Ectogenesis is perhaps the most fully realized expression of the old symbol of a human in a bottle or flask, which was introduced by alchemy and particularized in the creation of the homunculus (Figure 8.1). Intriguingly, Aldous Huxley's brother, the eminent biologist Julian Huxley, seems to have invoked the image when, aged four, he wrote a letter to his grandfather Thomas Henry Huxley after reading Charles Kingsley's *The Water Babies* and finding his famous relative mentioned therein:

Dear Grandpater have you seen a Water-baby? Did you put it in a bottle? Did it wonder if it could get out? Can I see it some day? Your loving JULIAN

The elder Huxley responded: 'I never could make sure about that Water Baby. I have seen Babies in water and Babies in bottles: but the baby in the water was not in a bottle and the Baby in the bottle was not in water.'

Indeed not. Some of us have seen those bottled babies too, preserved in formaldehyde in medical museums: fetuses dead from birth, some of them selected and pickled because they are specimens of developmental deformity. With their seemingly disproportionate skulls and blind eyes, they are an alien presence, visitors from legend or, as in the 2009 biographical movie of Charles Darwin, *Creation*, from the supernatural. Babies in bottles are not like other babies. They are harbingers; and let's not forget the obvious point that they are literally *in vitro*.

The original test-tube baby? The creation of the homunculus in Goethe's *Faust*.

Deformed fetuses preserved in bottles in the Vrolik Museum, Amsterdam. Image: Hans van den Bogaard.

Writing the future

The possibility of ectogenesis was highlighted by J. B. S. Haldane in his 1924 book *Daedalus*, the first of Kegan Paul's 'To-day and To-morrow' series. Haldane was only thirty-one when the book appeared, but he had begun it at an even more tender age, as a paper speculating on the future of scientific developments, written in 1912 when he was an undergraduate at Oxford. Haldane revised the text for a talk in 1923 to the Heretics, an anti-authoritarian Cambridge debating society. The presentation was heard by Charles Ogden, a scout for Kegan Paul, who encouraged Haldane to write it up for the forthcoming series. It established Haldane's name as a science communicator, and set the stage for a discussion about the role of technology in social reform and planning that was pursued in several other of the 'To-day and To-morrow' titles.

Haldane framed part of his text in the form of quotations from a retrospective essay on developments in science by a 'rather stupid' undergraduate 150 years in the future. Ectogenesis, Haldane's narrator says, was introduced in 1951 to combat the plummeting birth rate

in developed countries. By 1968, France was producing 60,000 ectogenetic childen a year. And by 2073 less than thirty per cent of children
were still 'born of woman'. 'We can take an ovary from a woman',
his narrator says,

and keep it growing in a suitable fluid for as long as twenty years, producing
a fresh ovum each month, of which ninety per cent can be fertilized, and
the embryos grown successfully for nine months, and then brought out into
the air.

Haldane clearly considered this vision plausible, pointing out that
experiments on the *in vitro* fertilization of rabbit eggs had already laid
the foundations.

What seems surprising now is not that some other commentators
doubted the alleged benefits of ectogenesis but that these doubts were
generally expressed in such mild terms. While in 1940 George Orwell
suggested that ectogenesis would be felt by many people 'to be *in itself*
blasphemous', most criticisms focused on the presumed context in
which it would be conducted: there was a common belief at the time
that ectogenesis would necessarily be coupled to some form of centralized state control of conception and child-rearing. It is true that in
Halcyon, or The Future of Monogamy (1929), the English feminist writer
Vera Brittain suggested (in a futuristic 'retrospective' modelled on
Haldane's) that 'natural methods of reproduction' would reappear after
a period of ectogenesis because of the psychological problems that
became apparent in children 'made' this way. (By that time, she argued,
medical advances would have reduced or eliminated the trials of pregnancy and the pains of childbirth.) But this concern was predicated on
the assumption that ectogenetic children would be deprived of parental
affection *after* birth, rather than a consideration of the physical and
psychological harms done during *ex utero* gestation.

In his book *The World in 2030 AD*, the Conservative peer the Earl of
Birkenhead (Frederick Edwin Smith, a former Lord Chancellor)
expressed reservations based on the same premise.* Birkenhead was

---

* This was not a part of the 'To-day and To-morrow' series but very much a response
to it. Indeed, Haldane felt that Birkenhead's book explored his own ideas without
properly acknowledging their provenance, leading him to make accusations of plagiarism in a mildly satirical tone.

no knee-jerk Luddite, being generally optimistic about the potential of science to improve the human condition. But in ectogenesis he saw only the spectre of state control and social engineering. His comments reveal how the potential of ectogenesis was often assessed within a highly deterministic view of genetics: he claimed that it could furnish a government with the means to (if it should so desire) 'leaven the population with fifty thousand irresponsible, if gifted, mural painters'. Or they might make 'strong, healthy creatures, swift and ductile in intricate drudgery, yet lacking ambition' – human robots, 'the exact human counterpart of the worker bee'. It is easy to ridicule such thoughts now, but they were widely shared at the time and would have conditioned the way ectogenetic fantasies such as *Brave New World* were received: specifically, the technology was perceived as a means by which states might mass-produce a population to precise specifications.

Birkenhead's reaction also invoked an older theme: an ectogenetic child, he said, would be 'monstrous' – not necessarily in physical form, but *because of the nature of its origin*. He considered that this origin imparts a social and moral judgement of character: no decent school, he said, would ever accept such a child. In other words, the anthropoetic monster again stands proxy for the mob, the unspeakable, amoral and unpredictable masses.

The crystallographer J. Desmond Bernal, a Marxist with a strong interest in the social applications of science, used ectogenesis as the launching point for a considerably more fantastical and speculative future in *The World, the Flesh and the Devil: An Enquiry into the Future of the Three Enemies of the Rational Soul* (1929). As the title implies, Bernal argued that the human body is one of the impediments to advancement of the human mind. Although he welcomed Haldane's future of ectogenetic procreation, he was disappointed that the beings so produced would still suffer the limitations of flesh and blood. He therefore posited it as merely the first step towards a Utopia of post-human cyborgs with machine bodies created by surgical techniques. The brain would be almost all that remained of the 'flesh', housed in a rigid protective cylinder. It would then be possible to wire these disembodied brains together to permit direct transference of thought. Over time, this might lead to a merging of human consciousness and an erasure of individuality and mortality:

Consciousness itself might end or vanish in a humanity that has become completely etherealized, losing the close-knit organism, becoming masses of atoms in space communicating by radiation, and ultimately perhaps resolving itself entirely into light.

Bernal recognized that to achieve such a goal we would need to overcome the 'distaste and hatred which mechanization has already brought into being'. But he was not so naive as to imagine that everyone would welcome his vision, and admitted that he shared the instinctive aversion to such a transformation of the body that many readers would feel. Equally chilling was Bernal's prescription for how the cyborg society might be governed: it would, he said, be best done by scientists and their institutions. 'Scientific corporations', he wrote, 'might well become almost independent states and be enabled to undertake their largest experiments without consulting the outside world.' Transformed scientists living in space might even turn the rest of the world into 'a human zoo, a zoo so intelligently managed that its inhabitants are not aware that they are there merely for the purposes of observation and experiment.' Quite how Bernal felt about this possibility is not clear; I suspect the response of most of us is easier to predict.

The best of humanity?

To Haldane and his contemporaries, ectogenesis had plenty to recommend it beyond the capacity to control population increase. It would also facilitate eugenic selection of the 'products': only the 'best' gametes would be used to create the next generation.

A eugenic (literally 'good genes') attitude to human breeding is very ancient, for it is a natural corollary of class consciousness: the aristocracy has always sought to procure 'good' marriages and has worried about the fecundity of the rabble. In 548 BC, the Greek poet Theognis of Megara wrote:

One would not dream of buying cattle without thoroughly examining them, or a horse without knowing whether he came of good stock; yet we see an excellent citizen being given to wife some wretched woman, daughter of a worthless father . . . Fortune mixes the races, and the odious adulteration is bastardizing the species.

And Plato's *Republic* attributes to Socrates the sentiment that 'If we want to prevent the human race from degenerating, we shall take care to encourage union between the best specimens of both sexes, and to limit that of the worse.'

The necessity of eugenic selection was almost universally felt among scientists in the 1920s, and it provided the context within which anthropoetic technologies were debated. One of the most influential advocates was Julian Huxley, who served as vice president and then as president of the British Eugenics Society from 1937 until 1962. In the wake of Darwin's evolutionary theory, there was widespread fear among the intelligentsia that human society was being degraded by the higher rate of reproduction among 'degenerate' and 'inferior' classes, whose numbers, along with those of 'mental defectives' and other disabled people, were no longer held in check by the selective pressures that had guaranteed the persistence of 'superior' types in more straitened times. This argument was expounded by Darwin's cousin Francis Galton in his book *Hereditary Genius*, first published in 1869, in which he expressed concern 'at the yearly output by unfit parents of weakly children who are constitutionally incapable of growing up into serviceable citizens, and who are a serious encumbrance to the nation'. By selective breeding, he argued in the 1892 revised edition, we might gradually raise 'the present miserably low standard of the human race to one in which the Utopias in the dreamland of philanthropists may become practical possibilities'. Echoing Theognis' words, he wrote, 'If a twentieth part of the cost and pains were spent in measures from the improvement of the human race that is spent on the improvement of the breed of horses and cattle, what a galaxy of genius we might not [*sic*] create.'

If, however, we let population growth continue in laissez-faire fashion, Galton and the other proponents of eugenics feared that the human race would deteriorate inexorably. For Haldane and Huxley the solution lay with 'positive eugenics': boosting the reproductive rates of better stock. This conviction was shared by many politicians. As early as 1905, US president Theodore Roosevelt warned of the danger of 'racial suicide' unless the 'best' of society were to have more children. He urged 'good' mothers to do so as a matter of national duty. Needless to say, these women were to be found among the affluent and well-educated members of society: precisely those who were starting to suspect that life might and should have more to offer them than interminable child-rearing responsibilities.

More sinister, however, was the notion of negative eugenics: preventing reproduction among 'defectives'. Galton advocated 'the hindrance of marriages and the production of offspring by the exceptionally unfit'. And in 1905, the state of Indiana took this advice and Roosevelt's warning to heart by prohibiting marriage between 'mentally deficient people', people with 'transmissible diseases' and habitual drunkards. Two years later the state introduced still more drastic measures, enforcing sterilization of 'criminals, idiots and imbeciles'. Enthusiasm for eugenics in the United States was spearheaded by the Harvard biologist Charles Davenport, who championed it in his 1911 book *Heredity in Relation to Eugenics*. Davenport's advocacy had a distinctly racist flavour: he argued that interbreeding of blacks and whites led to genetic decline. By 1936, thirty-five states had sterilization programmes, some of which persisted even after negative eugenics had been tarnished by its association with Nazi science. It is estimated that around 60,000 people were sterilized under these measures. Similar laws were enacted in Sweden, Denmark and Finland. The compulsory sterilization introduced in Nazi Germany in 1934 soon became state-sponsored euthanasia: doctors were authorized to kill patients with mental illnesses, homosexuals and alcoholics. It was then only a short step to the elimination of entire 'diseased races'.

Ectogenesis and other reproductive technologies were most often discussed in the pre-war years as enablers of eugenics. According to Haldane's fictitious student in *Daedalus*, 'Had it not been for ectogenesis there can be little doubt that civilization would have collapsed within a measurable time owing to the greater fertility of the less desirable members of the population in almost all countries.' Even the reactionary Birkenhead conceded that ectogenesis might be the 'only form of salvation which science and ingenuity can suggest' to avert social decline. But Haldane, Huxley and other biologists deplored some of the crude attitudes to eugenics that were guiding government policies. Haldane recognized that negative-eugenics programmes were racist, sexist and class-based, and he argued that the notion of 'best' might be relative in any case: different races were, he said, probably adapted to different environments. In his 1935 book *Out of the Night: A Biologist's Vision of the Future*, the American biologist Hermann

Muller, a former student of Julian Huxley's who was awarded the 1946 Nobel Prize for his work on the propensity of X-rays to cause genetic mutations, called mainstream eugenics 'a hopelessly perverted movement . . . lending a false appearance of scientific basis to advocates of race and class prejudice, defenders of vested interests of church and state, Fascists, Hitlerites and reactionaries generally.' 'The Nazi racial theory', Huxley agreed in 1942, 'is a mere rationalization of Germanic nationalism on the one hand and anti-Semitism on the other.'

These biologists, mostly of a left-wing persuasion, believed that eugenics should be accompanied by general social reform: they accepted that antisocial activities in deprived communities had environmental causes too. Huxley insisted that differences in height or intelligence between poor and wealthy classes were most probably a result of poor nutrition, not poor genes. 'Until we equalize nutrition, or at least nutritional opportunity,' he wrote, 'we have no scientific or other right to assert the constitutional inferiority of any groups or classes because they are inferior in visible characters.' The society Huxley wanted to bring about sounds admirable: one that rewards altruism, cooperation, sensitiveness, sympathetic enthusiasm, rather than egoism, low cunning and insensitivity.

But make no mistake: Huxley was concerned only that socio-economic factors should not mask 'bad genes' and thus undermine the effort to weed them out. 'Once the full implications of evolutionary biology are grasped', he said, 'eugenics will inevitably become part of the religion of the future, or of whatever complex of sentiments may in the future take the place of organized religion.' And despite the appearance of even-handedness, he clearly harboured suspicions about the fitness of the lower classes:

The upper economic classes are presumably slightly better endowed with ability – at least with ability to succeed in our social system – yet are not reproducing fast enough to replace themselves . . . We must therefore try to remedy this state of affairs, by pious exhortation and appeals to patriotism, or by the more tangible methods of family allowances, cheaper education, or income-tax rebates for children. The lowest strata, allegedly less well endowed genetically, are reproducing relatively too fast. Therefore birth-control methods must be taught them; they must not have too easy access to relief or hospital treatment lest the removal of the last check on natural selection should make it too easy for children to be reproduced or to survive; long unemployment should be a ground for sterilization, or at least relief should be contingent upon no further children being brought into the world.

Eugenics could not succeed, Huxley insisted, without 'control of the human germ-plasm'. He spoke at times as though the object of his attention were not individual human beings but some kind of generalized genetic material that was to be praised or condemned. He deplored the existence of 'nests of defective germ-plasm inspissated by assortative mating and inbreeding.'

For Huxley, ectogenesis could make eugenic selection faster, easier and more flexible. He argued that one of the gravest obstacles to his eugenic Utopia was 'the prevailing individualist attitude to marriage, and the conception . . . of the subordination of personal love to procreation'. These things, he said, prevent us from grasping

the implications of the recent advances in science and technique which now make it possible to separate the individual from the social side of sex and reproduction. Yet it is precisely and solely this separation that would make real eugenics practicable, by allowing a rate of progress yielding tangible encouragement in a reasonable time, generation by generation. The recent invention of efficient methods on the one hand of birth-control and on the other of artificial insemination have brought man to a stage at which the separation of sexual and reproductive functions could be used for eugenic purposes . . . It is now open to man and woman to consummate the sexual function with those they love, but to fulfil the reproductive function with those whom on perhaps quite other grounds they admire.

Muller's *Out of the Night* describes a scientific Utopia ushered by eugenics and ectogenesis, as well as other techniques for improving genetic stock: fertilizing human eggs without sperm (much as Jacques Loeb had done for sea urchins), or egg transplants to surrogate mothers so that 'females possessing characters particularly excellent' might produce children without being burdened by pregnancy or the need to raise them. Muller took practical steps towards his goal, initiating a 'Germinal Choice' programme in which people would relinquish their selfish desire to propagate their own genes and submit to centralized control of repro-duction. The 'best' women would (perhaps with their husbands' assist-ance) pick sperm from a pool of the 'best' men, according to information about the donor kept in detailed dossiers. Muller even made plans to avoid paternity lawsuits (this was the United States, remember) by recom-mending that frozen sperm be kept for twenty years, or until the donors had died, before use. Julian Huxley backed this bizarre scheme, and so did the British biologist Francis Crick. Sometimes, the accusation that scientists are apt to pursue their personal fantasies detached from ethical or humane considerations is not unwarranted.

What's a woman to do?

For Goethe's Wagner, separating reproduction from sex by anthro-poeia was seen as an act of liberation. And that is how both Haldane and Huxley regarded ectogenesis.

In the early twentieth century, childbirth was typically more painful and certainly more physically dangerous (the discovery of antibiotics lay some years away) than it is today. In *Chronos, or the Future of the Family* (1930) by Maurice Eden Paul, son of the To-day and To-morrow series publisher, it is still described as 'Eve's curse'.* If the effortful process of bearing a child was reflected in the term 'labour', ectogen-esis captured the progressive technological spirit in being 'labour-saving'.

But childbirth was also seen by some women, and also by men sympathetic to the feminist cause, as much more than a danger and discomfort: it was an impediment to the social advancement of women. Bearing the next generation was considered not only a conjugal duty but also a social obligation, particularly after the decimation of Europe's young men in the First World War. And eugenic enthusiasms meant that this pressure was felt all the more keenly by precisely those women who were best placed to see beyond it.

This was why ectogenesis was welcomed by many progressive thinkers as an emancipating technology for women – and condemned by

---

* This is a reflection of the conservative stance of Paul's book, in which birth control and the sexual and economic freedoms becoming available to women are denounced as the cause of the decline of the family. It is an early example of how any changes in reproductive norms are considered to threaten traditional family structures, which are, however, miraculously still with us.

conservatives for the same reason. 'There is no sexual reformer', said Eden Paul, 'but must wish that woman could be freed from the slavery of child-bearing, and that our offspring might come into the world out of a broken eggshell.' But to Paul the liberating potential of ectogenesis was precisely what was wrong with it. For Birkenhead, ectogenesis would have a 'shattering' effect on society because of its implications for marriage structures.* His antagonism was rooted in deeply sexist doubts about the value of giving women more freedom in the first place. 'The man of creative genius is more valuable to society and to the state than any woman', he wrote; women will fall 'just short of the best work done by men', no matter how hard they try.

Ectogenesis meanwhile alarmed some women by threatening to steal from them the privilege of motherhood. It was equally disturbing to the gender security of some men, for the supply of eggs is relatively limited but the supply of sperm is plentiful: a single individual of good breeding stock could provide enough sperm to spawn a battalion of babies. And so this speculative reproductive technology became seen as a means to make one sex or the other redundant. The English writer and social critic Anthony Ludovici unveiled this paranoid fantasy in all its ludicrous glory in his *Lysistrata, or Women's Future and Future Women* (1924). Ludovici's central fear was that feminism, in which he felt women were gratifying 'their vanity in every kind of sterile pursuit that gave them the appearance of being important', was in the end going to strip women of their 'eagerness to express physical passion' with red-blooded chaps like him. As a result, he said, women will come to regard procreation as a clinical affair to be conducted with the minimum of fuss and personal involvement, so that artificial insemination would become the preferred mode. 'When once artificial impregnation is an every-day occurrence,' he wrote, 'a Parliament of women will doubtless pass legislation to make it illegal for any man to procreate a child naturally, if it is the wife's desire to

---

* For social and religious conservatives, society and gender roles are disturbingly fragile to the shocks of new reproductive technologies. When in 2004 Japanese researchers induced parthenogenesis in mice by transferring some chromosomes from one mouse egg into another from a different female, professor of public health sciences Nancy L. Jones echoed Birkenhead with the claim that 'if [this] mode of creation were to be extrapolated to humans, the very basis of our society would be shattered'.

have one by the intermediation of science.' The hapless future man was doomed to a life of joyless masturbation when his wife demanded more sperm.

But it will be far worse, Ludovici predicted, once work on tissue culturing by the likes of the American biologist Alexis Carrel (see p. 263) reaches a stage that allows 'the fertilized ovum [to] be matured outside the female body'. With unconscious allusions to the medieval homunculus technologies, he envisaged that ectogenesis will first happen by 'enlisting the cow or the ass into our service', so that the embryo 'will mature very much as chickens now do in incubators'. Then, in an amusing conflation of chimeric myth and trans-generational contempt, he foresees that this will lead to 'intensified bovinity or asisinity' in the people so produced. The conservative's fear of socialist engineering enters the fray too: 'public centres will be provided where the Borough Council will undertake to "grow" children for the destitute and the poor'.

In the end, says Ludovici, men will be considered a mere source of 'fertilizer', superfluous above a ratio of about one to every 200 women. 'The legislature will establish laws to guarantee that this minimum should not be surpassed', he went on,

and in a very short while it will be a mere matter of routine to proceed to an annual slaughter of males who have either outlived their prime or else have failed to fulfil the promise of their youth in meekness, general emasculateness, and stupidity.

This gives the game away, of course. As is clear from the outset, Ludovici's real fear is the rise of female power and independence, which he interprets as an encroaching feminization of culture. Once again anthropoeia merely supplies the lens through which broader social and cultural concerns can be refracted.*

---

* Scientists are now becoming wise to these tropes of gender annihilation. In the Japanese work on mouse parthenogenesis mentioned in the previous footnote, the resulting embryo was transferred to a third female mouse for gestation, leading it to be described in media reports as having three mothers and no father. Developmental biologist Azim Surani predicted that 'The popular press will say that men are redundant.' They certainly did so five years later when British scientists found a way to produce sperm-like cells from stem cells. 'The end of men?' asked the *Daily Mirror*, while the *Daily Mail* fretted that we are 'on the brink of a society without any need for men'.

Ludovici's bizarre vision was criticized by the feminist and socialist Dora Russell in *Hypatia, or Women and Knowledge* (1925), and by the Australian sexologist and sexual reformer Norman Haire in *Hymen, or the Future of Marriage* (1927). Haire did, however, share Ludovici's belief that animals might be used as gestational surrogates for human fetuses, and the shadow of eugenic planning hangs over his vision: he believed that surrogacy could enable children to be reared from women 'particularly suited to parenthood' in the event of their accidental death.

Ectogenesis remains to some later feminists a potentially emancipating technology. It was embraced by Shulamith Firestone in *The Dialectic of Sex* (1970), now considered one of the foundational texts of contemporary feminism. Firestone argued that the female role in reproduction was the fundamental reason for the social inequality of the sexes, and that this imbalance would never be addressed unless some other means of childbearing were to become possible. 'I submit, then,' she wrote,

that the first demand for any alternative system must be . . . the freeing of women from the tyranny of their reproductive biology by every means available, and the diffusion of the childbearing and childrearing role to the society as a whole, men as well as women.

This did not mean, she stressed, simply that there should be more social provision of childcare and that men should play a greater part in it. Rather, pregnancy itself needed to be abolished. Firestone's revulsion towards pregnancy – she called it 'barbaric' and compared childbirth to 'shitting a pumpkin' – guarantees that her position does not reflect the experiences and feelings of many women; but like all extreme views it usefully serves to locate one pole of the debate. Not only would ectogenesis remove the burden of motherhood, she said, but it would actually *benefit* the child by freeing them from the possessive mothering that pregnancy tended (in her view) to encourage. Ectogenetic children would be loved 'for their own sake', and not on account of the physical and personal sacrifices the mother had made. One might argue that this seems an extreme solution to the occasional feelings of victimhood in post-partum mothers, and that Firestone seems to be generalizing from a very particular narrative of what it is to be a mother. But her views contrast strikingly with the more common concern that ectogenetic severance of the mother–child bond *in utero* will do it harm in later life.

Early science-fiction writers, particularly women, were intrigued by the implications of ectogenesis, and their explorations of its benefits and consequences anticipated some of the discussions among scientists and social critics. Firestone's disgust at pregnancy is prefigured in Lilith Lorraine's 1930 novel *Into the 28th Century*, where she speaks of the 'horror' of childbirth before ectogenesis eliminated it. The fear, meanwhile, that ectogenesis will lead to a woman-dominated world was expressed in August Anson's *When Woman Reigns* (1938), where this has come to pass not through any Machiavellian scheme to do away with men but simply because liberation from childbearing means that women are no longer held back in society. And Sophie Wenzel Ellis's 1930 short story 'Creatures of the Light', although in most respects an undistinguished 'mad scientist' tale written for the pulp science-fiction magazine *Astounding Stories*, shows that ectogenesis and eugenics were already being blended in popular culture with the Frankenstein template. In this tale the hunchbacked and 'Teutonic' Dr Mundson is using a Life Ray to speed up the physical and mental development of babies produced from the eggs and seed of 'perfect couples' and grown in jars called 'Leyden jar mothers', harking back to the Leiden jars used in the late eighteenth century to 'bottle' electricity. He plans by this means to create a super-race:

He lifted one of the black velvet curtains that lined an entire side of the laboratory and thereby disclosed a globular jar of glass and metal, connected by wires to a dynamo. Above the jar was a Life Ray projector. Lilith slid aside a metal portion of the jar, disclosing through the glass underneath the squirming, kicking body of a baby, resting on a bed of soft, spongy substance, to which it was connected by the navel cord.

'The Leyden jar mother,' said Dr Mundson. 'It is the dream of us scientists realized. The human mother's body does nothing but nourish and protect her unborn child, a job which science can do better. And so, in New Eden, we take the young embryo and place it in the Leyden jar mother, where the Life Ray, electricity, and chemical food shortens the period of gestation to a few days.'

At that moment a bell under the Leyden jar began to ring. Dr Mundson uncovered the jar and lifted out the child, a beautiful, perfectly formed boy, who began to cry lustily.

'Here is one baby who'll never be kissed,' he said. 'He'll be nourished chemically, and, at the end of the week, will no longer be a baby. If you are patient, you can actually see the processes of development taking place under the Life Ray, for babies develop very fast.'

Northwood buried his face in his hands. 'Lord! This is awful. No child-hood; no mother to mould his mind! No parents to watch over him, to give him their tender care!'

Behind the leaden prose are perennial themes: electrical animation of life, the speeding up of a 'natural' process, the demise of the family, the baby in a bottle. And this, remember, predates *Brave New World* by two years.

Ectogenesis is used to create human clones in Pamela Sargent's 1976 science-fiction novel *Cloned Lives* (see p. 294). The 'artificial womb', she explains, 'was used only in cases of grave need, for women who could not survive a normal pregnancy or who bore premature chil-dren'. In an echo of the theological debate over Adam's physiology, the clones' genetic father Paul is relieved to find that there is an arti-ficial umbilical cord attached to the fetus, so 'at least they'll have navels'. Paul reminds himself that, once 'born', these children will be like any other; the wider world, it transpires, is not so sure.

Margaret Atwood's *The Handmaid's Tale* (1985) belongs in this genre even though it does not make explicit use of ectogenesis, for the 'handmaids' of the title are reduced to the status of 'living wombs', kept only for reproductive purposes: they free the wives of high-ranking officials from the burden of childbearing. This practice also recalls the ambitions of Haldane and the ectogeneticists to separate sexual pleasure from procreation. Yet although *The Handmaid's Tale* has several targets – among them, totalitarianism, religious funda-mentalism, anti-pornography feminism, and the oppression of women – it does not obviously offer a critique of reproductive technology. Indeed, it seems almost to present a *failure* of that technology: had artificial wombs been available, the handmaids (who are far more troublesome) would not have been needed. And in taking its lead from *Brave New World* by setting the narrative within a totalitarian, militaristic society regimented by a strict caste system, it is a throw-back to an old style of dystopia, more appropriate to the 1930s than the 1980s.

Machine surrogacy

The artificial womb is taking much longer to arrive than Haldane and his peers anticipated, but the objective of full-term gestation by artificial means is still explicitly on some scientists' research agendas. Japanese physician Yosinori Kumabara at Juntendou University in Tokyo has sustained goat fetuses for up to three weeks in plastic containers containing amniotic fluid, with a tube connected to the umbilical cord that provides nutrients. In 1997 he told reporters, 'If I have the time and money for experiments, maybe within ten years we will have made the move from animal to humans.' That hasn't yet happened; even if it does, the experiment will little resemble the 'test-tube uterus' of popular imagination (see the image below). Meanwhile, inter-species gestation has already been demonstrated, with ibex kids gestated within and born from the uteruses of goats. Some researchers believe that the first child gestated outside the human womb will indeed be sustained in that of a cow.

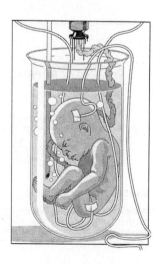

Ectogenesis as envisaged in an image from the Australian *Sunday Mail*.

One of the main challenges is that a uterus does much more than provide a protective environment for the fetus. It is in constant biological interaction with the growing organism and the placenta supporting it. According to reproductive biologist Roger Gosden, a former student of IVF pioneer Robert Edwards, 'Something much more sophisticated than tubes and pumps will be needed to replace the biochemical dialogue between mother and fetus for full development.' The placenta is crucial: it serves as a kind of multi-purpose organ while the fetus is developing its own.

Considerations of this kind motivate Gosden's colleague Hung-Ching Liu of the Weill Medical College at Cornell University to attempt to make an artificial uterus by tissue culture, for possible transplantation into women who lack a functional womb. Liu and her colleagues grow cells from the lining of the uterus on an artificial, biodegradable plastic scaffold. They have found that spare embryos from IVF will implant in the tissue and remain there for six days, at which point the experiments are stopped – this is too early to tell whether embryos can begin to develop placental tissue. When asked by a reporter whether it is 'science fiction to say maybe in the far future you could have a real breathing embryo and have a child in the laboratory', Liu replied simply, 'That's my final goal.' There are rumours not only that artificial uteruses have already been developed in China but that plans are afoot to transplant them to men. The newspaper *China Today* has reported the list of alleged requirements for 'potential male moms': 'A strong desire to have a child of their own genes and hereditary features; payment of a 200,000 yuan ($24,000) surgery fee; being possessed of a courageous spirit, and trust in science.' Well, indeed.

Should we welcome these developments or fear them? Australian bioethicist Peter Singer believes that women will be helped by the option of ectogenesis, while Aline Ferreira, a specialist in the cultural representation of reproductive technologies, has argued that 'the introduction of ectogenesis might be the decisive event paving the road to substantial operative changes to women's lives in the contemporary world'. In contrast to this optimism, Rosemarie Tong, a bioethicist at the University of North Carolina, believes that 'artificial wombs could lead to a commodification of the whole process of pregnancy . . . it may lead to a view of the growing child as a "thing"'. According to Christine Rosen, an editor of the magazine *New Atlantis* which takes a conservative stance on reproductive technologies, 'at stake in this debate is the very meaning of human pregnancy: the meaning of the mother–child relationship, the nature of the female body, and the significance of being born, not "made"'.

Misgivings about the enterprise are understandable when we consider the callow manner in which ectogenesis has occasionally been presented. In his 1971 survey of new reproductive technologies, science writer David Rorvik unwittingly presented the beneficiaries as naive, complacent, consumerist gulls straight out of a 1950s advertisement:

It seems apparent that many women . . . would avail themselves of the services of the test-tube mother [the artificial womb] as soon as they were proffered, eagerly forgoing the rigors of pregnancy by dropping off at the baby factory their two- or three-day-old embryos for predestination, gestation and decanting. Along the nine-month way, they could drop in at the laboratory to check on their babies' progress, perhaps ordering a few changes here and there. The 'coming out' party could well take on a whole new meaning as proud parents and their friends gather at the mouth of glass-and-steel wombs for the 'birth' of their babies, once they've reached term.

One assumes that the party will be liberally supplied with soma.

It is no surprise that Jeremy Rifkin, a veteran critic of biotechnological innovation, should oppose ectogenesis, but his objections are revealing. When he asks 'What kind of child will we produce from a liquid medium inside a plastic box?', the question is valid enough. But he goes on to say:

How will the elimination of pregnancy affect the concept of parental responsibility? Will parents feel less attached to their offspring? Will it undermine the sense of generational continuity that is so essential for reproducing and maintaining historical continuity and civilised life?

Will ectogenesis, in other words, bring about the end of the family, and then of civilization? One might imagine that Rifkin could have sought some answers (they are reassuring) from the wealth of research on children who are *not* the biological offspring of their parents. That he did not suggests his real concerns are more tied up with concepts of 'naturalness' than with actual human experience.

What of the argument that ectogenesis would deprive women of their procreative prerogative altogether? Rifkin puts that concern in these terms:

Other feminists view the artificial womb as the final marginalisation of women, robbing them of their primary role as progenitor of the species. The artificial womb, they argue, becomes the quintessential expression of male dominance, a way to create a mechanical substitute of the female womb. Armed with the artificial womb, asexual cloning technology and stem cells to produce all the extra body parts they need, men could free themselves, once and for all, from their dependency on women.

Australian sociologist Robyn Rowland shares this view, predicated on the notion that the ability to conceive a child is the only bargaining chip with which women can purchase any social or personal power. If the entire process of conception and gestation can be done *ex utero*, she says, 'We may find ourselves without a product of any kind with which to bargain. We have to ask, if that last power is taken and controlled by men, what role is envisaged for women in the new world? Will women become obsolete?' Setting aside this terribly bleak view of gender relations (and its implication that women who cannot conceive already have no power whatsoever), we can now begin to recognize this fear of gender obsolescence for what it is: one of the central myths of anthropoeia.*

The counterclaim of feminists like Firestone is that the provision of ectogenesis will offer woman greater reproductive *choice*. Certainly, it is hard to see why it should be prohibited while surrogacy is permitted; it avoids the charge of inequality levied against surrogacy (that is, the idea of rented wombs in the developing world or in poor communities), as well as the legalities and anguish of reneged contracts.

Yet it would be irresponsible to suggest that the decision should be left solely to the mother or the parents, for they are not of course the only parties in the process. It remains far from clear that a child gestated wholly *ex utero* would not be damaged physically or emotionally from

---

* The headline of a report in the UK's *Observer* newspaper on Hung-Ching Liu's work on artificial wombs managed the sleight of hand of doing away with both sexes: 'Men redundant? Now we don't need women either.'

the experience. And when there is the risk of harm to a child, the state may legitimately intervene. (If it does not prevent pregnant mothers from actions that might harm the developing child, this is because of the legal and practical obstacles to doing so, not because such actions are deemed an acceptable expression of personal choice.)

We should be wary of privileging the mother–infant emotional bond that may develop during pregnancy – not only is this not every woman's experience, but it is clear that strong and secure psychological attachment can develop between a child and a (for want of a better term) non-biological mother. We should certainly be wary of suggestions that ectogenesis will inevitably sever the mother–child bond simply because it violates 'human nature', an argument seemingly made by the conservative political journalist Charles Krauthammer speaking on George W. Bush's Council on Bioethics (of which we will hear more later):

Why do we want the embryo to be housed in its mother? One of the reasons is that it creates an innate connection between the child and the mother, and the mother becomes uniquely protective and attached. That's human nature.

Nonetheless, all else being equal, it might turn out to be preferable for a child to be gestated inside the mother's womb rather than not. My (ungrounded) suspicion is that this is so. But that is not necessarily the choice that ectogenesis would speak to – one might imagine its use being restricted, for example, only to embryos that would otherwise not survive, or would otherwise not even be conceived, for example because the mother's womb is incapable of harbouring an embryo. What is more, this kind of objection is rarely couched in terms of what is *preferable*, but in terms of absolute alternatives – in particular the presence or absence of a mother–child bond – that are demonstrably ficticious and survive that obvious fact only by drawing on the powerfully seductive rhetoric of 'human nature'.

It is not immediately clear how the risk of harm to the child could ever be tested without actually doing the experiment. One might have said the same about IVF, except that in that case a certain amount of information on the physical risks was available from studies of non-human animals. While such findings did not necessarily translate across species, there were no obvious reasons to suspect they would not.

And the risk of psychological damage as a direct result of having been merely *conceived* outside the body never sounded plausible. But for ectogenesis, where many of the strongest concerns are psychological rather than physical, animal models are silent.

This would appear to create an ethical impasse: we cannot know without doing, and we should not do without knowing. But there is a way the barrier might be overcome. In 1984, Peter Singer and bio-ethicist Deanne Wells forecast that ectogenesis would become a reality essentially by accident: as embryos created by IVF are sustained in a viable condition for ever longer periods before being returned to the womb, and as new technologies made it possible to keep premature babies alive from an ever earlier stage, eventually the two limits would meet. This, said Singer and Wells, would eliminate the principal ethical objection – that the procedure could inflict psychological and physical damage on the fetus – by default, since no one could claim that attempting to keep a premature baby alive was ethically wrong. And as such babies spent an ever shorter time in the womb, we would gradually learn about the possible consequences for development without inflicting avoidable harm on the child.

But since 1984 the gap separating IVF embryos from premature babies has not significantly narrowed; in fact, Roger Gosden says that the scientific progress 'has slowed to a crawl'. However, ectogenesis of animals is approaching feasibility. Developmental biologist Lynne Selwood at the University of Melbourne has, for example, cultured embryos of a mouse-like marsupial called the stripe-faced dunnart (with a gestation period of eleven days) to within one day of full term. Selwood's interest in such reproductive technology stems from its potential value for marsupial conservation.

What if ectogenesis were found to cause no or negligible risk to the child? Should we then condone it? Are we simply at the comparable stage to the early research on IVF, when it was widely opposed on the grounds of vague discomfort with the 'unnaturalness' of the procedure? In cases where the only alternative to ectogenetic growth of a child is the equally fraught option of surrogacy – where, for example, a woman is unable to gestate a child herself – it is far from clear that a potential child is better off unborn than being subjected to the risks of such an origin. (The ethical and philosophical problems with this kind of counterfactual argument are discussed further in Chapter 11.)

We may feel uneasy about the use of ectogenesis as a matter of mere convenience, but we need to recognize at least that women have been among the strongest advocates of that option (while granting that the roots of the problems a woman might wish to avoid through ectogenesis are not located in pregnancy and child-rearing as such but in the society in which this happens). And some fertility researchers suspect that ectogenesis will, in making the growth of the child directly visible, actually 'enhance rather than diminish the value we place on the fetus' – arguably, ultrasound imaging has already had an effect of that kind.

Others will never be convinced that there is a place for an artificial womb. To the extent that this is a visceral response, it should not be rejected out of hand as a triumph of sentiment over reason: visceral responses may have their reasons too. But when we look closely at the arguments, what we often find are the unexamined assumptions to which this book is addressed. Christine Rosen says:

Even the phrase 'artificial womb' appears at odds with itself: 'artificial' conjures images of chemical sweeteners, synthetic fabrics, second-best imitations, while 'womb' still retains its mystery and its gravity.

She means here to imply that there is something awry in this juxtaposition of words. But the assertion demands that we swallow whole the mythology of the 'artificial'. 'There is something about being born of a human being – rather than a cow or an incubator – that fundamentally makes us human', says Rosen. It sounds right, until you notice the corollary that a child gestated in a cow's uterus or an incubator is somehow *fundamentally not human*. Isn't this where we always end up: by denying the humanity of those who were not made 'like us'? If something is lost in the process of ectogenesis, it will most certainly not be humanness.

Ectogenesis creates genuine ethical, social and medical real dilemmas. All I should like humbly to suggest is that visions of *Brave New World*, gender annihilation, subhuman monsters and unnatural progeny should no longer provide the framework for debating them. Aldous Huxley's book created one of the most powerful myths of the twentieth century. A myth is precisely what it should remain.

9 Opening the Bottle

Now, it appears to many, the sexual act ... has been dragged out and 'manipulated' on the laboratory bench, and therefore degraded in some way.
Gerald Leach (1970)

We are entering an age of human manufacture ... That has never happened in human history. We now have in our hands the technology where we can make and create kinds of human life, variants of human life never before imagined.
Charles Krauthammer (2002)

The truth of J. B. S. Haldane's dictum that every biological innovation is seen initially as a perversion was never more apparent than when, in 1969, Robert Edwards and Patrick Steptoe announced the first successful fertilization of a human egg by sperm outside the human body – in a 'test tube', as it were (although test tubes were never involved). A BBC television series in the wake of the report warned of the possibility of 'mass-producing people without the advent of a mother at all'. A *New York Times* article three years later described progress towards *in vitro* fertilization under the title 'The Frankenstein myth becomes reality', and highlighted the comment that 'We have the awful knowledge to make exact copies of human beings', placed alongside a picture of Hitler.

Reactions of this sort prompted an editorial in *Nature*, which had published Edwards and Steptoe's results, saying that 'these are not perverted men in white coats doing nasty experiments on human beings, but reasonable scientists carrying out perfectly justifiable research.' Yet merely by conjuring up this negative image, the defence arguably only helped to perpetuate it.

And when, in what might be (contentiously) considered the first case of genuine anthropoeia, the 'test-tube baby' Louise Brown was born in Oldham hospital in July 1978, the same set of responses was rehearsed. *Time* magazine announced that '[Aldous] Huxley's prophetic vision had become reality', and to cover all bases the magazine's editorial threw in references to Frankenstein and Faust's homunculus. The media response was a queasy mix of celebration at the implications for treating infertility and hand-wringing about the breeding of 'superbabies' and Master Races. If all this sounds hyperbolic today, we have nevertheless learnt little from it: each new development in reproductive technology elicits the same stock imagery, as we shall see.

Assisted conception is in its broadest sense a very old technology – it is hard to know when artificial insemination was first practised, since it would doubtless have been deemed illicit for several reasons.* But IVF served to focus the cultural myths that have grown up around it, which are based largely on notions of a 'natural order' of procreation and the moral and physical dangers of violating it. Sensationalization of assisted reproduction, whether by association with monstrosity (the danger of deformity in babies) or miracles (as in 'miracle babies'), then acts as a defence against the ways in which it is deemed to threaten norms of family, sex, and the functions of male and female bodies. A quarter of a century after Louise Brown's birth, American professor of women's studies Karyn Valerius said that

assisted reproduction continues to provoke anxiety that technological intervention in procreation might compromise the humanity of humans, and monstrosity continues to figure prominently in the mediascape inhabited by high-tech assisted reproduction . . . Frankenstein's monster, that icon of scientific hubris, remains an absent presence lurking behind the various public

---

* Even when artificial insemination was introduced for cattle breeding in the 1930s, bishops objected to its 'unnaturalness'. It was officially sanctioned for use in humans from the early twentieth century, but was considered 'adulterous' when donor sperm was used, and until the 1960s the children born this way were deemed illegitimate even if the mother was married.

culture representations of assisted reproduction as unnatural conception with potentially hazardous results ... Alternately celebratory, incredulous, and fearful, much of the US public discourse on assisted reproduction tropes on literary, popular, and religious traditions of monstrosity to position assisted reproduction as a fantastic and unprecedented departure from natural procreation.

Assisted conception, she adds, 'destabilizes any simple equation of the biological with the natural.'

Perhaps most significant is the way that IVF and other techniques of assisted conception have literally exposed the hitherto hidden aspects of human reproduction: the formation and growth of an embryo. In doing so, they have turned the embryo and fetus into fetish objects, repositories for a host of preconceptions and prejudices about personhood, sex and family. According to feminist writer Donna Haraway, the fetus acts now in US culture as 'a kind of metonym, seed crystal or icon for configurations of person, family, nation, origin, choice, life, and future'. The test-tube baby carries a lot on its little shoulders.

Who is to blame?

Underpinning social attitudes to the 'baby-making' technology of *in vitro* fertilization is the unacknowledged fact that infertility has long been seen as a moral issue. This is clear in the biblical story of Rachel and Leah. Jacob loves his cousin Rachel, but through the duplicity of her father Laban he finds himself married to her older sister Leah. 'It is not our custom', Laban explains, 'to give the younger daughter in marriage before the older one.' In this polygamous culture, Jacob may marry Rachel too, and he later does so. But the Lord gives children only to Leah, so that Jacob might come to love her too. Poor Rachel is condemned by God to be infertile: 'Give me children, or I'll die!' she implores. Aware of her sister's jealousy, Leah reminds her bitterly that she has stolen her husband's love. These family dynamics are manipulated capriciously by God, who confers or withholds children as he wishes: Rachel's 'barrenness' is recognized by all as a sign of divine disfavour, although she seems to have done nothing to warrant it except to have been Jacob's first love. Even she admits that her situation is cause for shame; when God finally relents and

Rachel bears a son, Joseph, she declares, 'God has taken away my disgrace.'

In some obscure moral calculus, then, Rachel is regarded as having 'deserved' her plight. In the libertarian environment of Restoration England, this assessment took a curiously inverted form: women who were infertile (or apparently so) brought it upon themselves because they were averse to sex. As a manual of midwifery from 1694 put it, 'Sterility happens likewise, from the Womans Disgust ... or her dullness and insensibility therein.' In many cultures, it has always been seen as a woman's duty to bear her husband children – and if she did not, this was occasion not for pity but for shame, punishment, expulsion, even murder.

In other words, not only is infertility a moral judgement but this judgement has almost invariably fallen on women. That men might suffer from infertility has been denied throughout history. The notorious witch-hunting manual of the late fifteenth century, the *Malleus maleficarum*, blames male problems with conception on witches, who could make men impotent, perform sterilization and castration, and even make men's penises 'disappear'.

While both of these themes are still explicit today in cultures that make male infertility taboo and female infertility a curse, in the West they have simply been reformulated. Media stories of women fretting over their ticking biological clock are written by women for women, barely betraying the fact that men might have a role in baby-making: this is a women-only issue. And the stigmatizing of women who do not, for whatever reason, bear children, has always been insisted upon by both sexes. Arguably it has been the condemnation of other women that has cut deepest, which is not to deny the patriarchal origins of the morality of infertility but reminds us that its acceptance tends to be universal.

The development of IVF over the past four decades has played out within this context. Those who seek to use these procedures have often been portrayed as selfish: they only have themselves to blame for leaving things too late, they divert precious medical funding better spent in other ways, they are too egotistical to adopt. And so this way of making babies is tainted from the outset not just because it is 'unnatural' but also because the motivation is deemed to be morally dubious – or even, if this does not stretch the point too far (I do not think it does), because it seeks to evade the 'divine' chastisement that infertility represents.

Conception in the looking glass

Robert Edwards meandered almost by chance into the study of mammalian fertilization. He began his career reluctantly studying agriculture at the University of Bangor in North Wales, but switched to zoology, in which context he developed an interest in embryology. He was intrigued by one question in particular: 'Why does only one spermatozoon enter an egg?' After taking a postgraduate course in genetics at Edinburgh University, he decided, almost on impulse and despite his undistinguished academic credentials, to apply for a research position there with the famous embryologist and geneticist Conrad Hal Waddington. To his surprise Waddington gave him a post, and he began to research fertilization of mouse eggs *in vitro*.

One could hardly work on mammals without wondering how the findings might apply to humans. Edwards became aware that an American biologist named Gregory Pincus had already thought about that three decades earlier. Pincus was an expert on the hormonal signals involved in the process of 'ripening' that prepares eggs for fertilization: knowledge that he eventually used to develop the contraceptive pill. In 1934 Pincus claimed to have fertilized rabbit eggs with sperm *in vitro* and implanted them in a rabbit uterus to bring about gestation and birth. It later transpired that these eggs were not actually fertilized, but had developed parthenogenetically, much like Jacques Loeb's sea-urchin eggs. Yet Pincus's claim grabbed headlines in its time, prompting the narratives that later became obligatory. It is scarcely surprising that he should have been compared (in the *New York Times*) to Aldous Huxley's Bokanovsky, given that *Brave New World* had been published just two years previously: the newspaper claimed that Pincus's work might lead to embryo factories in which 'ninety-six human beings grow where only one grew before'. The article also explained that human *in vitro* fertilization would serve the same ends then being forecast for ectogenesis: facilitating eugenic selection, emancipating women, and separating love and sex from the imperatives of procreation.*

---

* Pincus's work on the contraceptive pill in the 1960s genuinely achieved the latter two objectives, but we might note that eugenics was still on the agenda then: the birth-control advocate Margaret Sanger, who enlisted Pincus for the research and helped fund it, was a committed eugenicist.

On the other hand, *Collier's Magazine* saw in the use of disembodied sperm a threat to the male: an article on Pincus's work in 1937 was called 'No father to guide them'. Linking this 'goofy' experiment to the work of Jacques Loeb ('the hugely famous Portuguese Jew') on parthenogenesis,* the journalist claimed that 'Man's value would shrink, the mythical land of the Amazons would then come to life. A world where woman would be self-sufficient; man's value precisely zero.'

In the 1940s Pincus claimed to have fertilized human eggs in a Petri dish, but this was never verified. Similar contentions, likewise unsupported by convincing evidence, were made in 1944 by the American gynaecologist John Rock, who (despite his Catholicism) led the clinical trials for Pincus's contraceptive pill. In what might now be seen as something of an own goal, Rock and his colleague Arthur Hertig cited *Brave New World* as a source of inspiration.

Regardless of what he actually achieved, Pincus's claims encouraged other efforts. In the 1950s the Chinese-American biologist Min Chueh Chang presented persuasive evidence for live births of rabbits conceived by IVF. And work was continuing on human eggs: Daniele Petrucci at the University of Bologna alleged that he had grown a fertilized human egg into an embryo that he cultured for twenty-nine days (some reports put it at an even less credible fifty-nine days) before destroying it because it became deformed. In 1961 he said he was ready to transfer an IVF embryo into a woman's uterus. Again, these assertions were never substantiated.

There was always controversy about such research, but in attempting to follow up Rock's work the American biologist Landrum Shettles of Columbia University fanned the flames with apparent glee. He too claimed in the 1950s to have fertilized a human egg and to have sustained it for several days. When Shettles attended a fertility conference in Italy in 1954, the Pope referred pointedly to those who 'take the Lord's work' into their own hands. But no one would have mistaken Shettles for a saint in the first place. By all accounts he was not the sort of person to instil confidence in science's wisdom to intervene in human

---

* This connection was perceived more widely; the *New York Times* asserted that 'Not since Professor Jacques Loeb hatched fatherless sea-urchin larvae . . . has so striking a success been achieved.'

procreation. With the puerile machismo of that era, he bragged that he would happily develop an egg fertilized *in vitro* into an adult, because 'There's nothing I'd like better than to grow a beautiful lab assistant.' (Here once again we have the image of the sex-doll replicant.) Writer Robin Marantz Henig says that Shettles 'seemed to have almost an obsession with human eggs'. He published an entire volume of photographs of eggs in the early stages of development, called (in anachronistic style) *Ovum Humanum*. 'His manner was strange', says Henig, 'because he didn't look people in the eye, he was awkward. He couldn't quite live in this world.' Despite (or because of?) being the father of seven children, Shettles is said to have spent most nights in his workplace, Columbia Presbyterian Hospital in New York. He later courted more controversy by promoting a method that allegedly enabled couples to determine the sex of their child by timing their intercourse appropriately. We will hear more of Landrum Shettles shortly.

All this meant that Edwards was by no means venturing into uncharted territory. Nevertheless, the idea of conducting IVF experiments on humans remained unpopular even among many scientists, as Edwards discovered soon enough. After a brief stint in California, he moved to the National Institute for Medical Research in Mill Hill, north London, where the director Charles Harrington became aware of the experiments Edwards was conducting with human ova. 'I don't want any human eggs fertilized here', Harrington growled. Edwards' departmental head at Mill Hill, Audrey Smith, shared those reservations but seemed reluctant or unable to articulate them. It is unethical, she declared; when Edwards asked her why, all she could say was: 'Because it is.'

To be sure, there were ethical questions aplenty with Edwards' egg-procurement strategy, although this is scarcely what the critics had in mind. He would simply go begging to surgeons and gynaecologists until he found ones sympathetic to his aims – the eggs came mostly from patients undergoing ovarian surgery at Edgware General Hospital. There were at that time no ethical guidelines for this sort of research, and Edwards did things that today would be considered grossly unacceptable.* The egg donors were not asked for their consent – they did

---

* Edwards was at least respectfully humble about his work. In contrast, when Landrum Shettles was conducting his IVF experiments in the 1950s, he bragged that he simply 'poached' his eggs (it's not clear the pun was intended).

not even know about the research – and Edwards would use his own sperm to attempt fertilization. It was inevitable that the work lacked proper controls, since Edwards was working in virtual isolation, with no precedent, and almost no one knew he was doing it.

Yet this all plays into the mythical image of the lone scientist dabbling with life by robbing graveyards and charnel houses for human parts. And Edwards unwittingly contributed to that perception, for his account of the work conducted after leaving Mill Hill for Cambridge conjures up visions of Paracelsus, secluded from the rest of humanity in his alchemical laboratory on a lonely quest to make a being in a flask of putrefied sperm:

Sometimes, at night, I could hear below, across the way, rock music coming from the Regal Cinema or, more distantly, the harmonious sound of a Cambridge University choir. That evening, as I collected my own semen, removed the excess fluid and added my spermatozoa to the eggs in the culture fluid, there was no music or noise at all. Just a tap dripping into the sink.

Creation or procreation?

The press first became aware of Edwards' attempts at *in vitro* fertilization when he reported preliminary results in the *Lancet* in 1965. With what it seemed to regard as perceptive originality, the *Sunday Times* proclaimed them 'reminiscent of Aldous Huxley's *Brave New World*'. Soon, the news report claimed ambiguously but ominously, 'births may be by proxy'. It was a taste of what was to come, and left Edwards wary of journalists.

Needless to say, *in vitro* fertilization to make an embryo was only the initial stage of actually making babies this way. Recognizing that he needed help from a surgeon in order to extract eggs from patients at the right stage and to replace them in the Fallopian tubes after fertilization, Edwards teamed up with gynaecologist Patrick Steptoe at Oldham hospital. Steptoe was a specialist in the technique of laparoscopy, in which a small telescope-like instrument allowed a surgeon to see inside the abdomen during surgery. He was also a rarity among clinicians: a man who not only understood what Edwards was attempting but sympathized with his motives and believed it was possible. Having treated many women for fertility problems, Steptoe was very familiar with the anguish it caused, and he hoped IVF might overcome infertility caused by blockages of the Fallopian tubes, which prevented eggs descending from the ovaries.

In their landmark paper in *Nature* in 1969, Edwards and Steptoe, along with Edwards' doctoral student Barry Bavister, reported that a human egg exposed to sperm within a culture medium of essential nutrients in a Petri dish had shown the telltale signs of fertilization. The researchers had needed to experiment endlessly and frustratingly with different culture fluids and growth conditions, and their eventual success was by any standards a dramatic finding. But it was not nearly as dramatic as the press was to insist, with headlines almost identical to those that greeted Jacques Loeb's artificial parthenogenesis: 'Life is created in test tube'.

The 'test-tube baby' was immediately granted linguistic currency, and led to inevitable confusion about whether the procedure actually involved a form of ectogenesis in glass – an idea perpetuated in popular images of IVF, in which the 'being' in the vessel is rarely a fertilized egg but instead a full-term infant or even a young child. The image of sperm and egg united in a glass vessel has alchemical associations: it is the Chymical Wedding under a microscope, a process that, like the alchemist's ministrations, merely assists what could happen naturally. Or so the 'alchemists' say; but does the suspicion of unnaturalness still cling to it? The test tube was now the symbolic crucible of the chemist, where miraculous or baleful substances were concocted. Well-motivated efforts by scientists to do away with the term 'test-tube baby' altogether – the molecular biologist and Nobel laureate Joshua Lederberg said it was like calling surgery 'anatomical manipulation' – were doomed to fail because such images are never arbitrary but arise for deep-seated reasons.

This elision of *in vitro* fertilization and gestation was often explicit. A British television report on the *Nature* paper claimed that 'the scientists are now in a position not only to replace the fertilized egg in the mother's body, but to continue to develop the fertilized egg artificially and perhaps even to produce a human baby without using the mother's body at all'. They were, as we have seen, in a position to do no such thing, nor had any intention of trying. Meanwhile, the *Guardian* newspaper deemed it necessary to point out that this was not after all the *de novo* synthesis of a precursor to a person: 'Life, or at least the

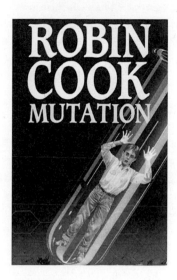

Popular images of IVF tend to show a full-term infant or even a grown child in a 'test tube'. Left © Monkey Business Images/Shutterstock.

potentiality of life, resides in the oocyte* and the spermatozoa; the artificial creation of either of these is so far beyond the present horizon of knowledge that it can be discounted as a serious possibility.' The need for this disclaimer indicates that the confusion existed, and it is unlikely that the *Guardian* dispelled it.

Moreover, that newspaper's efforts to banish misconceptions about what was being 'made' in the experiments of Edwards and Steptoe were somewhat undermined by accompanying a description of the work with a cartoon that combined all the archetypal anthropoetic themes with the Oedipus myth. It showed a white-coated scientist making a baby in a test tube that grows to monstrous size and is finally seen holding the pleading scientist imprisoned inside a test tube of his own. The mix'n'match imagery makes essentially no contact at all with the aims, results, implications or even the ethics of the *Nature* paper (unless one grants the very loose metaphor that the work of Edwards and Steptoe was going to grow out of all proportion, with dire consequences). Yet

---

* 'Oocyte' can be loosely considered synonymous with 'egg' or 'ovum', although in fact it refers to an immature egg, which has not yet gone through the ripening stages.

somehow readers will have imagined they knew what was being implied, and why. This appropriation of vague associations was not uncommon: *Time* called Edwards' Cambridge laboratory 'Orwell's baby farm', apparently a totalitarian fantasy brewed up from a mixture of *Brave New World, Nineteen Eighty-Four* and *Animal Farm*.

Not all the coverage was quite so hyperbolic, but an article in *Life* magazine illustrated many of the fixations that began at once to crystallize. The main cover image was of a fetus with recognizably human features – eyes, ears, nose, hands and feet – attached by an umbilical cord to a mass of tissue and yet apparently resting on a flat, artificial surface and visible through a round glass porthole. The implication here is again that gestation, not merely fertilization, will proceed *in vitro*. Yet this effect is somewhat offset by a photo of a mother holding her baby, who, while certainly not newborn, is apparently quite normal. And the cover words highlight what, for *Life* at least, were the key questions:

When new methods of human reproduction become available –
- Can traditional family life survive?
- Will marital infidelity increase?
- Will children and parents still love each other?
- Would you be willing to have a 'test-tube' baby?

How *Life* covered the test-tube baby.

I will consider shortly the concerns about family structures. And it seems I lack the imagination to see how IVF might increase infidelity ('Fancy a romantic liaison at the fertility clinic, darling? We won't have sex, but I'll masturbate in a cupboard and you'll have a laparoscopy'). It was no coincidence that IVF was developed against a backdrop of sexual liberation and uncertainties about the relationship of sex to procreation – for these things were ushered in by the contraceptive pill, the requisite scientific advances for which (mostly concerning the use of hormones to regulate the female reproductive cycle) were not wholly independent. But the resulting climate of unease about gender roles, marriage and family values meant that IVF was never going to be judged simply as a medical solution to infertility for couples, which is how Steptoe and Edwards envisaged it.

The doubts expressed in *Life* (and elsewhere) about the formation of an affective bond between an IVF child and his or her parents seem to suggest that this bond is associated with the moment of conception – as though, if this event took place outside the body, a connection would fail to develop. Put like that, it is exposed as a more or less occult belief, hard to comprehend or even to formulate in rational terms. But it is a belief that was apparently widely held. A poll of 1,600 American adults conducted for *Life* indicated that thirty-nine per cent of women and forty-five per cent of men doubted that an '*in vitro* child would feel love for family'. Notice that it is the feelings of the child, not of the parents, that are deemed to be affected: IVF children would, apparently, be cold, indifferent, *soulless* creatures. While we should be wary of reading too much into an anecdotal incident, there is probably a symptom of broader attitudes in the note left attached to the car of Californian fertility doctor Jeffrey Steinberg when he first began offering an IVF service: 'Test tube babies have no souls.'

The accompanying article in *Life* shows that, from the outset, the reception of IVF was dictated by imagined futures – or rather, one must say, reimagined, for the fantasies will now start to have a familiar, even clichéd, ring to them. The article (adapted from the book *The Second Genesis* by *Life*'s science editor Albert Rosenfeld) began by saying that 'Radical new techniques in biology promise (or threaten) a near-future world of reproduction that is artificially assisted or even sexless and of babies grown in glass wombs in the laboratory'. Rosenfeld went on to talk about an *'in vitro* world', in which it was implied that IVF was merely the first step towards *in vitro* ectogenesis, which in turn would lead to the severance of children from their biological parents – to be raised, perhaps, by 'volunteer parents' or in a 'state nursery'. Reproductive technology, in other words, was portrayed as having a deterministic social trajectory ordained by the gravitational pull of Aldous Huxley's dystopia.

An editorial in *The Times* in response to Edwards' and Steptoe's work meanwhile fumed about impoverished nations 'breeding a race of intellectual giants', while the newspaper's science correspondent foresaw state-controlled human cloning:

Ultimately we could have the know-how to breed these groups of human beings – called 'clones' after the Greek word for a throng [*sic*] – to produce a cohort of super-astronauts or dustmen, soldiers and senators, each with identical physical and mental characteristics most suited to the job they have to do.

All these speculations resolutely refused to locate themselves in the society that produced them. The social and moral transformations that would be necessary before any government or institution could contemplate breeding a cohort of identical dustmen make the scientific problems seem a relatively minor part of the challenge. This is a recurring characteristic of dystopian warnings against reproductive technologies: they build the dystopia in advance.

Evidently, even professional science communicators were by no means immune to the unconscious assumptions and expectations that IVF aroused. The title of the 1971 book by *Time*'s scientific and medical reporter David Rorvik, *Brave New Baby*, established his point of reference from the outset, and he took care to ensure that the reader's vision should never stray too far from Huxley's scenario. He called the scientists working on this new technique the 'predestinators of Aldous Huxley's prophetic fiction materializ[ing] in the form of white-frocked lab technicians who tend test-tube babies by day and play cards and go to the movies like everyone else at night' – the assertion of normality inviting us to be surprised that these scientist-priests who 'tend' their flocks of fetuses are just like us.

Rorvik reminds the reader that these men have been called 'monsters', and he insists on the Faustian/Promethean aspect of their work:

When man assumes – some might say *presumes* – powers previously held to be the exclusive province of God and nature, society will be forced, at the very least, to provide new definitions of God and man, perhaps to conclude, in a flash of irresistible hubris, that they are one and the same, or even that the latter has eclipsed the former.

One can't entirely blame Rorvik for this simplistic view when we consider what some scientists were saying. According to Rorvik, the reproductive biologist E. S. E. Hafez at Washington State University painted the following picture:

Hafez . . . foresees the day, perhaps only ten or fifteen years hence, when a wife can stroll through a special kind of market and select her baby from a wide selection of one-day-old *frozen* embryos, guaranteed free of all genetic defects and described, as to sex, eye color, probable IQ, and so on, in details on the label. A color picture of what the grown-up product is likely to resemble, he says, could also be included on the outside of the packet.

As this lurid image recognizes, an IVF embryo can be placed in *any* womb, not necessarily that of the genetic mother. This was felt to imply that all manner of 'unnatural' family structures would be created – or rather, groupings that did not, in the popular view, qualify as a family at all. Hafez's scenario also shows that the technology was thought to herald the age of the designer baby, with features and abilities made to order, and that it was expected rapidly to progress from a medical to a consumer technology: babies would be bought and sold in the marketplace, a 'product' packaged and labelled like frozen peas. None of these expectations is pure fantasy, but they tended to be spun out in a social vacuum, devoid of any consideration of what a society will permit or deem proper. The message is that technology alone will determine the course it takes. Most often, the context for such flights of imagination is pulp science fiction – for as Rorvik goes on to say,

Ultimately, Dr Hafez believes, entire colonies of men and animals can be contained in small, frozen packages and launched to distant planets, along with a few fully grown scientists and mother surrogates.

In this chillingly Baconian fantasy, the scientists are in total control while the 'mother surrogates' are basically just walking wombs. That Rorvik sees in this scheme nothing but 'Buck Rogers charm' should perhaps not surprise us coming from someone who could sing the praises of eugenic surrogacy thus:

It is a technique that will be used not only to improve the quality of man, but to give the barren woman the fulfilling experience of childbirth, and the wealthy woman the opportunity to avoid the rigors of pregnancy and still have children of her own.

Yet if Rorvik was an especially callow commentator, it was by journalists like him that the public were told what IVF ultimately implied. The real issue, however, both then and now, is not that the science is simplistically misrepresented, but *what forms that misrepresentation takes*.

Edwards and Steptoe were frustrated by such fanciful reporting. There is no question, they insisted in a press conference, of 'mass-producing test-tube babies'. People become alarmed, said Edwards,

because their imaginations have already been dramatically doom-lit and gaudily coloured by science-fiction fantasies and visions – fantasies of horror and disaster, and visions of white-coated, heartless men, breeding and rearing embryos in the laboratory to bring forth Frankenstein genetic monsters.

Their reaction, like that of many scientists researching human fertilization and development today, was simply to lament these 'science-fiction' stories, to imagine that if Mary Shelley and Aldous Huxley had never committed them to paper then all would be well. That complaint, while understandable, fails to perceive the root of the problem. As we've seen, neither Shelley nor Huxley *invented* their stories.

'Miracle' baby

The *in vitro* fertilization described by Edwards and Steptoe in 1969 shocked the world, but it was only the first step along a profoundly uncertain path. Would the fertilized ovum implant and develop normally if returned to the womb? In those early days IVF was even more invasive and painful for women than it is today: it requires a programme of hormone treatments to control the cycle of ovulation, and surgical intervention to extract the eggs and put back some that fertilize. No one could be certain that it would work at all, nor what, if implantation happened, the consequences would be. Yet as soon as Edwards and Steptoe announced that they were seeking volunteers at Oldham hospital, they were inundated with requests.

The press, now fully aware of the unfolding story, continued doggedly to follow the trail with a mixture of apocalyptic spleen and prying prurience. While journalists tried to track down couples 'waiting for a test-tube baby', a BBC documentary on the developments began with footage of the Hiroshima bomb, lest we should forget the perils of scientific discovery and technical innovation. The British tabloid the *Sun* paraded its trademark disdain for nuance by demanding that we should 'Ban the test tube baby'.

That would have pleased many Christians. In an article in the prestigious *Journal of the American Medical Association* in 1972, the American Methodist theologian Paul Ramsey called work on IVF 'unethical medical experimentation on possible future human beings', which was therefore 'subject to absolute moral prohibition'. The Catholic journal *America*, believing that some eggs would be fertilized by IVF solely for research purposes, had still harsher words:

The spirit of Frankenstein did not die with the Third Reich. His blood brothers regard a human being as just another expendable microbe, provided it is legally defenceless, physically helpless, and tiny enough to ride on the stage of the microscope . . . To produce a human being, holding it captive like a genie in a bottle, and doom it to inevitable death is to exercise an irresponsible dominion that cannot be justified by any appeal to the common welfare of mankind or to the advancement of scientific goals.

If this were not bad enough, the scientific community continued to look askance. An application by Edwards and Steptoe for a grant from the Medical Research Council was declined, partly on the grounds that not enough was known about the safety of the process – an issue that of course could not be addressed unless further research was supported.*

Public opinion, while mixed, could not be transparently read from polls. As is typical for advances in biomedical science, people tended to be more receptive when they were told of the potential benefits of the technique rather than simply given a description of it. The poll for *Life* in 1969 had shown that, while only nineteen per cent of people approved of artificial insemination on the basis of a technical account, nearly twice that many approved when informed that it might offer a couple their only chance to have a child. All the same, this poll revealed that a majority opposed assisted conception in general, and that just one in four people approved of IVF. Opinion shifted as the technology evolved, perhaps simply because of a growing awareness of what it comprised and why it was being developed: a Gallup poll conducted in the United States in 1978 found that sixty per cent of people regarded IVF favourably as a solution to infertility for married couples. Yet in an Australian poll in the early 1980s the most common reason given for opposition to the technique was that it was 'unnatural'. Even the earlier Gallup poll tacitly acknowledged that 'unnaturalness' was the most likely reason for objection, phrasing its question as follows:

Some people oppose this type of operation because they feel it is 'not natural.' Other people favor it because it allows a husband and wife to have a child they could not otherwise have. Which point of view comes closer to your own?

Whatever the public felt, in the United States work on human IVF was considered too politically troublesome to broach. The National Institutes of Health imposed an unofficial moratorium on funding. But Landrum Shettles in New York had not abandoned his dream of making a baby this way. Throughout the 1960s he had been trying to grow an embryo to the stage where it might be transferred to a womb, despite being warned by his superiors that the research was considered

---

* This decision of the MRC was, however, complex, and not motivated primarily by a conservative aversion to the nature of the research itself. The issue has been explored in detail in Johnson et al. (2010).

impermissible. He was surprised and alarmed to learn from the 1969 *Nature* paper how far Steptoe and Edwards had progressed, but his real sense of urgency came from the landmark *Roe* vs *Wade* decision of the US Supreme Court in 1973, which established a legal right to abortion. In the backlash, attention came to focus on the excess embryos that would be created in IVF but not replaced in the womb. Shettles feared that a ban was imminent, and he was desperate to press on before that happened. He found a couple in Florida, Doris and John Del Zio, who agreed to give IVF a try. But he told no one at his workplace, Columbia Presbyterian Hospital, of his plan.

Shortly after Shettles had mixed Doris's egg with John's sperm and left them so that fertilization might proceed, the experiment was discovered by the departmental head Raymond Vande Wiele, who was vehemently opposed to such work. Worried both that it would besmirch the reputation of the hospital and that it might lead to the birth of (in his words) a 'monstrosity', Vande Wiele ordered the sample to be taken out of the incubator and destroyed, and he more or less sacked Shettles on the spot. In 1978 the Del Zios filed a suit against the hospital for $1.5 million. But the incident helped to bring about a temporary prohibition from 1974 on the implantation in the womb of a human ovum fertilized *in vitro*.

In the early 1970s Edwards and Steptoe attempted many times to develop a pregnancy through IVF, always without success and sometimes plagued by perilous complications such as ectopic implantation of the embryo in the Fallopian tubes. In the autumn of 1976, Lesley Brown, who was infertile owing to a tube blockage, was referred to Steptoe. She and her husband John agreed to try IVF, and the single egg that Steptoe extracted was successfully fertilized and produced a pregnancy. As the birth approached in the summer of 1978, the press attention became so intrusive – journalists disguised themselves as staff to gain access inside Oldham hospital – that Steptoe had to ask for guards to be put on the door of Mrs Brown's room. (The request was refused.) Newspaper editorials agonizing about the threat IVF posed to human welfare and dignity thus sat alongside editors' utter disregard for the welfare and dignity of its first beneficiary.

But everyone loves a baby. Louise Brown's birth by Caesarean section was greeted with oddly joyful headlines which seemed almost to imply that we should celebrate the dawn of the era so balefully forecast beforehand. 'She was born at 11.47 pm with a lusty yell', wrote *Newsweek*, 'and it was a cry round the brave new world.'

Could a baby conceived in a 'test tube' be wholly human? The child-ish suspicion lingered that she had been concocted from chemical reagents – if not literally so, then, you might say, figuratively. US chat-show host Johnny Carson joked:

How would you like to be the world's first test-tube baby? What would you do on Father's Day? Do you send a card to the Dupont [chemicals] Corporation? I understand that after the baby was conceived in the labora-tory, a pair of beakers smoked a cigarette and stared at the ceiling.

John Brown recalled the attitudes with West Country bluntness: 'When they say "test-tube baby", everybody had the impression that she was going to be nine feet tall and a quarter of an inch wide. You know, something kind of out of a comic strip.' There was a prevailing sense of cognitive dissonance: how could this ordinary baby, a child of ordinary parents, be created from the technologies of dystopia?

Views from the pulpit

Robert Edwards would be fully justified now in considering his goal, and the persistence with which he strove towards it, vindicated by the Nobel prize for physiology and medicine awarded to him in 2010. (Steptoe would undoubtedly have shared it if he was still alive; he died in 1988.) But even now this belated recognition was not welcomed by everyone: Ignacio Carrasco de Paula, head of the Pontifical Academy for Life, called the award 'completely out of order'. Although Carrasco stressed that he was voicing a personal opinion, the Catholic Church continues to oppose IVF today. Here is what the Church Catechism has to say on the matter:

Techniques involving only the married couple (homologous artificial insem-ination and fertilization) are perhaps less reprehensible, yet remain morally unacceptable. They dissociate the sexual act from the procreative act. The act which brings the child into existence is no longer an act by which two persons give themselves to one another, but one that 'entrusts the life and identity of the embryo into the power of doctors and biologists and establishes the domination of technology over the origin and destiny of the human person. Such a relationship of domination is in itself contrary to the dignity and equality that must be common to parents and children.' 'Under the moral aspect procreation is deprived of its proper perfection when it is not willed as the fruit of the conjugal act, that is to say, of the specific act of the spouses' union . . . Only respect for the link between the meanings of the conjugal act and respect for the unity of the human being make possible procreation in conformity with the dignity of the person.'

There are two issues here. First, the passage insists that procreation can only be properly brought about by sexual intercourse. Why so, given that there is no statement to this effect in the Bible? The assertion originates in the theology of natural law, which insists that all things have a 'natural end' (see p. 24). The natural end of sex is procreation, and it is for this reason that artificial contraception is prohibited by the Church.* The objection to IVF seems to invoke this reasoning in reverse: since the natural end of sex is procreation, the natural beginning of procreation is sex. Whether or not we regard this prescription as anything but arbitrary (it is after all merely an assertion of what is normally the case), the point is that it involves concepts of naturalness informed by the tradition of natural law. And the argument given in the Catechism is that, unless this concept of naturalness is respected, the result of procreation will be *deprived of its proper perfection*. The fruit will be imperfect.

Let's be clear about what this aspect of the Catholic prohibition implies. 'Sex' here does not refer to the biological definition of sexual reproduction in terms of the combination of gametes – which indeed does have the 'purpose' (insofar as we can use that treacherous term in a Darwinian framework) of diluting deleterious genetic mutations. That is to say, sex does not here mean the confluence of egg and sperm. Nor does it take into account the intentions of the participants in the sexual union – whether or not they wish to produce a baby. Rather, sex means 'the conjugal act', a euphemism for the matter of which bits of the body go where: the crude mechanics of the process.

---

* 'Natural contraception', meaning the use of the (inefficacious) rhythm method, is tolerated on the grounds that family planning is desirable. But this makes no sense: in effect, it implies that it is permissible to try to evade what God intends for sexual intercourse provided that you do not try too hard.

That point is emphasized by the condemnation issued by Pope Pius XII in 1949 against artificial insemination. Even though in this case conception takes place in its 'natural' location, the procedure is forbidden because it involves an 'act against nature', namely masturbation to produce sperm. (Pius gave cautious approval of artificial insemination between married couples if semen is collected by syringe after vaginal intercourse and then deposited at the entrance of the cervical canal. In other words, so long as the ritual of intercourse is observed, it doesn't much matter what happens after.) Again, no account is taken here of intention – whether, say, it is done by a loving married couple wishing to have a child. No, it is the physical act that determines the moral judgement. This notion of sex seems peculiarly, even brutally, mechanistic: what is 'natural' is decided on the basis of the sex organs being in a particular configuration at a particular time.

This is made even more clear by a version of IVF that *has* been sanctioned by the Catholic Church, called gamete intra-Fallopian transfer or (auspiciously) GIFT. Here the woman's eggs are collected as in IVF and mixed with sperm *in vitro*. This mixture is then immediately transferred back to the woman's Fallopian tubes, so that fertilization can occur inside the body. One claimed benefit of GIFT is that the embryo can begin its earliest development in 'natural surroundings' rather than in an 'artificial environment'. It is not clear that in the very first stages an embryo cares in the slightest about this distinction, and indeed GIFT both is more invasive than standard IVF and makes it impossible to select the embryo of best apparent quality from several prepared *in vitro*. But the Church will permit it – provided that the sperm is collected using a condom (a perforated, leaky one, mind) in sexual intercourse and not by masturbation – because everything then seems to be happening in its 'natural' place, with just a momentary sleight of hand involving a Petri dish. This obsession with the 'proper' mechanics, notwithstanding the lengths that are necessary here to achieve it, speaks of a deeply strange attitude towards the relation between sex and procreation, not to mention the bizarre and, I should have thought, highly disrespectful notion of a God who watches as if with clipboard in hand (but ready to avert his eyes at the crucial point) to tick off each step when it happens as it 'ought'.

The second significant aspect of the prohibition against IVF in the Catholic Catechism is that it makes a judgement about the moral character of technology. Here it is again: IVF 'entrusts the life and identity of the embryo into the power of doctors and biologists and establishes the domination of technology over the origin and destiny of the human person'. The problem with this view is that, without those 'doctors' and this 'technology', there is no human person, no embryo, to be 'dominated', no 'origin and destiny' to speak of. This makes the pronouncement logically incoherent in the first place: there is in general no choice to be had between making the child 'naturally' and making it 'technologically'. When such a choice exists, it is hard to imagine (except for medical reasons such as the possibility to screen for genetic disease) why anyone would opt for the discomfort, the inconvenience and typically the financial cost of IVF. Among the shards of this fractured logic, however, is a remarkable assertion: *technology is inevitably corrupting*. There seems no room here for the possibility that the doctors and the technology might be working towards the good, namely the provision of a much-wanted child. Rather, technology creates an indelible stain on the 'origin and destiny' of any person in whose conception it intervenes.

The same implication is found in John Paul II's condemnation of all artificial procreation, which he dismissed as a 'technological procedure which is devoid of human value and subject to the dictates of science and technology' (see p. 4). It is reflected in Pius XII's judgement on artificial insemination, which can be read as a pre-emptive view on IVF too:

To reduce the shared life of a married couple and the act of married love to a mere organic activity for transmitting semen would be like turning the domestic home, the sanctuary of the family, into a biological laboratory.

This comment is profoundly revealing about attitudes to anthropoeia in the age of biotechnology. What it shows us is the horror felt at the intrusion of biological processes into procreation. Put that way, it sounds absurd – what is procreation if not a biological process? But that is not generally how we experience it. Because we never previously witnessed the fundamental biological events, we never had to dwell on them. It is one thing to know that each of us began as a cluster of cells, quite another to see it. The explicitness of that vision is unsettling, and for good reason: intuition can never connect that abstract, microscopic mass to a human being. Even the shrimp-like fetus has something of the human about it, and ultrasound images of the gestating baby fill us with empathy. But cells? One defence against the visualization of our origin is, as we will see, to cram all the attributes of a fully developed person into these membrane-bounded sacs of genes and cytoplasm. It is a very poor fit.

Pius XII's distaste at the explicit unveiling of the biology of human conception is quite normal and understandable. The problem is that it is received not as a 'gut response' but as an authoritative theological pronouncement, despite the fact that Christian theology managed to get by for centuries after its inception with never a mention of 'biological laboratories'. Notice also that the imagery is purely rhetorical. Of course artificial insemination does not really turn the home into a biological laboratory – many people have achieved it with the simplest of domestic equipment in their own homes, with not a lab-coated scientist in sight. What Pius XII really wants to imply here is that assisted conception makes procreation a matter of ugly biological fact – and worse than that, for there is no doubt that we are meant to regard the 'biological laboratory' as a sinister place in which perverse and unnatural things happen. If he had said merely that the act of married lovemaking had become a matter of 'biology', the effect would have been far weaker – we would be inclined to respond that of course biology is involved. But the laboratory is a place of manipulation, of synthesis, of *art*, a place where nature is traduced.

The Catholic opposition to IVF stands incongruously alongside the Church's determination otherwise to remove all obstacles to procreation. As Peter Singer and Deanne Wells put it, 'In no other case does the Catholic Church espouse the view that if pregnancy represents a risk to the child of that pregnancy, then measures guaranteed to avoid the pregnancy should be taken'. Why should it be so wrong to keep eggs and sperm apart in one instance, but to bring them together in another? Again, the answer seems to be that this is not a question of eggs and sperm, which are in the end only unwelcome biological intruders into an act that is being required to channel an awful lot of moral clutter. Sexual intercourse and procreation are, the Catechism insists, utterly inseparable: each demands the other. Yet why, when no one is necessarily harmed by their separation, should this equation be insisted upon so strenuously? The answer comes down to a tautological (and thus impregnable) idea of 'naturalness', coupled to the implication that 'natural' is a moral virtue. That this is a pure expression of Hume's naturalistic fallacy is clear in the objection (here apropos human cloning) by Christian bioethicist Gilbert Meilaender: 'To say that a child ought to spring from the sexual union of a man and a woman (indeed, more precisely . . . from the union of husband and wife) is . . . to utter a moral rule.'

Some Christian spokespeople attacked IVF from a different angle. One Catholic priest objected to the psychological damage that might be inflicted on IVF children: 'They could become complete egoists, knowing that, in their generation, they stand apart from the rest of mankind.' The arbitrary, wholly speculative nature of this assertion is painfully evident in retrospect, as is the flimsiness of the suggestion that a method of alleviating infertility should be cast aside because of the danger that children conceived this way would end up boasting about it. (Others were more concerned that IVF would impose a character-stunting stigma, not a badge of superiority.) More telling is the suggestion that IVF children would 'stand apart from the rest of mankind' – that is, they would not quite be a part of the same species.

Not all Christian views on IVF were negative. The Protestant theologian Joseph Fletcher offered one of the most eloquent defences of the technique yet voiced, not least because it focuses for once on the *intentions* that lie behind the procedures:

Any attempt to set up an antinomy between natural and biologic reproduction, on the one hand, and artificial or designed reproduction, on the other, is absurd. The real choice is between accidental or random reproduction and rationally willed or chosen reproduction . . . laboratory reproduction is radically human compared to conception by ordinary sexual intercourse. It is willed, chosen, purposed and controlled, and surely those are among the traits that distinguish *Homo sapiens* from others in the animal genus, from the primates down.

Will the family survive?

It is a curious fact that the field of bioethics, which developed in response to the possibilities of modern biological technology in the late twentieth century, for all its academic mien has little more by way of entry requirements than the possession (for whatever reason) of strong opinions. Many religious moralists have become 'bioethicists' simply by expressing views on biomedical science. As a result, what sits under the rubric of bioethics is a decidedly mixed bag.

In 1971, the American journal *Science* published an article by biochemist turned bioethicist Leon Kass called 'The new biology: what price relieving man's estate?'. While IVF provided only a part of the context of Kass's article, his comments were motivated largely by the capabilities that Edwards and Steptoe's early work had then revealed. 'Those who welcome and those who fear the advent of "human engineering" ground their hopes and fears in the same prospect', wrote Kass, '*that man can for the first time recreate himself.*' The gendering here, while a habitual reflex at that time, is not incidental, for it accentuates the purported 'unnaturalness' of this state of affairs: man begets man. In the literal sense Kass's statement is patently false (humans have always been able to recreate themselves), but it is clear what he means: we can reproduce *artificially*.

Kass invested this capability with a remarkable attribute. To him, a spermatozoon and an ovum united in a Petri dish represented the first step towards the removal of the individual from reproduction. 'The complete depersonalization of procreation', he said,

shall be, in itself, seriously dehumanizing, no matter how optimum the product . . . Would not the laboratory production of human beings no longer be *human* procreation? Could it keep human parenthood human?

Kass remains (intentionally?) vague in his terms. Is IVF already 'the laboratory production of human beings', since after all it results in human beings and involves a kind of laboratory? If so, does this mean that an IVF baby would no longer be the result of human procreation? That is, at least, one way to read this remark.

Kass worries that without 'biological parenthood' – by which he means not just a genetic investment in one's children but a 'natural' process of conception and gestation – we may lack the naturally selected mechanism that 'fosters and supports in parents an adequate concern for and commitment to their children'. This is so astonishingly ignorant that one has to suspect it stemmed from some unvoiced fantasy. There is not, nor was there in 1971, any firm evidence that adopted children or stepchildren are less fostered and supported by their parents, despite in this case even the absence of genetic kinship. Nor is there now any evidence that this is so for children born of IVF.

Nonetheless, Kass felt that 'transfer of procreation to the laboratory will no doubt weaken what is presently for many people the best remaining justification and support for the existence of marriage and the family'. If it later surprised Kass that the overwhelming demand for IVF came from married couples wanting to start a family, it surprised no one else. Might his views stem from the fact that all fictional dystopias (and utopias) created by artificial reproduction seem to lack a family structure? Certainly, he raised the inevitable spectre of *Brave New World* with its 'dehumanized' citizens. Or might he in fact doubt that a 'real' family can be produced by 'unnatural' procreation?

That was the position taken by Paul Ramsey: IVF threatened parenthood because he chose a definition of parenthood designed specifically to exclude non-traditional means of conception. In his view, parenthood is not the state of raising a child within a family, but the 'God-given' act of conceiving a child by bodily sexual intercourse within marriage. This position, which excludes millions of parents (including some married Christian couples with biological children) from that designation, is a perfect example of how religious opposition to assisted conception is apt to proceed from rigid (but ultimately capricious) doctrines that take no account of intentions, effects, or actual relationships between people.

The worry that IVF would undermine the traditional nuclear family, and perhaps, as a consequence, society as a whole, was more widely voiced once the technique became a clinical reality. In a debate in the UK House of Lords in 1982, the Conservative Lord Campbell of Croy warned that

without safeguards and serious study of safeguards, this new technique could imperil the dignity of the human race, threaten the welfare of children, and destroy the sanctity of family life.

Notice that it is not the family per se that is deemed to be threatened here, for after all IVF was a means of *creating* families. It is the *sanctity* of the family: a particular idea of what a family is. It is hard to interpret Lord Campbell's remark in any other way than as a warning against permitting families to be made *in the wrong way* – an idea that would seem again to have roots in a concept of natural law.

Leon Kass served on the short-lived Ethics Advisory Board which, in the two years of its existence from 1978 to 1980, made recommendations about IVF to the US Department of Health, Education and Welfare. In its key paper on this topic, Kass wrote that the logic used to justify IVF 'knows no boundary', so that 'it will be difficult to forestall dangerous present and future applications of this research and its logical extensions'. This is the classic slippery-slope argument, which assumes that regulation will be impossible, leaving prohibition as the only option. 'Once the genies let the babies into the bottle, it may be impossible to get them out again', Kass wrote. While his main justification for this turn of phrase was perhaps the pleasure of having thought it up, it is not incidental that it brings to the technology a whiff of demonic magic, nor that it harks back to Aldous Huxley's 'babies in bottles' (let's remind ourselves once again that there are no bottled babies in IVF). There seems little to distinguish the Catholic position from that which Kass took in the wake of Louise Brown's birth: 'this blind assertion of will against our bodily nature – in contradiction of the meaning of the human generation it seeks to control – can only lead to self degradation and dehumanization'. Here again is the assertion that the process is 'against nature', which implies that there is a meaning to human generation that is bound up with the bodily mechanics of how (and not the question of why) it happens.

In expanding on his fears of a slippery slope in 1972, Kass was more explicit about his views on what is and is not unnatural in the options that IVF seemed to present:

Is it a right to have your own biological child? Is the inability to conceive a disease? Whose disease is it? Does infertility demand treatment wherever it is found? In women over seventy? In virgin girls? In men? Can these persons claim either a natural desire or natural right to have a child, which the new technologies might or must provide them? Does infertility demand treatment by any and all available means? By artificial insemination? By *in vitro* fertilization? By extracorporeal gestation? By parthenogenesis? By cloning – i.e. 'xeroxing' of existing individuals by asexual reproduction?

He adds, in the manner of a schoolteacher admonishing an errant pupil to go away and think about the bad things they have just said, 'Those who seek to submerge the distinction between *natural* and *unnatural* means would do well to ponder these questions.' The implication is that some telling point has been made. But Kass has merely presented a list of questions to which one might calmly respond, each in turn, 'Yes', 'No', or perhaps 'That's a tough call; it depends on the circumstances.' The escalation of the degree of intervention (ending with several that are not feasible at present) invites us to conclude that all must be wrong, apparently because they are all 'unnatural'.

Test-tube monsters

In the early days of IVF, not all the criticism came from religion and its surrogates. James Watson, the co-discoverer of DNA's structure and in the early 1970s the director of the Cold Spring Harbor Laboratory for research on genetics, attacked the work of Edwards and Steptoe at a 1971 conference in Washington DC on 'fabricating babies'. It is plain enough today that Watson is not one to let facts cloud his judgement, but at that time he was one of the most powerful voices in molecular biology – so when he spoke up at the meeting, people listened. And what he said to Edwards, also a participant of the meeting, was this:

You can only go ahead with your work if you accept the necessity of infanticide. There are going to be a lot of mistakes. What are we going to do about the mistakes? We have to think about some things we refuse to think about.

Watson has long been obsessed with the notion of genetic 'perfection', speculating that it might be possible and acceptable to screen for and to eliminate genes for homosexuality and 'stupidity'. Visions of IVF babies with severe genetic abnormalities crowded his thoughts, not in the light of any carefully considered scientific assessment of the available evidence (from animal studies) but because of private fantasies informed, as much as those of the lay public, by old myths of monstrosity.

He was not alone. The molecular biologist Max Perutz, another Nobel laureate for his work on the structure of protein molecules, warned that

Even if only a single abnormal baby is born and has to be kept alive as an invalid for the rest of its life, Dr Edwards would have a terrible guilt upon his shoulders. The idea that this might happen on a larger scale – a new thalidomide catastrophe – is horrifying.

The thalidomide affair weighed heavily on the minds of many scientists at that time, even though a decade had passed since the link between this drug (administered to pregnant women to alleviate morning sickness) and birth defects had been established.* But the connection between that specific chemical intervention and the general practice of uniting sperm and egg in a 'test tube' is far from obvious, and represents the kind of leap of logic one might expect to see in a newspaper editorial and not to hear from a Nobel laureate. Perutz also assumes without basis that it would be a straightforward matter to distinguish a birth defect caused by the IVF procedure from one resulting from other factors (for how would we otherwise know that the 'guilt' was Edwards'?).

Kass also played the risk card at the 1971 Washington symposium, saying, 'It doesn't matter how many times the baby is tested while in the mother's womb, they will never be certain the baby won't be born without a defect.' This is of course an unanswerable charge: whatever you do, it asserts, you'll never be certain. The accusation could scarcely have carried much force if it did not draw strength from the suspicion that IVF babies would be monstrously deformed. Here that fear has become almost an article of faith: no matter how normal they look, something will be wrong. Others argued that, even if these babies looked and seemed normal at birth, we could never be certain that something wouldn't go disastrously wrong as they grew older. When the origin is 'unnatural', they implied, the seed of imperfection must be lodged somewhere and will bloom horribly sooner or later.

It was entirely proper that concerns should have been raised about the safety of the IVF process for the development of the child. Indeed, it now seems that standard IVF does result in a slightly higher incidence of birth defects than is observed in babies conceived by sexual intercourse: figures vary, but an increase of around thirty per cent is commonly cited. But the absolute risks are low in both cases: around one to three per cent for 'normal' conception, up to four per cent or

---

* Thalidomide is recognized now as a 'teratogen', an agent that induces abnormalities during fetal development – or literally, a 'monster-maker'.

so for IVF.* And the most common defects are not like the visible abnormalities associated with thalidomide, let alone severe impairments of the kind that would result in Perutz's 'invalids' requiring extended life support, but gastrointestinal and cardiovascular problems that often have straightforward surgical solutions. As Edwards pointed out at the time, most serious growth abnormalities in a developing embryo will result in early termination of its own accord.

How one regards these risk factors is a matter for debate. But it would be very hard to argue that, because the absolute risk of defects (the vast majority of which are not life-threatening and do not impair mental development) seems to be raised by at most one per cent from a low initial baseline, IVF is somehow unethical. Nobody now does so. And notice that Watson and Perutz did not, in those early days, call merely for caution and more research; they pronounced grave warnings about legions of grotesque deformities.

There is nothing to distinguish the response of these eminent scientists from the gut reactions of the general public. As Steptoe put it in 1980:

Many, if they consider the matter at all, thought such a procedure [laboratory fertilization of a human ovum] undesirable: the notion of initiating human life *in vitro* being repugnant to them, unacceptable in principle. Some were fearful of the physical consequences of fertilization *in vitro*. They asked 'Supposing the baby born is abnormal, a cyclops say, or some other monster?'

He pointed out that some would even find such an outcome gratifying as a warning against the perceived hubris, saying that Paul Ramsey had declared he 'half-hoped that the first child born through such a method would be impaired and be publicly displayed!'

---

* The figures must be gathered carefully. Since infertility problems may sometimes have a genetic component, any fertility treatment might carry a raised risk because it potentially passes on these genetic abnormalities. In line with that consideration, fertility treatments that do not involve IVF seem to pose a risk midway between that of IVF and that in the general population. And because birth defects are more common for multiple births, and the latter are more common with IVF, this also distorts the risk due solely to the *in vitro* process. Moreover, the technique called intracytoplasmic sperm injection (ICSI), which selects a spermatozoon and injects it into an egg rather than allowing the sperm to enter of its own accord, might be expected to remove some of the biological barriers to inappropriate unions. But so far, the risk from ICSI does not appear to be significantly higher than that of standard IVF.

Some objections to IVF seem resolutely secular and grounded in its real-world consequences. That does not necessarily make them any more sound. Theologian Maura Ryan argues that IVF can prolong the difficulty of coping with childlessness: couples commit to one fruitless cycle after another, rather than getting on with their lives. This is possible – and all the more likely when IVF happens in an unregulated, competitive commercial arena. But to argue that people who wish to try IVF should not be allowed to do so because they might repeat it addictively if it does not succeed seems an insufferably paternalistic denial of opportunity that does not at all address the moral calculus of withholding a treatment from some because it won't work for all. Ryan adds that assisted conception constrains other options for resolving childlessness, such as adoption – which encourages the common view that these options are equivalent, a pernicious notion in terms of both opportunities and challenges for would-be parents. She cites the objection of some feminists that 'scientist-controlled, technological reproduction' requires women to hand over their 'life power' to the medical establishment – specifically, that it 'necessitates interference from the state and its institutions, especially medical science professionals'. Yet there is hardly a birth in the industrialized world that does not incur such 'interference'.

Bioethicist Laurie Zoloth criticizes 'advanced reproductive technology' on the basis that it portrays infertility as a disease which must be 'cured'. And Leon Kass complains that IVF is in any case no 'cure' at all, because it does not address the underlying biological problem. As we'll see in the next chapter, there are valid concerns about the way IVF has become a commercial business with a product to be sold. But it is characteristic of critics like Zoloth that they duck the unpalatable corollaries of their criticisms. Should we, then, ban IVF and tell those who suffer from infertility that they must simply learn to live with it? Or might we merely want to constrain and monitor how it is done? At the same time Zoloth herself pathologizes infertility to a grotesque degree, saying that 'the hunger of the infertile is ravenous, desperate' – with the implication that it is also dangerous and lacking all moral restraint. Kass's complaint here is merely bizarre: most women with blocked Fallopian tubes, or men with a very low sperm count, do not particularly care about their condition, nor suffer from it, except for the critical fact that it prevents them from having children.

Objections to IVF are sometimes dismissed as 'symbolic', meaning that they articulate preconceptions and doctrines (the 'sacredness of life's mystery', say) that are wholly subjective. Maura Ryan argues that they should not be ignored on that account, for symbolism is intrinsic to human experience:

It is only when we recognize our anxieties about what we are doing when we fertilize human eggs *in vitro* or reproduce asexually [that is, without sexual intercourse] as acknowledgements that we are close to, and may be violating, some of the deepest and most basic of human matters – procreation, kinship, marital intimacy – and our unwillingness to pay disrespect to human embryos as a powerful statement of our societal commitment to the value of human life generally that we come to the important questions: What practices, what acceptable harms, are ultimately consistent with such a commitment? What research goals are we willing to sacrifice to sustain it?

In other words, our fears and concerns are telling us something important and worth heeding, a position Kass has called the 'wisdom of repugnance'. I suspect Ryan and Kass are right to say that instinctive repugnance and unease have something to tell us, but wrong in what they believe that message to be. Look again at Ryan's statement (and don't be distracted by the sleight of hand which makes the combination of gametes 'asexual', in contradiction to biology). She is proposing that we may be 'violating . . . procreation', notwithstanding the evident fact that when IVF works we are procreating. The unspoken text is again that we are not procreating *in the right way* – we are doing it unnaturally, and therefore in a morally reprehensible manner.

The important point here is to recognize the mode and motivation of the arguments opposed to IVF. Behind the flimsy, illogical, censorious or merely vague condemnations seem to lurk unarticulated ideas about naturalness and its violation, expressed by one critic as 'the integrity of sexuality, reproduction, and parenthood'. While it is tempting to react with either laughter or despair at the medieval objections of one of the interviewees in *Life*'s 1969 poll on artificial conception, 'mother of three' Marjorie Foster of Toledo, Ohio, I find something oddly admirable in the way it accesses so directly the mythical ingredients that hide beneath a veneer of sagacity in the remarks of some academics, both then and now: 'I just wouldn't feel the child was mine. It might sprout horns or wings or something,' Angels and demons, indeed.

Men only

The fantasies of gender supremacy that we encountered earlier became attached seamlessly to IVF. The Feminist International Network on New Reproductive Technologies (now called FINRRAGE), formed in 1984, opposes all reproductive technologies as a global conspiracy of male 'technodocs' that aims either to turn women into experimental guinea pigs or to eliminate them and usher in 'the moment when men will play God and be the sole procreator of the human species'. The FINRRAGE manifesto is called *Test-Tube Women*, and its mission is to infiltrate the patriarchal medical establishment, learn its impenetrable language, and reveal its evil plans to the world. To these now-familiar myths and conspiracies, FINRRAGE adds one with a twentieth-century Marxist flavour, claiming that reproductive technologies are attempts to stimulate capitalist over-consumption. In the mid-1980s, radical feminists attacked IVF centres: a group in Germany called Rote Zora (Red Zora) bombed clinics and stole documents, and in 1987 members of the group Gen-Archiv, which seeks to ban all reproductive technologies, were arrested and charged with terrorist activities.

These developments don't in any sense represent 'the feminist' position on assisted conception – as social historian Naomi Pfeffer has pointed out, FINRRAGE seemed to consult the views of all women except those who suffer from infertility.* But it is clear enough now that the science-fiction narratives on which these radical positions depend are not arbitrarily constructed. How otherwise might we explain the same kind of sinister global conspiracy (in that case, a

---

* A 1990 study by feminist Renate Klein in which she sought views on IVF from women for whom it had not worked does not undermine that accusation. Klein believes that IVF has very little to do with helping infertile people, but is instead an 'inherently eugenic' means to secure 'a constant source of experimental subjects for research be it investigating their hormonal cycles or as egg donors' (to make embryos for research). So it is perhaps not surprising that she chose to mention only the extreme negative responses from the women she questioned, and to suggest that any woman involved in IVF who does not share her conclusions is akin to the women who continue to love and refuse to leave their abusive partners. While it is dismaying to see such distorted accounts given a patina of academic respectability, and while on the other hand one can acknowledge the fact that the IVF business sometimes does sell its successes and marginalize its failures, 'studies' like this are useful for what they inadvertently tell us about the myths that feed them.

corporation seeking to *engineer* infertility in order to profit from IVF) recurring in the thriller *Vital Signs* by Robin Cook, a writer not known for his affiliation to radical feminism?

Assisted conception has also become burdened by myths about strange procreation. For example, media fascination with (extremely rare) instances of confusion between IVF embryos – particularly when these are transracial – draws on legends of changelings. There are genuine ethical conundrums posed by the options that IVF makes possible for shifting and blurring conventional distinctions of race and parenthood, as for example when in 1993 a black woman chose to bear a white child conceived from her husband's sperm and the donated egg of a white woman, in the belief that the child would have a better life if it was untainted by racism. But when faced with such unfamiliar moral terrain, we will intuitively fall back on the resources provided by the oldest stories, for better or worse. Most often – as in that particular case – such narratives present themselves in terms of an alleged transgression of nature. As feminist writer José van Dyck says, the hurdle that these new technologies face

is not that they are defying nature, but that they are defying culture. They profoundly challenge our preconceptions of a society that seems 'naturally' divided along axes of gender, race, class, age and physical or psychological condition. They also challenge traditional social structures, such as the nuclear family, identifiable races or ethnic groups.

As another example of how IVF plays out at the mythic level, consider the much-publicized legal wrangles over frozen embryos created by couples who subsequently separate. In the typical scenario, the man wants the embryos destroyed, the women wants them implanted. The court battles become in a sense a matter of one's right to one's gametes, recalling the belief in demonic succubae who arrive in the night to steal male seed spilled outside the womb and, as Paracelsus warned, 'carry it to a place where they may hatch it out'. This comes to pass, Paracelsus insisted, because the sperm has not found its 'natural' place. That's the trouble with ways of making our babies which do not demand that we be physically present at the conception: they still seem like a form of witchcraft, and from them our ancient nightmares come bodying forth.

10 Anthropoeia in the Marketplace

An embryo is not a little baby in the freezer.
Ruth Deech, Chair, HFEA (1996)

As long as [assisted] conception remains a furtive trade – a
business cloaked in the garb of science – it will remain vulnerable
to both the excesses of its fringes and the attacks of its critics.
Debora Spar, *The Baby Business* (2006)

Robert Edwards and Patrick Steptoe launched a small but energetic
niche industry devoted to the creation of people through technological
intervention. No longer was anthropoeia a matter of covert manipulations in secret laboratories; it had found a mass market. The British
team's success was soon repeated, first by fertility specialist Carl Wood
in Australia, who had been working on IVF since the early 1970s.
Private IVF clinics proliferated, the first of them established by Edwards
and Steptoe at Bourn Hall, near Cambridge. Only ten years after Louise
Brown's birth, *Nature* could say that 'infertility therapy has become
big business'.

You might want at this point to protest that it is sleight of hand to
conflate the sober business of assisted conception with what I have
called the anthropoetic tradition, which embraces automata and moving
statues, the homunculus and Frankenstein's creature. But I hope you
might now appreciate why this association is not only permissible but
essential. If we were simply to complain that the constant reference
to these legendary sources in the media was nothing more than an
expression of ignorance about what the technology involves, we would

fail to grasp the true nature of the anxieties that assisted conception and reproductive medicine stir up. There is, without doubt, a world of difference between the idea of assembling an oversized humanoid from the limbs of dead bodies, and fertilizing an egg in a Petri dish. But we should recognize, first, that continuity between the two is ensured by our evolving view of what life is and how it comes about (a topic that I consider further in Chapter 12); and second, that the central issue in both cases is that human life is seen to be initiated by *art*, by means of human ingenuity rather than merely human biology. The being that results would not exist without having been in some sense fabricated.

However, assisted conception, and particularly *in vitro* fertilization, adds a new element to the procedure that does not really have a precedent in the anthropoetic tradition. For IVF does not exactly *make people* – it fertilizes eggs to make embryos. That it might become routine, *easy*, to fabricate people was a possibility anticipated in Čapek's *R.U.R.* and to a certain extent in *Brave New World*. But in those cases anthropoeia was an all-or-nothing affair: either you made a person or you did not. With IVF, we cannot agree about precisely what has been created, except to say that some of these created entities may become people. We lack a coherent and consensual framework even for debating the issue.

The tension inherent in this ambiguity was acutely felt once IVF clinics began producing more human embryos than they could use for an attempted pregnancy. A woman might, under the regimen of hormonal drugs, produce just one viable egg, or twenty of them; the average number is around twelve, of which most might become fertilized but only a few returned to the womb.* Among the others, some might show signs of being unlikely to develop beyond the few-cell stage. But apparently healthy embryos that are not transferred might be frozen for possible use in another cycle, or discarded, or voluntarily donated for research.

For scientists studying human development, and especially those

---

* Multiple births, with attendant complications for the mother and children, pose the biggest health risk from fertility treatment. In 1996, a British woman became pregnant with octuplets after a course of fertility drugs (not IVF). All of the babies died. From 2004, the UK Human Fertilization and Embryology Authority ruled that no more than two embryos may be transferred to women under forty, and no more than three for women over forty. It was more recently developing a strategy for 'elective' single-embryo transfer for women with a high chance of conceiving.

who wanted to improve the safety and reliability of IVF, this has been a boon: they could conduct research on the 'spare' embryos. But doesn't that reduce these embryos to the status of raw material for scientific exploitation? As IVF clinics multiplied, anti-abortion groups identified a new target. Carl Wood's colleague Alan Trounson recalls that in Australia,

There were walls daubed with whitewash about 'Wood and Trounson: mass murderers', and newspaper articles which were highly offensive. And even television programmes run – one of them by the BBC – which interposed us with Mengele and the experiments of Nazi Germany.

Because every person who has ever lived began as an embryo, there is a strong intuitive impulse to regard embryo creation as the mass production of human beings, or, at least, of potential people. This is not, as we will see, necessarily the right way to regard the situation in *biological* terms, but one might argue that biology should not be the sole arbiter. Yet what else do we have to draw upon? What social scientist and ethicist Audrey Chapman has said about human cloning applies equally to the prospect offered by IVF: 'None of the traditions involved in the debate could simply reapply past deliberations on human nature and the human role in creation to answer the unprecedented dilemmas posed.' By and large, however, we have done just that: we have felt compelled to situate our views with reference to religious and secular myths about the artificial creation of people.

Because of the incendiary controversy over the status of the embryo, some governments have refused to regulate assisted conception, since they know that whatever they do will arouse fierce condemnation. This has left biomedical anthropoeia exposed to the tender mercies of market forces – a predicament that no legend has foreseen, and one that bodes ill for future technologies that assist in the creation of people.

Awkward questions

However politically unwelcome the dilemmas of IVF were after the birth of Louise Brown, they could not simply be ignored. So both the British and the US governments did what most governments do under such circumstances: they appointed a committee. In the United States this was the Ethics Advisory Board (EAB), which concluded that there was no obvious ethical barrier to federal funding of IVF and embryo research, provided that there were clear scientific aims. But the issue was too volatile for the government to act on that advice, and so the EAB's recommendations were quietly ignored. Although the law required that any federal money granted for human embryo research be approved by the EAB, after 1980 there *was* no EAB: it was dissolved, and neither Ronald Reagan nor George H. W. Bush were eager to appoint a successor. So public funding of IVF became subject to a de facto moratorium.

It wasn't until 1993 that the National Institutes of Health created a replacement body, the Human Embryo Research Panel (HERP), which again advised (limited) federal funding of embryo research according to strict guidelines. Like the EAB, however, HERP refused to grasp the nettle of how to think about the moral status of the embryo. And again inaction was deemed the prudent political course, while in any event Congress refused to countenance any such funding. The impasse became a formal restriction in 2001 when President George W. Bush banned the use of federal funds for research on embryonic stem cells taken from human embryos after that date. (Removing stem cells from human embryos results in the embryo's destruction. These versatile cells, discussed in the next chapter, now provide one of the major focuses of embryo research.) Although the ban was lifted by Barack Obama in 2009, the legislative and regulatory situation for federal support of embryo research more generally remains to be clarified, and still lacks any coherent and transparent framework.

So from its inception, IVF was permitted, and flourished, in the American private sector largely without any regulation, while the research that was needed to make it safer and more effective was rendered impossible in the public sphere. As bioethicist Paul Lauritzen put it in 2001, 'We have a situation where society as a whole has simply decided not to take responsibility.' Another bioethicist, Carol Tauer, pointedly remarked during George W. Bush's administration that:

It may seem strange that a federal government so bent on outlawing any use of federal funds for research involving IVF is at the same time so uninterested in regulating these procedures and research on them in the private sector. This stance could stem from a libertarian view of private enterprise . . . or from a political calculation.

Of course, it could also have been a mixture of both.

US politicians avert their eyes from the shambolic consequences, whereby, for example, American IVF clinics can claim pretty much what they like: one in California has promised (impossibly) 'a baby, guaranteed'. Addressing HERP during a consultation in 1994, reproductive biologist Jonathan Van Blerkom said that IVF and embryo research in private clinics lacked proper scientific and ethical controls and was often shoddy. As a result, IVF was inefficient and cost more than it needed to, and carried health risks to women and babies that might have been avoided with better research. Some felt that the honest course would be for IVF either to be supported by properly funded research or to cease entirely. Tauer suspects that if the choice were presented to the American public in this fashion, they would think quite differently about the ethical issues.

In the United Kingdom, there was at first a similar political reluctance to intervene. In 1978 an advisory group of the Medical Research Council concluded that embryo research could be ethical so long as it had clear and valuable scientific objectives. But the group had no regulatory power, and four more years passed before the government appointed a committee, headed by the moral philosopher Mary Warnock, to draw up guidelines for appropriate regulation.

The Warnock committee again took a permissive stance on research but recommended that no IVF embryo be kept alive for longer than fourteen days – the point at which a feature called a 'primitive streak' appears. This is a furrow-like structure which indicates the axis of bilateral symmetry of the developing body, and its appearance corresponds more or less to the stage at which twinning of the embryo can no longer happen. This was taken by the committee as some kind of marker for the inception of a human form – crudely speaking, as an emblem of personhood. The arbitrary basis of the limit was criticized by scientists, although it was not clear that any other choice would have been less so. Indeed, unless it was going to recommend either prohibition or absolute liberalization of research, the Warnock Report was bound to upset most parties. Scientists saw it as too restrictive, while anti-abortion groups such as the Society for the Protection of Unborn Children actively opposed it.

The government simply continued to prevaricate, and into the vacuum in 1984 came the Unborn Children (Protection) Bill, drafted by the Conservative MP Enoch Powell, which sought to impose a complete ban on embryo research. The bill had the support of a majority within the House of Commons, but its opponents prevented it from being passed on a technicality, using delaying tactics to ensure that the Commons debate ran out of time. Meanwhile, the government decided that what was now needed was a public consultation exercise, which was launched at the end of 1986. A regulatory framework was finally proposed in a White Paper of 1987.

There was little public discussion during this time of why embryo research might be needed or what it might do. Instead, the debate was dominated by the familiar imagery. The *Sun* newspaper accompanied a report on the White Paper with a movie still from *Frankenstein*, and even the White Paper itself was compelled to acknowledge that

one of the greatest causes of public disquiet has been the perceived possibility that newly developed techniques will allow *the artificial creation of human beings* with certain predetermined characteristics through modification of an early embryo's genetic structure [my italics].

Some newspapers failed to recognize their own reflection in that 'perceived possibility', and instead took it as a statement of intent. Under the headline 'Clamp on Frankenstein scientists', *Today* reported that 'Scientists are to be banned from creating superbeings in the laboratory' – with the obvious implication that, without such a ban, they would go right ahead.

The level of political discussion was not much higher than that in the tabloid press. In 1984, Baroness Masham remarked in the House of Lords that 'the further we get away from nature, the more problems we shall have'. The baroness's response shows how opposition to embryo research often expressed itself as inarticulate repugnance – a primitive fear of 'violating nature' – for which images of Frankenstein's monster, Nazi scientists, super-races and cloned armies supplied a set of off-the-shelf references. Masham's remarks betray an almost stream-of-consciousness terror: 'I once watched a science-fiction film about a man who was made in a laboratory. He escaped and did all kinds of terrible things. It makes me think: is this the start of something which could lead to a Hitler theory of only the perfect human beings being brought into our society?' Another speaker commented that 'If we take the rectification of genetic defects to its logical conclusion, one day we shall live in a society in which medical developments applied to *in vitro* fertilization will be so advanced that . . . one will be able to book an embryological configuration – tall, dark, handsome, short, intelligent, athletic, shrewd or perhaps even a Frankenstein monster if that is what one wants.' Others used *Brave New World* as a vague shorthand for ill-defined fears about the 'debasement of human generation'. The Conservative MP Trevor Skeet warned of children grown by ectogenesis with predetermined characteristics, and raged against those who defy the law to satisfy their own 'thirst for knowledge' (which seemed to rather cast doubt on the value of a ban in the first place). In general, the opponents of embryo research rooted their opposition in imaginary (but not at all imaginative) futures, allowing the other camp to denounce them as sheer science fiction. Neither side appreciated that they were debating mythical fears.

Another speaker in the Lords debate, Lady Saltoun, presents an interesting case. Initially she voiced her concerns in the same vague, ominous language as Baroness Masham:

I do not know when the soul enters the body and I do not believe anyone else knows, either . . . Until we know beyond all shadow of doubt, I think that the answer must be no to all experimentation on human embryos . . . If we leave that door open any longer we shall never get it shut again . . . And then, sooner than we realise – because of the speed of scientific development – we shall have experiments in inter-species breeding and genetic engineering with a view to creating a master race . , . not to mention a growing disregard for the sanctity of human life.

Yet we should not be too hard on Lady Saltoun, even if she seems to consider the timing of ensoulment of embryos a matter that science will settle, for she was one of the few willing to find out more about what human embryo research planned to achieve. And it became plain to her that the 'modification of an early embryo's genetic structure' mentioned in the White Paper did not involve the creation of a super-race, but rather the detection and elimination of inherited genetic disease. Shortly before the bill that stemmed from the White Paper was debated in Parliament, the fertility experts Robert Winston and Alan Handyside in London announced that they had successfully screened embryos produced by IVF to determine which were male and which female, in order to avoid the risk of a sex-related genetic disease. So *this* was what embryo research was all about!

Lady Saltoun decided by 1988 that embryo research might be a good idea after all. She changed her mind twice more before voting in its favour in 1990, but that prevarication can be fairly considered a reflection of her willingness to engage with the issues. The biomedical community had finally awoken to the need to explain the potential benefits of their work, but the campaign focused almost exclusively on the kind of genetic screening that Winston and Handyside pioneered, barely mentioning the possibilities that embryo research might offer for its original *raison d'être*: the treatment of infertility. Political and public opinion was swayed largely by the apparent potential of embryo selection to avert the kinds of 'monsters' that previously the technique had been expected to produce: as one peer in the House of Lords commented, 'such research can make an important contribution to preventing the creation of grossly deformed and mentally handicapped babies'. In fact, screening was focused largely on 'invisible' but life-threatening conditions such as cystic fibrosis, haemophilia and muscular dystrophy – but public attention was still fixated on the making and unmaking of monsters.

What won the day for the Human Fertilisation and Embryology (HFE) Bill were personal stories of how the work could diminish suffering. 'Wife who found joy backs embryo research' reported the *Daily Mail* (stressing the marital status of the happy mother). Yet as the sociologist of science Michael Mulkay points out, the narrative used to support embryo research tended to be as simplistic as its counterpart – the benefits of genetic screening (and perhaps gene therapy) were presented with no analysis of what was necessary to enable them or what social change might follow. 'Genetic risk' is a clear enough concept for cystic fibrosis, but what if 'risk' were deemed to be associated with ginger hair or darker skin? And there was little discussion of what these technologies implied for women; in the overwhelmingly male British Parliament, the debate was all about embryos.

As these medical applications became better appreciated, the debate sought less recourse in mythical tropes. 'Science-fiction imagery was almost entirely absent from the press during the final stages of public debate [1989-90] over embryo research', says Mulkay. (Like many commentators, Mulkay regards *Frankenstein* and *Brave New World* as 'science fiction', as though they had sprung pristine from their authors' imaginations.) Increasingly it was the pro-research lobby that perpetuated such imagery by erecting it as a straw man; *New Scientist*, for instance, felt it necessary to deny that 'any embryo researcher has tried to produce a monster with bolts in its neck, horns on its head or a pointed tail'.

All the same, opponents to the bill made what currency they could of the traditional Faustian theme. The *Daily Telegraph* spoke of the 'overweening arrogance of our perverted science', while the equally reactionary *Evening Standard* confirmed the view that Josef Mengele was now the modern Victor Frankenstein, at the same time using the old trick of turning an early embryo into a 'child' (or even, cutely, a 'little one'):

There is only one minor distinction between the life of a human being who has been born and the life of a recently conceived child in the womb or the test tube: the former is visibly human and the latter, though no less human, is invisible. This distinction is one of appearances only. It is the sole basis for regarding Dr Mengele, whose obsessive desire to extend the frontiers of medical knowledge led him to repudiate the humanity of the victims whom he slaughtered, as more obviously culpable than the medical experimenter of today, whose desire to discover ways of preventing genetic disease is honourable but whose failure to respect the humanity of the little ones upon whom they wish to experiment is, to say the least, regrettable.

Notice that this line of argument now requires that the conservative position *deny* the distinction of 'unnaturalness' previously attached to IVF embryos: to 'save' them, they must be made just like the 'rest of us'.

The HFE Act of 1990 made embryo research subject to a licensing procedure administrated by a body called the Human Fertilisation and Embryology Authority (HFEA). It stated that 'No person shall bring about the creation of an embryo, or keep or use an embryo, except in pursuance of a licence.' It forbade, among other things, the transfer of non-human embryos to a woman's womb, or of a human embryo to an animal, or the maintenance of a human embryo *in vitro* beyond the appearance of the primitive streak or an absolute limit of fourteen days. The HFEA was charged with the responsibility to regulate all research and manipulation of human embryos, whether for IVF or for scientific experimentation.

Until its impending dissolution by the UK's current coalition government, the HFEA has licensed and monitored all British fertility clinics and academic research involving embryos, according to laws determined by the parliamentary process. In the rapidly changing field of embryo research, this legal framework is recognized to require constant revision and extension: the latest amendments to the HFE Act were made in 2008, which, for example, banned sex selection of embryos for non-medical reasons and established restrictions on the creation of mixed human–animal embryos (see p. 288). The HFEA has become widely lauded as a model of how regulation of assisted conception may be conducted (which makes its imminent demise all the more an act of egregious political vandalism). But that is not, of course, the same as saying that the questions have been settled to everyone's satisfaction.

When life begins

And in truth they never will be, for the really hard questions about making embryos are not scientific but philosophical ones, and therefore inevitably dependent on one's religious and cultural context. But if they do not have absolute answers, we can nonetheless hope to evaluate and perhaps exclude some of the answers that have been offered so far.

Intervention in the process of conception and procreation surely demands some way of deciding whether and when a 'person' has been 'made'. That need has been highlighted by controversies about abortion during the past hundred years, and it was already implicit in older debates about the legalities of induced miscarriage. The question is usually phrased as a matter of 'when life begins', but this is a poor way to pose it, because conception merely passes the baton: from parent to offspring, there is never a moment when life is absent. The issue is rather: at what point is a *new* being formed, and at what point can we regard it as a person? *In vitro* fertilization brings this matter literally into focus, for the fertilized egg is then no longer a hidden, perhaps even an unrecognized, entity subject to the uncertain

influences of the uterine environment but is there under the microscope, demanding an answer to the question: is this a person, or not?

Pope John Paul II asserted that the position of the Catholic Church
was and always has been very clear:

Throughout Christianity's two-thousand-year history, the same doctrine has
been constantly taught by the Fathers of the Church and by her Pastors and
Doctors . . . the Church has always taught and continues to teach that the
result of human procreation, from the first moment of its existence, must
be guaranteed that unconditional respect which is morally due to the human
being.

This forms the basis of the Church's opposition to abortion, human
embryo research, and any other procedure that results in the destruction of embryos. But is the Pope's statement consistent with the biological evidence?

Some people will consider that an irrelevant question, insisting that
holy scripture carries more authority than the laboratory. Yet it is not
obvious what Christian scripture *does* say on the issue, and Church
authorities in the past were by no means as unanimous as John Paul
II asserts. The Bible says little about the process of conception beyond
a few ambiguous hints. As one speaker commented during the British
parliamentary debate on embryo research in the 1980s, 'Christians have
a rather difficult time because it is no good looking in the New
Testament . . . because we shall not find in it what our Lord said about
embryos. It is a question of the development of Christian doctrine,
and undoubtedly there is more than one view taken in the Church as
a whole.'

The biblical view of conception is, needless to say, grounded in
a very different notion of human biology from that of today. When
Job says to God, 'Did you not pour me out like milk and curdle me
like cheese, clothe me with skin and flesh and knit me together with
bones and sinews?' he is repeating an image used in the ancient
world: Aristotle spoke of the formation of a human being as initially
akin to the curdling of milk. A living organism was thought to begin
as formless chaos, onto which shape and organization are imposed.
In Hebrew tradition, the forming of the body is God's work. He is
described as a craftsman, knitting and weaving the tissues – Psalm

139 says, 'For it was you who formed my inward parts; you knitted me together in my mother's womb.' For Aristotle, unformed matter is the female contribution to the new person, on which the male seed acts like the rennet that curdles cheese. But is the milk already a human person? Is the cheese?

We saw earlier that Aristotle believed human beings have a three-part soul: the nutritive soul that plants possess, the sensitive soul shared with other animals, and the rational soul possessed by humans alone. He suggested that the body acquires each type of soul when it grows the organ needed to exhibit the associated characteristic. Thus, the appearance of a human soul is a gradual affair. Until that has happened, the embryo is not specifically human:

it is not the fact that when an animal is formed at the same moment a human being, or a horse, or any other particular sort of animal is formed, because the end or completion is formed last of all, and that which is peculiar to each thing is at the end of its process of formation.

Aristotle is unclear, however, about when the rational soul appears. Since he says that this, the defining stamp of humanity, is not contingent on the possession of any particular organ, but comes from 'outside', the theologian David Albert Jones argues that it may be present right from the earliest stages of embryogenesis. But that is supposition – Aristotle himself simply did not specify the timing. In any event, he asserted that the human form is complete only when the child is first felt to kick in the womb. So while it is possible (but not at all clear) that Aristotelian biology regards the embryo as being in some sense human at conception, it nevertheless makes a distinction between the fully formed human and the partially formed one.

The reason why we should care about what Aristotle said in this regard is that it informed both religious and secular views on embryogenesis until the Enlightenment. The distinction between the fetus before and after movement ('quickening') is felt to begin in the womb was important for establishing legal and theological positions on abortion and penalties for causing miscarriages. Was this homicide, or a somewhat less serious crime if the fetus was not deemed fully human? The ancient Hebrew law is laid down in Exodus 21:22–25:

If men who are fighting hit a pregnant woman and she gives birth prematurely but there is no serious injury, the offender must be fined whatever the woman's husband demands and the court allows. But if there is serious injury, you are to take life for life, eye for eye, tooth for tooth, hand for hand, foot for foot, burn for burn, wound for wound, bruise for bruise.

This is highly ambiguous. For one thing, 'serious injury' applies to either mother or child. Does 'giving birth prematurely' apply only to late-stage pregnancy, or to early miscarriage? The status of the early embryo is not obviously determined here.

A set of Irish canons in the seventh to the ninth centuries distinguished the penances for abortion according to whether the embryo was 'unformed like water', formed in flesh but without a soul, or fully ensouled. In thirteenth-century England, abortion was considered homicide 'if the fetus is already formed and animated and especially if animated', but not otherwise. Thomas Aquinas proclaimed that abortion was a very serious sin, but less so than murder. Aquinas shared the Aristotelian opinion that full humanity appears in the fetus only by degrees – he asserted that it is acquired by the male fetus after six weeks and the female after thirteen, which is when the fetus gains a soul. Even in the nineteenth century, English law against abortion distinguished the seriousness of the offence according to the quickening criterion. Although doctors at this time began to question its biological relevance, the distinction was abandoned in the United States only around 1880 (when all abortion became illegal in most states).

Aristotelian developmental biology, for all its historical significance, is of course wrong. The seventeenth-century English physician William Harvey challenged the view that embryos are created from the coagulation of fluids, saying instead that 'all living things begin as an egg'. Yet when Harvey investigated the wombs of deer immediately after intercourse, he found no sign either of fluids or of eggs. Nor did an embryo seem to appear until several weeks later. Harvey guessed correctly that the eggs must be present all along, but are of microscopic size. He agreed with Aristotle, however, that embryonic growth involves a progression from lower to higher forms of life, saying that around the fourth day after fertilization the human egg changes from being vegetable-like to animal-like. Thus the gradualist view of embryogenesis and 'personhood' survived the transition from the 'curdling' to the 'fertilized egg' picture.

Weighing the embryo

A qualitative distinction between human embryos at different stages of development is precisely what opponents of embryo research seek to challenge, and what advocates have sought to maintain. In the British debate that followed publication of the Warnock Report in 1984, supporters of embryo research took to labelling the early embryo (less than two weeks old) a 'pre-embryo'. The distinction was scientifically arbitrary, however, and was widely criticized – *Nature* called it a 'cop-out'.*

The Warnock Report itself called the embryo 'a potential human being', but did not specify how that is distinct from an 'actual' one, nor when one becomes the other. 'Although the questions of when life or personhood begin appear to be questions of fact susceptible of straightforward answers', the report said,

we hold that the answers to such questions are in fact complex amalgams of factual and moral judgements. Instead of trying to answer these questions directly we have gone straight to the question of *how it is right to treat the embryo*.

The report was right to say that the matter is very complex, but does not explain how one can possibly judge 'how it is right to treat the embryo' without resolving it.

Indeed, the report did not even resolve how one should treat human embryos in practical terms, let alone philosophically and ethically. It asserted that they deserve 'special respect', but without saying what

---

* Attempts to win the argument by linguistic sleight of hand have been more prevalent on the other side of the fence, for example in the way anti-abortion groups label themselves 'pro-life'. And the theologian David Albert Jones uses the word 'embryo' when the biology is being discussed but shifts without comment to 'child' when it comes to ethics. 'Christian compassion must embrace *both* mother and child', he says. No one is going to condone killing of a child – but is the week-old embryo a 'child'?

that implied. Did it mean that researchers should be dutifully solemn while tipping them down the sink? One could hardly avoid the suspicion that this phrase was inserted in an attempt to preserve the sensibilities of the onlooker rather than to recognize anything intrinsic to the embryo. As bioethicist Bonnie Steinbock puts it, 'Unless we can give a convincing account of "special respect", the suspicion will remain in many minds that this phrase merely allows us to kill embryos and not feel so bad about it.' Perhaps there are reasons why we should *not* feel so bad about it, but if so then they need to be made explicit. 'Special respect' might reasonably imply that we should not treat these entities frivolously, but as with the treatment of holy books, one person may consider acceptable what another will find shocking and obscene.

The Warnock judgement was echoed by the US Human Embryo Research Panel in 1994, which called for the status of the embryo to be given 'serious moral consideration'. Again this says nothing about what uses are appropriate – one might argue that we already grant laboratory animals and human cadavers 'serious moral consideration' while dissecting them. Bioethicist Daniel Callahan points out that the HERP judgement was silent on how 'a moral calculus [was] to be constructed to do the necessary balancing of the claims of research vs claims of the embryo'.

Can this be done? Most attempts to define 'humanness' have built on the idea that it stems from the possession of certain *capacities*, or of *potentials*, or *genetic uniqueness*. Each of these candidates can be made to sound somewhat plausible, but each also has ambiguities, and the criteria they invoke are not mutually consistent.

A definition of personhood based on capacities – saying that a person is a being capable of particular actions and thoughts – clearly offers no succour to those who demand that the embryo is a person from conception, given that an embryo consisting of just a few cells has no more capabilities than a slime mould. Yet if we accept John Locke's definition of a person as a 'thinking, intelligent being, that has reason and reflection', it isn't obvious that we can grant this status to a newborn infant either – it depends on what is meant by terms like intelligence and reflection. If the bar is raised too high, the result is that one might end up philosophically condoning infanticide, or at least making it less serious than the killing of an adult. And is a person in a coma no longer a person? Or a person with serious and irreversible brain damage?

A related definition derives from the notion of *interests*. Bonnie Steinbock argues that, because the embryo cannot be said to have interests in terms of possessing a stake in its own existence, it cannot be said to have a 'life' in the personal sense even if it clearly has 'life' in a biological sense. 'An embryo is not deprived of its life by being killed', she says. 'In an important sense, it does not have a life to lose.' But this argument seems rather circular. We don't consider that a bacterium has in this sense a 'life to lose', but some would argue that it is precisely the 'humanness' of the embryo that distinguishes it from a bacterium. One might say that, just as in the case of a comatose person, the embryo does not have to be aware of 'interests' in order to possess them.

The most common argument for granting an embryo the same rights as a fully developed human is that it has the *potential* to become one.* But what does it mean to say that an embryo comprised of a few cells has the potential to become a human? We might argue that an embryo left to itself within the womb will naturally develop into a human person. Of course, an embryo in an IVF Petri dish will certainly not do that of its own accord – it needs to be returned to the womb. But one could say that this would be only restoring nature to its 'normal' course after having disrupted it by technical intervention. It is true that it was this intervention that enabled the embryo to be created in the first place, but perhaps a consistent position on this argument from potential might stipulate that, once the egg is fertilized, we are morally obliged to take whatever actions are necessary to enable nature to take its course.

The problem is, however, that if nature were indeed to take its normal course, the most probable result is that the embryo will die. It's estimated that around sixty to seventy-five per cent of eggs fertilized *in utero* meet this fate, usually without the woman being aware of it. So the fate that we have designated as 'natural' is then not the

---

* A single embryo may divide and become two humans, but that doesn't substantially compromise this argument, since one can hardly argue that two humans should have fewer rights than one. On the other hand, doesn't it seem just a little odd to attribute 'personhood' to an entity that can be split in half (and which might do so spontaneously) and grow into two separate beings?

one that is most likely – in which case, in what sense can it be deemed natural? And how is that view changed if the chances of successful implantation and gestation can actually be enhanced by technology (such as hormone treatments)? Is that kind of intervention unnatural and therefore to be shunned, or is it, conversely, morally obligatory?

One of the relatively few contributions by a female politician to the debate over regulation of human embryo research in the UK during the early 1990s spoke tellingly on this issue:

A fertilized egg may try to develop in the [Fallopian] tubes and may never reach the womb. Its relationship to the mother is one of a malignant growth that will kill her if allowed to develop. So how can I agree with honourable Members who want me to regard such an entity as a human being?

What is more, embryos can become biologically incoherent yet continue to grow. One of the most disturbing manifestations of this is the teratoma – a benign growth found inside some babies that can contain all kinds of body parts and tissues in wild disarray: hair, teeth, eyes, arms, legs, nerves. A teratoma begins as an embryonic twin of the child, but the developmental process goes awry. It is in some ways more like a mature human than a newly formed embryo, because it has differentiated tissue types. But it makes little sense to regard it as a person, or even as a human being.

So any concept of 'what comes naturally' here can have little to do with what 'nature' would do, but is teleological, based on some assumption about intention. As we saw earlier, for the Catholic Church the definition of naturalness stems from the concept of 'natural order', which posits a view of what God has deemed 'good'. 'Natural' then becomes not what is likely, nor what would happen without intervention, but rather what has been divinely ordained and considered desirable. But what is 'desirable' is then decided by doctrine, not by the particulars of the human situation – not by whether, say, we would consider it desirable that a couple who cannot conceive a child by other means be enabled to do so by technology. For anyone who does not share that doctrine, the decision can never be anything but arbitrary: a claim to know what God intends.

One could argue nonetheless that, as a fertilized egg has the potential to develop into a human being (or more properly, as we have in general no means of being sure that it cannot – in practice there may be genetic or environmental factors that will inevitably prevent it), the ethical duty is to give this entity every opportunity to do so, even if that is not how things turn out. Isn't that at least a logically consistent position?

Biomedical science has now shown that it is not. For as we shall see in the next chapter, research on cloning and stem cells has demonstrated that probably every living cell in our bodies has the potential to make a new organism. It is likely soon to be possible to transfer the genetic material from any of our cells into an 'emptied' oocyte and trigger its development into an embryo. In fact, it seems conceivable that even this transfer process may not be required: biochemical agents may be found that return a mature body (somatic) cell to the stage at which it is capable of behaving like a fertilized egg (see p. 265). Even if we do not yet have the technological ability to carry out these processes in humans, there seems little doubt that the potential to grow into a human being is a property of every one of our cells.* The potentiality argument would then have to regard it as ethically wrong to discard any body tissue, or even to kill cells by scratching an itch. It collapses by *reductio ad absurdum*.

There's another objection to the potentiality argument. It is conceivable, although not yet achieved in humans, that we might make an embryo from cells that have been artificially genetically programmed to stop dividing after a fixed number of divisions – eight, say. The entity is then intrinsically incapable of becoming a baby. One might object that such manipulation would be horrific, like programming an embryo to die once it becomes a two-year-old child. Are the two things really equivalent, though? Imagine that we take a skin cell, add the 'genetic circuit' that provides the timed auto-destruct, and transfer the genetic material into an egg so that it starts to develop as a cloned embryo. This entity might be useful for studying the very early stages of embryo development, say. Doesn't there seem to be something odd about regarding such a modified skin cell with a prescribed eight-division lifespan as a human being? Even if we were to regard this as

---

* I can't help here thinking of the comment made apropos *Brave New World* by Charlotte Haldane, wife of J. B. S.: 'Biology is itself too surprising to be really amusing material for fiction.'

an unacceptable degree of 'tinkering with life', the objections would need to draw on a different source from concerns about the killing of a 'potential person'.

Some of these same arguments undermine any attempt to defend a definition of personhood in terms of genetic uniqueness, as a statement of the Catholic archbishops tried to do in 1980. That position runs aground in several other ways too. Clearly, genetic uniqueness does not constitute personal uniqueness, since identical twins share the same genome. And if we were to insist on conferring 'human' status on any entity that has the full complement of genes needed to make a person (or more than one), must we not also designate as 'human' an IVF Petri dish containing an egg and sperm that have not yet combined? You might want to insist (albeit on seemingly rather arbitrary grounds) that we should only accept the 'genetic' definition when the genes *have* combined: neither egg nor sperm alone possesses the full set of genes. Perhaps fusion of the gamete cells to make complete chromosomes is the right criterion. But in that case, a freshly fertilized egg is not a 'person', since the parental genes do not actually fuse until it has reached the two-cell stage. It would seem odd to make a profound philosophical distinction between genes separated by a few millimetres and a membrane, and genes separated by a few hundred nanometres and some cytoplasm. Besides, in the early phase of embryonic development, division of cells is controlled not by the genes in their nuclei but by the cytoplasm – in other words, the 'genetic identity' of the nascent organism is essentially irrelevant at this stage. The organism's genes are merely 'doing what they are told' by the maternal cytoplasm: the organism is not building itself under the autonomous direction of its self-defining genes. Indeed, the eventual activation of the embryonic genome is known as the 'maternal-to-zygote transition' (the zygote is technically the one-celled fertilized egg, but here it refers to the new organism as a whole, which at that point is called a blastocyst). In humans, this process begins at the eight-cell stage.*

---

* Let me point out for completeness that it is mistaken to consider a human being as an entity in which all the cells have human genes. Only around ten per cent of the cells in our bodies are genetically human – the rest are symbiotic bacteria. Without them, we would not be alive. So a being that is purely genetically human is an imaginary construct.

All this undermines the claim that the 'sacredness' of the embryo from the moment of conception finds support in biology. Leon Kass, for instance, has said that 'while the egg and sperm are alive as cells, something new and alive *in a different sense* comes into being with fertilization . . . there exists a new individual with its unique genetic identity, fully potent for self-initiated development into a mature human being'. And the Vatican has pronounced that 'Recent findings of human biological science . . . recognize that in the zygote (the cell produced when the nuclei of the two gametes have fused) . . . the biological identity of a new human individual is already constituted.' On the contrary, human biological science shows that there is no single stage at which that can be considered to happen.

In any event, to equate the genes with the person is biologically naive. For a start, the influence of genes may be moderated by environmental factors experienced during a person's life. And where is the 'person' in a stretch of DNA, most of which still has an unknown function, and some, perhaps most, of which has no biological function at all? Is it in the one per cent of genes that we do not share with chimpanzees? In arguing for the inviolability of the embryo in genetic terms, the Vatican is forced towards an indefensible and perhaps even impious genetic determinism: 'modern genetic science', it proclaims, 'has demonstrated that, from the first instant, the program is fixed as to what this living being will be'. Modern genetic science has, as we will see, often committed the sin of giving that impression, but only out of sloppiness and self-delusion. Once we ascribe personhood to our genetic constitution, we are liable to discover that it vanishes entirely, and all that remains are arrangements of atoms. It is no good looking for ethics within a molecular soup.

Some people consider the embryo inviolable not because of religious faith but because of what seems like innate moral intuition.* But that sort of intuition is a very personal affair. Most rational people agree that murder, theft and deceit are wrong, even if that does not always deter them. But on the ethical status of the embryo, nothing is obvious. Technology and biological science have brought us to the point where asking about the moral status of the embryo is akin to asking about

---

* There is in fact experimental evidence that moral intuitions precede judgements based on doctrine even for religious people.

the moral status of a drop of blood, a piece of skin, a fingernail paring. That's not to say that the human embryo is equivalent to these things, but the difference is one of attitude, not of science – not of fact. One can advance a good scientific argument that a human embryo *is* actually equivalent to other kinds of human tissue, except that it is not tissue that 'belongs to a person' – it is *all that exists* of this particular organism.

The problem, in the end, is that in evoking personhood we are trying to import an everyday concept into a science that cannot recognize it: our colloquial language here was never intended to apply to a series of biochemical processes stimulated by the fusion of gametes. Ancient prohibitions against killing and demands for human life to be respected were made in the context of our interactions with other people, not with other cells. It seems natural and entirely proper that these injunctions should be extended not just to babies but to unborn infants in the womb. But we cannot pretend that they were ever formulated with few-day-old embryos in mind, not just because sacred texts make no mention of such things but more obviously because they were written in ignorance of their existence.

It is not clear how, in creating moral and legal frameworks, we can negotiate this mismatch; but a forced marriage will not help. 'Personhood' is surely not an all-or-nothing affair, but a cluster of gradually acquired characters, stemming from such factors as memory, consciousness, sensation and history. When law professor Robert George asserts that 'a thing either is or is not a whole human being', he is pleading against hope that biology should respect our intuitions. When Eric Cohen, founder of the journal *New Atlantis*, implores us to object to the destruction of embryos by 'awaken[ing] a revulsion that is not naturally present', he is tacitly admitting that we don't intuit personhood in the embryo. When Leon Kass asserts that 'life and soul are irreducibly mysterious', it sounds more like a plea not to subject them to damaging scrutiny. Francis Fukuyama also quietly admitted defeat when he attempted in his book *Our Posthuman Future* (2002) to characterize what it is that makes us human: he could offer nothing but a label, 'factor X'. But factor X is not a material thing, nor even a well-defined concept. It is a suspicion, a myth – or perhaps, as molecular biologist Lee Silver suggests, a surrogate for the religious notion of soul.

'The scientific quest to understand the inner workings of life must be mediated by a foundational sentiment that life is a wondrous mystery, albeit a mystery amenable to our discovery', says bioethicist Courtney Campbell, thereby undermining the first phrase with the second – for a mystery is only that so long as you don't understand it. There is not a great deal of mystery left in the way a virus functions, nor even a bacterium (despite the fact that many of the scientific facts are still imperfectly understood). Likewise when Campbell insists that embryo research must recognize 'the fundamental wonder of life itself, encoded in the embryo', he again creates a self-cancelling statement from the confusion between life and human experience – why is this 'fundamental wonder of life' any less encoded in a parasitic worm? The problem with such sentiments is that they misdirect the awe that we experience in human life, thinking to locate it in 'life' defined as a process or a material, and not, as we should, in our humanity.

Filled with soul?

The Christian view of when personhood begins is bound up with the question of when a person gains a soul. Indeed, the two are generally considered to be synonymous. But what is this thing that they are supposed to gain – this entity that, as we have seen, has been commonly deemed to be lacking in artificially created beings? If the soul supplies the defining stamp of true humanity, we had better understand what is meant by it.

David Albert Jones says that for Christians 'the soul has been understood first and foremost as the principle of life' – a statement with so many possible meanings that it ends up having none. Aristotle's tripartite soul was more a characteristic of the organism than an incorporeal entity that passed into it. It is far from clear that this Aristotelian picture makes any connection with the Christian idea of a soul as something transcendent which survives after the death of the organism, for all that Thomas Aquinas strove to unite the two concepts. While the Aristotelian soul can conceivably mutate into a truly scientific concept – it might be understood as the means by which an organism achieves its organization and self-determining power – the Christian soul cannot. It cannot be measured, observed or analysed, notwithstanding the folk belief that souls have a measureable mass –

something of the order of twenty-one grams* – by which amount the dead body differs from the living one. One might then argue that science can say nothing about this article of faith.

But that isn't quite true. It seems reasonable to suppose that, even if the soul need not obey the laws of mechanics or thermodynamics, it should obey the laws of logic. It is one thing to say that the *nature* of the transcendental soul is a mystery, but if the *concept* is logically incoherent then there seems no compelling reason why it should be taken seriously.

Christian theology has generally held that each person has a unique soul which is united with the body during life. It is now the dominant view that the soul is created with the body, rather than pre-existing, although once made it is immortal. The important question for our purposes is when this union of body and soul happens, for that is generally deemed to be the moment that the 'body' becomes a human person.

Some Jewish scholars argue that God gives humans a soul 'at the first breath taken after birth', because God gave Adam a soul by 'breathing into his nostrils the breath of life'. In contrast, most early Christians thought that the soul appeared some time between conception and birth. They were not very specific, however, about when that moment arrives. According to the early fourth-century North African writer Lactantius, the soul 'is not introduced into the body after birth, as it appears to some philosophers [such as the Stoics], but immediately after conception, when the divine necessity has formed the offspring in the womb'. What 'immediately' implies is not clear, although Lactantius hints that this may not be until the human form has begun to develop. The Christian apologist Tertullian in the late second century took the view that the soul is not a preformed entity but that it grows, from the moment of conception, in a manner analogous to the growth of the body. This

---

* This popular myth arose from the work of a Massachusetts physician named Duncan MacDougall, who weighed human bodies just before and just after they had died. He reported his findings in the *Journal of the American Society for Psychical Research* and in *American Medicine* in 1907. 'One ounce of fact', he wrote, 'more or less will have more weight in demonstrating the truth of the reality of continued existence with the necessary basis of substance to rest upon, than all the hair splitting theories of theologians and metaphysicians combined.'

is the so-called traducian position, and it means that humanness is acquired gradually.

Several Christian authorities, including Thomas Aquinas, took a traducian view of ensoulment – it was, indeed, the official Catholic position until the late nineteenth century. Martin Luther also believed either that ensoulment happened only 'when the little body is prepared' – that is, when the human form becomes evident – or that it grew in a genuinely traducian manner. Even though the traducianism of the nineteenth-century Italian priest and philosopher Antonio Rosmini-Serbati was condemned in 1887, this condemnation was lifted in 2001 to prepare the way for his beatification in 2007. So whatever Pope John Paul II might assert about the Catholic Church's unanimity on the belief of ensoulment at conception, there is little doubt that its theologians have been and continue to be divided on the matter of how and when it occurs. The Declaration on Abortion drawn up at the US Catholic Conference of 1975 admitted as much.

If acquisition of a soul is delayed, this raised the idea – unacceptable to early Christian theologians – that Jesus might not have been 'fully human', with a complete human soul, since the moment of his conception. So they tended to insist that Jesus was ensouled from that instant. Some have deduced that this must be so for all other humans too. Why that should follow, however, when Jesus is exceptional in so many other ways is far from obvious. His divine nature and immaculate conception hardly seem to commend any generalization to the rest of humanity, let alone the fact that he was assumed to have been conceived as an omniscient homunculus, fully formed but in miniature (an embryonic or even a childlike Jesus being again too uncomfortable an image to be contemplated).

It should be said that not all Christians regard the soul as a Platonic 'thing' independent from the living body. The influential twentieth-century Jesuit scholar Karl Rahner rejected the idea of an other-worldly soul that animates the body, although in doing so he recognized that he diverged from the view of many practising Christians.* He was critical of a notion of the soul as 'an element within the totality of

---

* Rahner was also divergent in his permissive views on new technology, which indeed go further even than most scientists would allow: 'there is really nothing *possible* for man', he wrote, 'that he ought not to do.'

man which can be encountered immediately and in itself, and distinguished empirically and in test-tube [*sic!*] from the rest of him'. Rather, the soul is something that emerges through a person's relationship to the world. To establish that the soul is real, Rahner said, 'Theologians need only be able to affirm that human consciousness possesses that unlimited transcendentality in which there is present an openness, capable of legitimizing itself to the absolute reality of God.' Although the meaning here is characteristically elusive, it does seem plain that Rahner is linking the soul to human consciousness and self-knowledge: the soul, he says, obtains in 'the human person in the world'. If that is so, it is hard to see how it could have anything to do with a bundle of cells.*

Yet whatever else we might say about the concept of the soul, most theologians seem now to accept that it should not directly contradict modern biological understanding. How does it fare in that respect? The growth of identical twins from a single embryo does not in itself pose a serious problem for the belief that the embryo is human from conception, but it does real damage to the idea of a soul. If the soul is infused in the embryo at fertilization, where does the second soul come from when an embryo divides into two? At the very least, this makes for a highly interventionist God, who must quickly swoop down with another soul when that event occurs. David Albert Jones doesn't seem too troubled by this image, saying blithely that

As the new body is generated, God gives it a new soul . . . something similar could be said about twinning. As twinning results in a multiplication of human embryos, then God gives new souls appropriately.

Others might be more troubled by this notion of a God who, far from investing all creativity in the mystery of conception, hovers in case capricious biology might demand a second soul.

---

* Sadly, Rahner's writings offer little clarification of this interesting view. The theologian Terrance Klein explains that Rahner 'completely rejects the notion of a disembodied, unworlded soul surviving after death', only to add moments later that Rahner's view of the soul after death can be compared to the departed citizens of Grover's Corners in Thornton Wilder's play *Our Town*, watching the world with passive but keen interest. Perhaps the traditional view of the soul is just too tenacious after all.

And how does the original soul know in which of the twins it should remain? Perhaps, says Jones, twinning is predetermined, so that the original embryo is given two souls (forcing us to accept that two souls can exist in one 'body'). Or maybe twinning must be seen as the destruction of one individual and the simultaneous creation of two (which seems decidedly wasteful of souls). Other theologians have asserted that the soul is not infused until after the point at which twinning becomes impossible, around fourteen days after fertilization. That would seemingly negate objections to the use of early embryos in research, meanwhile positing an odd picture of God having to hold his horses in case biology springs a trick.

There is an opposite phenomenon to twinning, called chimerism, in which two separately fertilized eggs fuse into one embryo. If development continues, the baby may show no outward sign of having had such an origin. Where does the second soul go? Or does a chimeric baby possess two? What's more, some cells from the mother can become incorporated into the growing fetus. Do they bring a part of the mother's soul with them? Or can cells swap souls?

In the end Jones can't help but give the game away. 'The problem with twinning', he says,

seems less our inability to tell a 'soul story' and more the inability to judge between these stories. Until more is known empirically, it is difficult to know what sort of story to tell.

In other words, whatever biological research reveals, soul-believers will have a story ready for it. The explanation of ensoulment – this central tenet of Christian theology – will then be found not in scripture but in the biomedical literature. One might respond to Jones by saying that a 'story' is precisely what it will be.

Aside from the biological considerations, there is a far more obvious problem with the idea that an embryo receives a soul, whether at the moment of conception or shortly afterwards. Either way, since most embryos die before they can develop into human beings, the vast majority of souls in heaven will be those of 'people' who were never born. This difficulty occurred to St Augustine, who did not know the true extent of embryos that never survived but would have been aware of the frequency of infant death during or close to birth. He speculated that the resurrected bodies of such souls would appear as they would have looked if the beings had lived to maturity. But many embryos spontaneously abort in the early stages because they lack the capacity to grow any further: it is meaningless to imagine an adult form for them. 'This is not an easy question', Jones admits. One might equally conclude that it is not a question at all. Perhaps, however, this image of a heaven filled with embryo souls is after all an apt metaphor for the orthodox Catholic position on the status of embryos: it crowds real human beings out of the picture.

Mindful perhaps of these complications, the Catholic Church has never made an explicit official pronouncement about the timing of ensoulment. Instead, in 1974 the Holy See attempted to evade the issue by stating that even if ensoulment is delayed, the early embryo is still 'a *human* life, preparing and calling for a soul'. 'It suffices', the statement claimed, 'that the presence of the soul be *probable* (and the contrary will never be established), in order that the taking of life involves accepting the risk of killing a human being who is not only waiting for but already in possession of a soul.' This is, first, an admission that the whole business of souls is irrelevant for the purposes of debating the status of the embryo: even if the embryo is not yet deemed to have one, it is still sacrosanct because ensoulment works retrospectively, bestowing humanness from the 'first moment' the embryo existed. And second, the Church has no obligation to prove its claims philosophically, scientifically or theologically: it can merely assert that possession of a soul is 'probable', and rely on the fact that science can never disprove the existence of this incorporeal, non-physical entity. As a statement of bald superstition, this one takes some beating.

If belief in a soul is logically untenable, why is it so widely held? Admittedly, few people who believe in souls will have examined their faith along these narrow lines, and so will not have encountered the objections. But why did the idea take hold in the first place? Psychologist Jesse Bering has argued that a concept of the immortal soul is a corollary of a belief in life after death that, because it offers a psychological crutch and promotes peace of mind, has adaptive value in the Darwinian sense. In this view, we are hard-wired to be sympathetic to the idea of a soul. It's a plausible story but, like so many in evolutionary psychology, remains just that unless we can put it to the test.

On the other hand, Karl Rahner's view of the soul, teetering on the brink of what can be expressed in words, probably comes closest to articulating how many thoughtful Christians now regard it: as a representation of something transcendental in human experience that a materialist view of human biology will never adequately describe. Hard-line rationalists may scoff at such vagueness, but the testament of history and culture should make us hesitant to dismiss it out of hand.

Yet the notion of a soul as a countable thing, as a distinct entity that is either 'here' or 'there', can no longer be considered intellectually respectable, and can certainly play no part in discussions about what constitutes life or personhood, or how we should think about the status of the human embryo. It simply does not fit the picture. We cannot expect people of faith to relinquish the soul to the dictates of logic – but science cannot resolve that dilemma for them.

Aristotle's concept of soul is, in retrospect, evidently an attempt to bridge the gulf that made life so mysterious to natural philosophers in the ancient world. But the religious soul seems to have a different function, aside from offering the mental balm of potential immortality. In real terms, it 'does' nothing: it adds no material substance to the body, gives us no physical capabilities that we would otherwise lack, and apparently leaves no detectable trace on our behaviour. Rather, it supplies a kind of watermark of humanity. That is how it functions in the homunculus tradition, and in *Frankenstein* and in Čapek's robots. *It is a defence against the horror of materialism* – it protects us from the thought that we can be manufactured. No matter how well you do the job, it suggests, you can never make a real person because the soul cannot be created by human art. In the anthropoetic tradition, the soul serves as a tool of denial, a way of withholding humanness from the artificial being. And even if many might now deem it anachronistic to speak of souls, this unspoken act of denial remains: we will find ways to insist that we cannot be made.

Market forces

More than two million babies worldwide have now been born by IVF, and it would be tempting to suppose that, here at least, what once seemed unnatural and distasteful about 'manufacturing babies' has become thoroughly normalized. Where society continues to debate the process, the questions usually concern the problems of commercialization and the prioritizing of medical funds.

This situation shows that, in spite of the constant recourse to the imagery of *Brave New World* when each new development in assisted conception appears, Aldous Huxley got it wrong. Or rather, the myth he told was one specific to his age, not ours. To Huxley, artificial procreation was an instrument of state control and social engineering. But today it is not the all-powerful state that dominates the agenda; it is the free market. There is, as business academic Debora Spar has convincingly argued, now a market for making babies. Yet it is one that operates unacknowledged, for who wants to admit that babies are commodity items? As Spar says,

Because no one wants to define baby-making as a business, and because the endeavor touches deeply on the toughest of moral dilemmas, many governments around the world have either ignored the trade in children or simply prohibited it.

This means that either the market exists largely unfettered by laws, institutions and regulations (as in the United States), or that, when these do develop, they fail to recognize the true character of the thing they are supposed to regulate. 'We need to view reproductive medicine as an industry', says Spar, 'with all the commercial prospects and potential foibles that other industries display.' This is not necessarily an indictment – most of us do not consider industry to be intrinsically bad – but it does no good to ignore the reality out of moral squeamishness. We saw earlier how, because *in vitro* fertilization and the excess embryos it generated were considered too controversial to be addressed at a political level in the United States, prohibition of public funding sat alongside total permissiveness in the private sector. And as law professor Lori Andrews says, 'We need to give serious thought to whether our laissez-faire market mechanisms are the best determinants of how children should be brought into the world.'

For this is the reality of today's anthropoeia: making people is not magic, not demonology or perverted anatomy, nor industrialized totalitarianism, but commerce. 'It's no use being coy about the baby market or cloaking it in fairy-tale prose', says Spar. 'We are making babies now, for better or worse, in a very high-tech way.' 'Baby-making' is an unusual market, apparently shirking some of the standard rules of economics such as the cost relationship of supply and demand. But like any market, it will generate questions about economic concepts such as property rights: ownership of eggs, sperm and embryos. The very nature of this 'property' remains contentious: when a Chicago fertility clinic mistakenly discarded a couple's embryos in 2005, the clinic argued that this was a case of erroneous destruction of property, while the aggrieved couple sought to bring a charge of murder.

No market exists without advertising, product development and marketing. José van Dyck argues that the market for IVF, like the market for any technology, is never simply 'there' but needs to be created – a process, she says, 'which is ideological rather than technical':

> The creation of need is an intricate process of image-building and story-telling. In order to become accepted, technology needs a story, not only to sell itself, but also to justify the time, expense and efforts invested in a new invention ... Scientific projects are political and social arrangements, embedded in a society that gives rise to, or disapproves of, their legitimacy.

In this view, IVF doesn't simply get manipulated to fit a pre-existing set of ideas about anthropoeia; it creates its own stories. Van Dyck argues that 'the use of reproductive technologies reinforces the ideal that parents need to have a child they can call their own'. (It is implicit that she is talking here about genetically related children.) At first glance, it might seem harsh to imply that IVF creates such stories for the purpose of self-promotion. Isn't it a normal human instinct, not an invented narrative, to want a genetically related child? And when the procedure was being developed there were many people suffering from infertility who were clamouring to try it – no one, surely, needed to *create* a market for infertility treatments?

But van Dyck's claim warrants consideration. IVF has expanded far beyond the sole application that Edwards and Steptoe envisaged: circumventing the complications of Fallopian tube blockages. It can be used to enable genetic screening of embryos, to avoid the risk of inherited genetic disorders. It can be used (alongside sperm donation) by single women. Embryos might be created and then frozen for implantation when it is more 'convenient' for the (fully fertile) mother. With frozen sperm, it offers the possibility of a woman conceiving the child of a dead male partner, or of choosing a sperm donor on what might be considered eugenic grounds.

Each of these possibilities raises a host of ethical questions, and I do not propose to rehearse them here. Van Dyck's point is that there was an active process of story-construction in the development of IVF, focused on the arrival of the much-awaited healthy child, which does not necessarily reflect the way things often turn out. There was a determination to counter the fears inherent in the anthropoetic tradition by rendering the process 'normal', even 'natural', stressing that this is ordinary conception in which the *vitro* merely lends a helping hand. This perception was encouraged by images of IVF that show the moment of conception at a level that makes it indistinguishable from what happens inside a woman's body: as sperm entering egg. Thus, van Dyck says, IVF 'becomes incorporated in the myth of natural reproduction – the artificial smoothly aligned with the "natural"'. Sometimes IVF was even presented as a way of *improving* on nature – the old defence of the alchemists – as is evident in a *Time* story from 1991:

The beauty and power of IVF are that it allows doctors to take many key events in reproduction out of the body, where they are subject to the vagaries of human biology, and perform them in vitro.

The Christian bioethicist Joseph Fletcher has argued that there are benefits in conducting conception outside the womb, which is, he wrote, 'a dark and dangerous place, a hazardous environment'. Given that the human race has so far managed to sustain itself by procreating in this dark and dangerous place, one must suspect that there is indeed a new story being created here about how reproduction might (should?) happen. These narratives potentially establish an opposite pole, equally fictional and mythical, from that at which the technology is demonized as 'unnatural'. Rather than being hubristic, evil magicians trying to control life, the scientists become selfless saviours or seekers of pure knowledge. By extension, their patients become helpless victims needing technological succour: the 'desperate' would-be parents to whom media articles on IVF endlessly allude.

Playing God?

These very real concerns too rarely feature in a debate over reproductive technologies and embryo research that is dominated by the abstract, moralizing bioethics of what has been dubbed the 'theocon' position. When the advent of stem-cell research put human embryos back on the political agenda in the early twenty-first century, President George W. Bush was compelled to appoint another committee to provide ethical guidance – and knowing what he wanted to hear, he loaded it with conservatives. It was led by Leon Kass, and included Robert George, Charles Krauthammer and Gilbert Meilaender. As bioethicist Arthur Caplan commented when the members of the board were announced,

The ethics boat the president has launched is stacked with members who lean to the political right, who will rely on religious rather than secular principles to navigate its course and will do nothing to jostle any of the president's already espoused positions condemning stem-cell research, cloning and the creation of human embryos for research.

That intention was confirmed when, in early 2004, two of the committee members who advocated human embryo research – biologist Elizabeth Blackburn and ethicist William May – were abruptly dismissed and replaced by religious conservatives.

Cognitive scientist Steven Pinker has adroitly identified the 'sickness in theocon bioethics', which he says:

goes beyond imposing a Catholic agenda on a secular democracy and using 'dignity' to condemn anything that gives someone the creeps. Ever since the cloning of Dolly the sheep a decade ago, the panic sown by conservative bioethicists, amplified by a sensationalist press, has turned the public discussion of bioethics into a miasma of scientific illiteracy. *Brave New World*, a work of fiction, is treated as inerrant prophesy. Cloning is confused with resurrecting the dead or mass-producing babies. Longevity becomes 'immortality', improvement becomes 'perfection', the screening for disease genes becomes 'designer babies' or even 'reshaping the species'. The reality is that biomedical research is a Sisyphean struggle to eke small increments in health from a staggeringly complex, entropy-beset human body. It is not, and probably never will be, a runaway train. A major sin of theocon bioethics is exactly the one that it sees in biomedical research: overweening hubris. In every age, prophets foresee dystopias that never materialize, while failing to anticipate the real revolutions. Had there been a President's Council on Cyberethics in the 1960s, no doubt it would have decried the threat of the Internet, since it would inexorably lead to 1984, or to computers 'taking over' like HAL in 2001.

It is a depressingly impeccable statement of the situation, to which one might crucially add that the catalogue of distortions, exaggerations and apocalyptic extrapolations of the theocons is deemed credible only because, as we now see, it grows in fertile soil.

Time and again, the warning sounded by the theocon agenda is that by intervening in procreation we are 'playing God'. Paul Ramsey made artful play of this notion in his 1970 book *Fabricated Man*, saying that 'Men ought not to play God before they learn to be men, and after they have learned to be men they will not play God.' To the extent that 'playing God' is simply a modern synonym for the accusation of hubris,* this charge against anthropoeia is clearly very ancient. Like evocations of Frankenstein, the phrase 'playing God' is now no more than lazy, clichéd – and secular – shorthand, a way of expressing the vague threat that 'you'll be sorry'. It is telling that this notion of the man-making man becoming a god was introduced into the *Frankenstein* story not by Mary Shelley but by Hollywood (see p. 86). For 'playing God' was never itself a serious accusation levelled at the anthropoetic technologists of old – one could tempt God, offend him, trespass on his territory, but it would have been heretical seriously to entertain the idea that a person could *be* a god. As theologian Ted Peters has pointed out,

The phrase 'playing God' has very little cognitive value when looked at from the perspective of a theologian. Its primary role is that of a warning, such as the word 'stop'. In common parlance it has come to mean just that: stop.

---

* The two are not really synonyms, however, for hubris is an arrogant, excessive pride that sets out to *defy* the gods, not to usurp them.

And yet, Peters adds, 'although the phrase "playing God" is foreign to theologians and is not likely to appear in a theological glossary, some religious spokespersons employ the idea when referring to genetics'. It has, in fact, an analogous cognitive role to the word 'unnatural': it is a moral judgement that draws strength from hidden reservoirs while relying on these to remain out of sight.

Well then, what did President Bush's ethics committee do? It stared the debate's mythical elements in the face but failed to recognize them. When the committee met in 2002, its members had been assigned by Kass the task of reading Nathaniel Hawthorne's short story 'The Birth-mark'. This tells of a chemist named Aylmer who seeks to 'perfect' the beauty of his wife Georgiana by removing a small birthmark from her cheek. It turns out that this mark provides the vital link between her 'angelic spirit' and her 'mortal frame': when it is removed by Aylmer's chemical manipulations, she dies. The narrative is very clearly Faustian: by searching for biological perfection, Aylmer brings tragedy on himself.

Thus Kass established the tone he wanted the discussion to take. He claimed that 'The Birth-mark' deals 'with certain important driving forces behind the growth and appreciation of modern biology and medicine, our human aspiration to eliminate defects and to pursue some kind of perfection'. In the ensuing debate, Kass made much of our instinctive revulsion at Aylmer's aims. Indeed, we might deplore the (al)chemist's determination to eradicate the slightest blemish from his wife; we might also recognize some analogy with ambitions to genetically engineer or clone children to be 'better': more intelligent, more beautiful, more athletic. These things are widely condemned, for good scientific, ethical and social reasons. Why they should frame a general debate about biotechnology, embryo research and cloning either for reproductive or biomedical purposes was not at all clear, however, unless the quest for improvements in health is to be equated with a quest for human perfection.

But for Kass the story is a parable about attempting to eliminate *all* 'limitations' with which we are born – even mortality. While some of the board felt that Aylmer's actions were reprehensible because the birthmark was such a trivial flaw, and that the moral would be less clear if, say, Aylmer had been trying to remedy a serious physical deformity in his wife, Kass disagreed. 'I do not think the sign of the birthmark is superficial', he said.

What it means is deep and the attempt to go after the human condition to save it completely and to save it even from its mortality . . . means that there is something in the culture at large and something in medicine today, however modestly practiced, that almost says, 'Look, we will never stop until we can deal with mortality as such.' The question is, is that a worthy aspiration or is there something that necessarily gives rise to shuddering as a result of our efforts to do that?

This is, in other words, moralizing from the slippery-slope argument, which fails to acknowledge that we might agree to permit trivial interventions for trivial problems (braces for misaligned teeth) but only to allow hazardous interventions for life-threatening conditions. One board member argued that we do not always know how safe an intervention will be; this is true, but it does not mean we are wholly ignorant. We can feel fairly sure that orthodontic braces are not life-threatening.

And however we choose to interpret the symbolism of the birthmark, there is an obvious flaw in Hawthorne's moral: it requires him, rather than Aylmer, to play God. The fateful outcome of Aylmer's experiment results from the arbitrary imposition by the author of a supernatural connection between the birthmark and Georgiana's life. The tragic consequence is not so much due to Alymer's hubris as to Hawthorne's insistence that this is how the story must end. Hawthorne's tale does not demonstrate that a 'quest for perfection' will be ruinous; it merely describes a world engineered by its author to ensure that is the case. Would we feel so aghast at what Aylmer had done if his remedy had worked and the couple had lived happily ever after? After all, removing a birthmark today is often a simple matter, and while we might regret that it should be deemed desirable in any case, we do not legislate against it. Moreover, in order to express his distrust of science (a common position of late-eighteenth-century writers – see p. 64), Hawthorne loads the argument against the 'scientist' by making the imperfection one that is obviously trivial, so that the reader can easily interpret his efforts as perverse tinkering. (Aylmer's assistant even states explicitly that if Georgiana were his wife, the mark would not trouble *him*.)

Moreover, some of the scientists on the ethics board were understandably perplexed by the fact that Hawthorne's vision of a lone experimenter pursuing his course in secret was an eighteenth-century narrative that had nothing to do with modern science. And that is of course the case, for Hawthorne's scenario is very clearly a throwback to the trope of the obsessive medieval alchemist. Indeed, Hawthorne himself tells us that Aylmer's library was stocked with alchemical texts: he was writing with an eye on this older tradition, with all its Faustian and anthropoetic associations, and it is in that context that he alludes to aspects of the ancient art–nature such as the question of whether nature can be bettered.

The real problem was then that, as science historian William Newman says, Kass was asking his board to read 'The Birth-mark' in a historical vacuum, as an 'atemporal index of human repulsion to the hubris inherent in "the pursuit of perfection"'. In the end, Kass's assignment revealed more about the board members' preoccupations and assumptions than about an appropriate and contextualized moral reading of Hawthorne's tale. It was instructive as a barometer of the board's subsequent, largely censorious deliberations. But that may not have been quite what the committee's chair had intended.

11 Make Me Another

In the not-too-distant future, it will be looked at as kind of foolhardy to have a child by normal conception.
Gregory Stock (2000)

After fielding numerous inquiries about the possible cloning of Michael Jackson, Dr Brigitte Boisselier, head of Clonaid, today reaffirmed the company's policy of strictly respecting the privacy of each of its patients. 'Clonaid prides itself on never releasing the identity of the numerous individuals who have been cloned in the past six years,' Boisselier said. 'Even if that policy has been at the cost of my reputation, it's important for us that the celebrities and other interested parties contacting us know they won't be betrayed.'
Clonaid press release, 8 July 2009

B2: You said thing, these things
Salter: I said?
B2: you called them things. I think we'll find they're people.
Caryl Churchill, *A Number* (2002)

To make a homunculus, you needed at very least some human sperm. In the patriarchal age when sperm was considered to supply the soul and form of the body, adding the female *menstruum* to the mixture – the passive matrix of new life – was optional. Jacques Loeb reversed that view at the end of the nineteenth century by demonstrating that an egg could develop into an organism without ever coming into contact with sperm. One way or another, however, a gamete has been regarded as the minimum requirement for what we might call biological (as opposed to mechanical or chemical) anthropoeia. Contemporary techniques real and supposed, from ectogenesis to IVF, need both varieties.

But that may not be the case for much longer. The cloning of Dolly the sheep in 1996 did not exactly begin with an egg, but rather, with something more like an eggshell. Into this receptacle were injected Dolly's genes, taken from a cell derived from mammary-gland tissue of an adult ewe. Cloning of humans has already been attempted, and although it has never been shown to progress to the stage at which a cloned embryo implants in the womb, there is no known reason why that should not be possible. It will most probably happen within the next decade, although such predictions are notoriously hostage to fortune.

Whatever the social and ethical considerations of making human clones, cloning per se profoundly reconfigures the conventional picture of developmental biology. It forces us to think again about what our flesh represents: not just a mass of specialized cells that each has meaning only within the context of the others, but a matrix from which new beings could, in principle, spring. If that is unnerving, it is also a reminder that the origins of cloning lie partly in the science and technology of tissue culturing, in which cells are grown in an artificial environment. In fact, a tissue culture is itself a kind of clone, because the new cells are genetically identical to their progenitors. Cloning still remains closely associated with this field, being allied to stem-cell technologies that take advantage of the developmental versatility of the tissues of early-stage embryos. This association provides yet another lineage for modern anthropoeia – with, as we should now expect, its own legend-strewn landscape of hopes and fears.

Growing flesh

Julian Huxley did not have his brother's flair for fiction, but that didn't prevent him from having a stab at it. He wrote a short story called 'The Tissue-Culture King', which was published in the scholarly *Yale Review* in 1926 but found a more appropriate home a year later in the recently launched pulp-fiction magazine *Amazing Stories*. The story is in some ways stranger than anything in *Brave New World*, for it tells of a tribal leader who turns pieces of his own flesh into sacred, magical charms.

Huxley's tale uses the old literary trope of a lone traveller recounting a fantastical society encountered in some far-off place – the device used by Thomas More in *Utopia*, Jonathan Swift in *Gulliver's Travels*, and most resonantly in the present context, by Huxley's close friend H. G. Wells in *The Island of Dr Moreau*. Huxley's adventurer is taken captive by an isolated African tribe who, he discovers, are bewitched by the experiments of a stranded British biologist named Hascombe, 'lately research worker at Middlesex Hospital'. Hascombe has created an 'Institute of Religious Tissue Culture', where he capitalizes on the mystical protective power invested in the tribal king by culturing the king's tissues and dispensing them, like holy relics, to his subjects.

Huxley makes the old connection between transgressive biology and the generation of monsters. Like Dr Moreau, Hascombe has created an array of freaks, which he presents to the Africans as objects of worship: dwarfs, girls with moustaches, sexually mature children. There are biblical connotations in his plan to make a race of virgins from a kind of parthenogenesis based on the methods of Jacques Loeb.

Given Huxley's optimism, even naivety, about the potential of biological science to improve the human condition, it comes as a surprise to read the moral inserted clumsily at the end of the tale: 'It is the merest cant and twaddle to go on asserting, as most of our press and people continue to do, that increase of scientific knowledge and power must in itself be good.' It is hard, however, to imagine that Huxley would have been well pleased with the current reversal whereby the press and the public often suspect that any expansion of the power and scope of science must be inherently bad.

Like his brother, Julian Huxley based his fiction on what he saw happening in contemporary science. For tissue culture had become well established at the start of the century, and it seemed to imply that human flesh, if not the human person, could enjoy a kind of immortality.* What is more, this new technique revealed that human

---

* That tissue culture supplies immortality of a sort is implied in the very title of Rebecca Skloot's account of how cervical cancer cells taken from a black woman, Henrietta Lacks, in 1951 supplied a cell line called HeLa now routinely used in biomedical research: *The Immortal Life of Henrietta Lacks* (Macmillan, London, 2010).

body parts, and perhaps even the entire human body, could be *grown* in the laboratory.

Huxley's story was inspired by the work of an American surgeon named Alexis Carrel, a colleague of Loeb's at the Rockefeller Institute in New York. Carrel was a pioneer of organ transplantation and surgery – for which he won the 1912 Nobel Prize in medicine – and he was interested in how amputated organs might be sustained in a viable state. He found that body tissues could not only be kept 'alive' but could continue to grow for many days if kept in a culture medium containing all the nutrients that living cells need. By 1910 Carrel had cultured human thyroid and kidney tissue, and he was later successful with heart tissue: he showed that a chicken heart could be kept growing and pulsating *in vitro*. To Huxley this implied that 'even in mammals most of the parts of the body are potentially immortal'. His conviction in the feasibility of ectogenesis drew strength from this demonstration that tissues could be maintained outside the body.

Carrel's achievements became conflated with Loeb's work on parthenogenesis, to the point where it was rumoured that he too was on the verge of making artificial life. One press report awarded him Frankenstein-like powers: 'he could by taking various parts of different animals, build up a new living creature'. According to science writer Jon Turney, Carrel's work was 'incorporated into a myth of a collective scientific endeavour to create life'. This was clear when a journalist for the magazine *Titbits* described tissue cultivation being conducted at the Strangeways research hospital in Cambridge. Here, he assured readers, 'canned blood' might be used to make an artificial baby. 'Could you love a chemical baby?' the reporter asked rhetorically, before parading a royal flush of anthropoetic myths in a single sentence: 'Will the sexless, soulless creatures of chemistry conquer the true human beings?'

Growing and moulding human tissue *in vitro* by culturing cells within a supporting biodegradable polymer mesh is now a mature technology. As they grow and divide, the cells colonize the scaffold and produce a piece of tissue with a predefined shape. Gradually the scaffold is eroded, broken down by the cells' metabolism into innocuous chemical reagents such as carbon dioxide. This technique has been used to make 'artificial skin' for grafts, and by using more complex 'scaffolds' seeded with cells of different tissue types, entire organs such as livers or hearts might one day be grown. If seeded with the cells of the organ's intended recipient, the problem of immune rejection that complicates organ transplantation would be avoided. It is possible in principle to imagine growing a complete human body: a fully formed adult, say, created by clothing a plastic simulacrum in flesh. Humans grown from tissue culture were one of the prospects raised by *Life's* science editor Albert Rosenfeld in his 1969 article (p. 201). Quite aside from the improbability that such a being would spontaneously 'come alive' once every part is in place, however, the requisite structural complexity of the tissue looks impossibly daunting. On the other hand, many cell types, such as neurons, possess the ability to weave their own microstructures: the cells communicate by producing chemical signalling molecules which guide their neighbours into the right config-urations.

The finesse of this sculpting of living tissue may be refined by the use of printing technologies, like those devised for ink-jet printers, to deliver the cells to the required locations. Each nozzle of this 'bioprinter' would dispense a different cell type as if it were simply a primary ink. Under computer guidance, three-dimensional tissues could then be 'written' layer by layer, copying the arrangement found in cross-sections of real tissues and organs. In this way, organs can be 'sketched out' in rough, with the cells left to organize themselves into fine details such as the tiny capillaries of vascular systems. 'Living cells have a unique capacity to self-assemble into three-dimensional living human tissues', says Vladimir Mironov, a proponent of this method at the Medical University of South Carolina in Charleston. 'We just put them in the right place at the right cell density.' Mironov and others have shown that bioprinting works in principle: cells can be deposited in specified positions without being traumatized by the printing process.

Mironov remarks that 'if we are able to bioprint all human organs, then by logical extension we will also be able to print an entire human body'. He sees this as 'the realization of Pygmalion's dream': the sculpting of a living statue that, if desired, could conform to whatever model of beauty the sculptor prefers. This is nothing less than a contem-porary version of Čapek's fantastical 'robot factory' in which artificial people were assembled from tissues and other components spun and shaped on an industrial assembly line. Mironov admits he is 'not sure that humanity is ready for this' and doubts it will happen in his lifetime; but he insists that 'technology is probably not the biggest problem'. Most researchers, however, see a much more modest future for tissue engineering in regenerative medicine: to repair damaged body parts, for example by regrowing neurons to combat neurodegenerative diseases or to heal spinal-column injury.

All-purpose cells

These efforts in tissue engineering are increasingly exploiting the protean character of so-called stem cells, which have not 'differentiated' into specific tissue types such as liver, skin or heart. Stem cells are present in embryos in the early stages of growth, and are said to be 'pluripotent', possessing an ability to form any of the more than 200 tissue types in the human body. (Earlier still, the embryonic cells are 'totipotent', able to form the placenta too.) They gradually lose that versatility as development proceeds: 'multipotent' cells can, for example, give rise to a limited number of mature cell types, such as both white and red blood cells.

Stem-cell research has depended largely on the availability of embryonic tissue taken from the 'excess' embryos generated in IVF. Because the extraction of embryonic stem cells destroys the embryo, the research has been burdened with the questions about the moral status of the human embryo that I considered in the previous chapter. But these controversies may eventually be obviated by the development of methods that can make ordinary tissue cells revert to stem-cell status. Differentiation of cells happens not, as once suspected, by the loss of genes that are not required, but simply by their silencing – a phenomenon not fully understood, but apparently associated with the attachment of chemical labels to the respective stretches of DNA, a so-called epigenetic process of the sort discussed in the next chapter. In 2007 researchers in the United States and Japan discovered how to make 'induced pluripotent stem cells' from differentiated cells (in that case, connective-tissue-forming cells called fibroblasts) by using viruses to insert a handful of genes encoding transcription factors, proteins that bind to DNA and control gene activity. Other transcription factors have now been shown to convert one cell type directly to another: fibroblasts, for example, have been transformed this way into fully active neurons. In other words, we can make skin become brain simply by adding the right genetic 'spice'.

Thanks largely to human stem-cell research, tissue engineering is sometimes perceived to be poised on the brink of growing people. Jean Rostand already recognized this possibility, and its link with cloning, in 1959:

> If a biologist takes a tiny fragment of tissue from the freshly dead body [of a person], he could make from it one of those cultures which we know to be immortal, and there is no absolute reason why we should not imagine the perfected science of the future as capable of re-making from such a culture the complete person, strictly identical to the one who had furnished its principle . . . This would, in short, be *human propagation by cuttings*, capable of assuring the indefinite reproduction of the same individual – of a great man, for example.

Here Rostand falls prey to misconceived genetic determinism (see p. 284); but the basic point remains.

Because embryonic stem cells can develop into any tissue type, with a corresponding capacity to organize its own shape and microstructure, questions arise about the moral status of stem cells themselves. Is their disposal equivalent to that of any other body tissue – a mouth swab or a removed appendix, say? Or do they, in their potential to grow into any part of the human body, possess something of the moral standing of embryos themselves (whatever that might be)? These stem-cell cultures are imbued with more symbolic power than any differentiated flesh taken from Huxley's Tissue-Culture King, for not only are they a part of the 'king's' extended body but they seem imbued with every part of him. They are a kind of human essence made flesh.

These associations were implied (without, it must be said, any real insight into the implications) in a 2007 UK television drama series that presented itself as an 'update' of Mary Shelley's story. In this contemporary *Frankenstein*, the eponymous scientist, now modishly a woman named Victoria, is the head of a project to engineer a human heart from stem cells. When she modifies the process by secretly adding a drop of her dying son's blood, she triggers a change that produces not a lone organ but a proliferating 'chain' of them. In a twist inspired more by Hollywood's than by Shelley's version of Frankenstein's experiment, the equipment is struck by a freak lightning bolt and runs out of control. You can guess the rest.

The first clone

When Jacques Loeb triggered parthenogenesis in a sea urchin, he made a clone: a new organism with an identical genetic constitution to the one that made the egg.* In this way he induced a process that is rather common in nature, although unusual among vertebrates (fish and lizards do it) and virtually unknown in mammals. That is to say, it is virtually unknown for *new* mammalian organisms to be generated by spontaneous cloning. But cloning in the sense of the replication of genetically identical cells is the standard mechanism by which any multicellular organism develops: each of us is in essence a confederation of self-cloned cells.

Loeb was not actually the first person to clone a sea urchin. This was achieved in 1892 by the German biologist Hans Driesch, a former student of Ernst Haeckel's, working (as Loeb had recently done) at the Marine Biological Station in Naples. He discovered that he could separate the cells of developing sea-urchin embryos at the two- and four-cell stages simply by shaking them. He expected that each fragment would continue growing into precisely half of the mature sea urchin, but to his surprise he found that each developed into a complete (albeit small) embryo. This implied that cells at the very early stage of embryogenesis possess pluripotency, the ability to grow into all the component parts of the organism. It challenged the experiments of Wilhelm Roux, another of Haeckel's protégés, who had reported that

---

* There is in fact a fine distinction between parthenogenesis of an oocyte cell, which is the precursor to a mature egg, and of an egg itself. Oocytes, like all somatic (body) cells, have two sets of chromosomes (they are 'diploid'), whereas in the process of fertilization the chromosomes divide so that the resulting egg cells have only one set (they are 'haploid'), ready to fuse with those of a haploid spermatozoon. (Note how this implies that *directly* after fertilization the egg is not a viable organism with a full set of chromosomes, but contains too many, and those from each gamete are as yet uncombined.) Parthenogenesis of both oocytes and mature eggs may occur, and the resulting organisms are diploid and haploid respectively. So only if it happens from oocytes are the resulting organisms clones in the strict sense, with the same genetic make-up as the parent. Mammalian embryos cannot develop fully unless they are diploid.

when he destroyed one of the two cells in a frog embryo using a hot needle, the other cell developed into only half an embryo. (It turns out that Roux's result was due to the influence of the damaged cell, which remained attached to the pristine one.)

The hypothesis of pluripotency was supported by experiments performed in the early 1900s by the German biologist Hans Spemann. He transplanted tissue from one part of a young embryo to another part and found that it developed into the new tissue type appropriate for that location. Spemann also achieved cloning by mechanical means, using a noose made from the hair of his baby son to separate the cells in a two-cell salamander embryo. In 1928 Spemann demonstrated his skill in micromanipulating cells by moving the nucleus of one embryonic salamander cell (the compartment in which the chromosomes reside) into a newly fertilized salamander egg that had had its own chromosomes removed to make it 'enucleated'.* So the enucleated host cell is effectively given a new genetic programme: it becomes a clone of the cell from which the transferred nucleus came. Spemann found that the clone made this way by so-called 'nuclear transfer' would grow into a complete embryo. This indicates that even after several rounds of division the cells of an early embryo remain pluripotent. That nuclear transfer works at all is remarkable, for the nucleus is not (as schematic diagrams of cells sometimes imply) just a membrane sac floating in the fluid of the cytoplasm; rather, the cytoplasm itself is full of sub-microscopic structures linked to the nuclear membrane, so that pulling out the nucleus is likely to be traumatic. Dolly's creators have compared its disruptive effect to that of a brain transplant.

For how long does the versatility of embryonic cells persist, Spemann wondered? In 1938, three years after receiving a Nobel Prize, he speculated about a 'fantastical experiment' in which a fully differentiated tissue cell from an organism – a somatic cell – might be used as the donor of a nucleus for this type of cloning, a process now known as somatic-cell nuclear transfer (SCNT). But he lacked the technical capability to do it.

---

* The term is used only loosely here. Although most cells in higher organisms have a well-defined nucleus bounded by a membrane and containing the chromosomes, an oocyte does not. 'Enucleation' is then more properly a euphemism for removing the host chromosomes.

The centuries-old horticultural practice of grafting of shoots and branches on plants is another form of cloning, and this is in fact where the word comes from: *clon* means twig or branch in Greek. Cloning from single cells is quite another matter; but as plants seem, for some unknown reason, to retain pluripotent stem cells throughout their lives more often than animals do, the challenge is more easily met. The first plants to be artificially cloned from cell cultures were humble carrots, made in 1958. Applying the technique to high-end horticulture was potentially lucrative, and orchids were being cloned from single cells just two years later.

Cloning of higher animals had already been achieved by that time. Robert Briggs and Thomas King in Philadelphia built on Spemann's work by cloning tadpoles in 1955 using nuclear transfer. Amphibian eggs, particularly those of frogs, are well suited to this sort of surgery because they are large and peculiarly resilient. Briggs and King took the nuclei from embryonic cells that had undergone many rounds of division, when the embryos contained 8,000–16,000 cells, which showed that they still possessed pluripotency even at this stage.

In 1963 the Chinese embryologist Tong Dizhou used much the same principle to clone fish. And John Gurdon, a biologist at Oxford, began to work in the 1950s and 60s on nuclear transfer from somatic cells of the African clawed frog, taking differentiated cells from the frog gut as the chromosomal donors. He was able to grow some of the embryos made in this manner all the way to mature frogs. But even so, he was not able to produce adult frogs from fully adult donor cells.

Mammals were harder. In 1979 the Danish scientist Steen Willadsen at Cambridge made the first cloned sheep by the technique of splitting a two-cell embryo. Five years later he achieved it using the more challenging method of nuclear transfer from embryonic cells, and in 1985 he cloned a cow from embryo cells one week old, which had begun to differentiate. The German embryologist Karl Illmensee and his American colleague Peter Hoppe claimed to have achieved the first mammalian cloning – of a mouse – by nuclear transfer in 1981, but their work was never verified or replicated.

A *bona fide* cloned mouse was reported, however, in 1986, and in 1995 Ian Wilmut and Keith Campbell at Roslin Institute in Scotland, a government research laboratory, cloned two sheep named Megan and Morag from embryo cells that had been cultured for several weeks, so that they were unambiguously differentiated. (Stem cells will generally begin to develop spontaneously into fibroblasts if cultured for long enough.) Later that year the Roslin researchers made a quartet of sheep clones named Cedric, Cecil, Cyril and Tuppence from a culture of nine-day-old embryo cells. They used an electrical stimulus to trigger the fusion of the donor and enucleated oocyte cells, in effect causing nuclear transfer.

Hello, Dolly

In the light of this background, it seems surprising that the wider world had not realized how far cloning technology had progressed until the announcement of Dolly in 1997. What made Dolly special was that she was cloned by SCNT: by nuclear transfer from a fully differentiated mammary cell of a ewe that was long since dead. (Dolly was so named because, in a fit of schoolboy humour, the source of the parent cell put the researchers in mind of singer Dolly Parton.) Until the 1980s, most scientists still thought that differentiation of cells to make specialized tissues deactivated permanently all those genes that were not needed. Even experiments on nuclear transfer from embryonic cells a few weeks old gave no reason to believe that pluripotency of a cell's genome remained accessible even into adulthood. But the cloning of Dolly proved that the silenced genes could be reawakened by rather simple means – in this case, merely by 'starving' the cell of nutrients before transferring its nucleus. Again, Jean Rostand anticipated this possibility of using mature cells for cloning when he wrote that 'every speck of flesh [is] charged with seminal virtue'. 'In each element of a living creature', he said, 'there would reside the principle of the whole creature.'

Dolly's message was understood immediately. If cloning works only for embryonic cells, it is too late for you and me. But if what had been done for Dolly could be done for humans, then we can all be cloned.

Polly, a cloned sheep born in 1997, was in the eyes of Wilmut and Campbell a potentially more significant creature than Dolly, because she embodied one of their ultimate goals: she was genetically modified. Getting modified genes into cells is currently a difficult and uncontrolled, hit-and-miss affair. But if it is attempted within a culture

of many embryonic stem cells, the few successful hits can be identified by genetic screening methods. If these cells can then be grown
into embryos by nuclear transfer, one can make a genetically modified
animal. And with luck, the modifications will persist in the creature's
gametes, so that normal breeding techniques can then be used to make
a whole flock of modified animals – for example, with genes that
produce protein drugs in their milk.

Aside from biomedical applications, there are several reasons why
one might want to clone livestock. It would be a way of propagating
the best specimens, such as good milk bearers, without diluting their
genes by sexual reproduction. Laboratory animals such as mice and
rats with an identical genetic constitution would offer a standardized
population for research. And the ability to clone animals, and perhaps
maintain a stock of cloned frozen embryos, could be valuable for
conservation. These objectives, however, have been little noticed
outside the scientific media, because it has generally been assumed
that sheep were only a stepping stone to humans.

There are good medical reasons to clone human cells that have
nothing to do with cloning entire human beings – as we have seen,
for example, cloned somatic or stem cells might be used to generate
tissue cultures genetically matched to the donor for grafts and regeneration of organs and nerves. But such applications rarely feature in
the public debate over human cloning. What fascinates us is the
prospect of 'copying ourselves'.

The clone wars

Morag and Megan did not go unnoticed. The *Daily Mail* described
their cloning in characteristically Manichaean terms, asking if these
creatures were 'Monsters or miracles' (with the implication that they
must be one or the other). Other reports revived the 'extinction of
sex' theme, noting that no sperm were needed for the lambs' creation.
'When sheep are sheep and men are uneasy', quipped the *Daily
Telegraph*, while the *Observer* indulged in even more shameless punning:
'A clone again, naturally: Males don't figure in embryonic engineering,
but it's not all ova yet.'

But Dolly was the undisputed star. Her debut in a *Nature* paper by
Wilmut, Campbell and their colleagues instantly made front-page

news, and it was clear that there was more than science at stake. US president Bill Clinton claimed that cloning 'is a matter of morality and spirituality'. Leon Kass called it 'a serious evil'. And the old themes were revived with glorious predictability. *Time* magazine wrote of Wilmut that 'One doesn't expect Dr Frankenstein to show up in a wool sweater, baggy parka, soft British accent and the face of a bank clerk.' The German magazine *Der Spiegel* splashed the story on its cover, accompanied by an image of a battalion of Hitlers. 'Science on the way to cloned people', it insisted, calling it *Der Sündenfall* – the Fall of Man. The news even reached the murky netherworld of the sensationalist tabloids, where in the manner of medieval folk tales it stoked rumours of monsters stalking the earth. 'Dolly the cloned sheep kills a lamb – and EATS it!' declared the *Weekly World News* in a flight of pure fantasy.

Everyone assumed that human cloning would follow in short order, notwithstanding the fact that Wilmut and Campbell confessed to finding this prospect 'distasteful'. According to American bioethicist Dan Brock:

In the popular media, nightmare scenarios of laboratory mistakes resulting in monsters, the cloning of armies of Hitlers, the exploitative use of cloning for totalitarian ends as in Huxley's *Brave New World*, and the murderous replicants in the film *Blade Runner* all fed the public controversy and uneasiness.

Wilmut says he 'quickly came to dread the pleas from bereaved families, asking if we could clone their lost loved ones'. The old stereotypes of the artificial being also made themselves known – it was widely assumed, without any obvious rationale, that Dolly would be sterile. (She was not, and in fact the offspring of cloned animals are more likely to be healthy than the clones themselves.)

Cloning is not in any real sense a means to create artificial people – it simply involves a substitution of the genetic material in an ordinary egg. But in the popular arena that is a negligible distinction; all of the emblematic features, assumptions and behaviours of synthetic beings are transferred onto our perceptions of clones. The failed clones of Ripley in the movie *Alien Resurrection* (1997) are all hideously deformed monsters, an outcome that doubtless struck audiences as all the more likely because the beings are cross-species (human/alien) hybrids. The clones of the Clone Wars in the *Star Wars* series are, in their anonymous armoured ranks, mere ciphers for robots. Clones, like homunculi, are either sub- or supernormal, or both – there *must* be something wrong with them. When it was disclosed in 2008 that meat and milk from cloned animals might be entering the US food supply, critics voiced concerns about safety that they could barely articulate, being rooted more in unease about the 'unnaturalness' than in any obvious reason to anticipate health risks. When the first cloned human is born – and for better or worse this will probably happen within the lifetime of some of us – people will be as amazed at their normality as many were at Louise Brown.

It may surprise you to know that human cloning has perhaps been achieved already, after a fashion, in a reputable laboratory. In 1993 Jerry Hall and Robert Stillman, biologists at George Washington University Medical Center in Washington DC, announced that they had cloned seventeen human embryos produced by IVF. They had used the simple method of artificially induced twinning: they took embryos at the two- to eight-cell stage and divided them into individual cells, called blastomeres. These will not grow by themselves unless they are given a protective 'shell' called the zona pellucida, which all human embryos possess in the early stages. The embryo emerges from this shell about five days after fertilization, a process known as hatching. Hall and Stillman created artificial zona pellicudas for their blastomeres using an extract of seaweed, which forms a tough, biocompatible and flexible coating on the cells' surfaces. The blastomeres underwent several rounds of cell division, beginning to form embryos, but they did not reach a stage at which they could have implanted in a uterus.

The announcement unleashed a storm of controversy, among researchers as much as among the media, religious commentators and the broader public. Officials at George Washington University were alarmed to discover what was being perpetrated inside their institution, and they conducted an inquiry which charged Hall and Stillman with violating university policy by failing to gain full approval for the experiment before conducting it. Some people dismissed the claim on the spurious grounds that twinning is not really cloning. Others had technical objections of more substance. The embryos used for the experiment were 'polyploid': they contained an excess of chromosomes, because they were formed by the fertilization of an egg by two sperm. (This made them unsuitable for use in IVF, which is why they were available in the first place.) In consequence, there was no guarantee that all the cells in the initial embryo were genetically identical, and Hall and Stillman conducted no tests to establish if this was so. Nonetheless, the experiment showed that cloning by artificial twinning was possible in humans, and it precipitated a debate on human cloning, at least in the United States, well in advance of Dolly. Should we do it? Leon Kass voiced the widely held view that it should be banned absolutely, and that it involved 'Frankensteinian hubris' – a casual shorthand that, as I hope is now clear, does little more than stir the pot of legend.

Mounting a more reasoned assault on the ethics of human cloning is not easy. There are several possible objections, but most do not bear the weight of scrutiny. We might worry that (to take the flimsiest concern first) human cloning would divert research and medical funding from more urgent and important social needs. If that is so, it merely argues against the use of public funds, and is not an objection in principle. And the valid concern that human cloning might be exploited for commercial interest is an argument for its being regulated, as any commercial activity should be. Another common objection is that cloning might be used by governments or other groups for immoral ends, in the manner of *Brave New World*. We've probably seen enough now to recognize that this is largely a mythical worry, although it tends to dominate popular depictions of cloning.

Some objections warrant closer consideration. One is straightforward to appreciate: human cloning might not be safe. Cloned children might, say, have a high chance of suffering developmental defects. Perhaps their genes, if taken from mature body cells, would have deteriorated and might cause early degeneration, shortening the child's lifespan. Yet there are no strong signs from other cloned mammals that this would happen. Dolly died at the age of six, just half the typical life expectancy for a Finn Dorset sheep. But there was no clear reason to link this with the fact that she was cloned. Like many sheep, including several in the flock from which the genetic donor ewe came, she died of lung cancer caused by a virus. The Roslin researchers say that Dolly showed no signs of premature ageing.* After all, as is the case with IVF, most potential problems with the development of the embryo would become manifest by auto-termination at an early stage. It took Wilmut and colleagues 277 attempts before Dolly was born, but the other losses occurred primarily at the stage of the blastocyst – there was no grisly parade of mutant lambs before Dolly emerged intact.

There are other reasons why cloning could be biologically risky. For example, the process by which germline cells form the gametes, called meiosis, has the effect of purging some damaging mutations from the genome. It is conceivable that successive generations of clones, lacking this filter, might accumulate genetic flaws. And in 2000 Wilmut and Campbell said that 'fetuses produced by nuclear transfer are ten times more likely to die *in utero* than fetuses produced by normal sexual means, while cloned offspring are three times more likely to die soon after birth'. Other cloning experiments in mammals have incurred problems ranging from stillbirths, obesity and increased birth weight to immune dysfunctions and malformations of the liver, brain and urogenital tract. Some of these problems seem to be shared with other reproductive interventions such as cross-species embryo transfer. Only further research will show whether they are an inevitable result of cloning or are caused by the relatively crude methods currently used to achieve it.

In any event, the health risks of human cloning would need to be studied very carefully. But that is true for any new medical procedure. One could, with justification, assert that the health risks for human IVF were not explored as fully as they should have been before embryos were transferred to the womb. Yet any increased health risk would have to be significant (it's not immediately obvious how thresholds might be set) to warrant prohibition on these grounds alone. As we saw earlier, there is a slightly increased risk of complications for IVF compared to normal conception, but the procedure is not banned on that account. And as with IVF, risk perception for cloning is

---

* The first cloned mouse, named Cumulina, lived for slightly longer than the average lifespan of a mouse. And other experiments on cloned cows and mice have found that their telomeres – the 'end caps' of the chromosomes, which become eroded as we age – are actually longer than average, rather than shorter as seemed to be the (much publicized) case for Dolly.

gratuitously influenced by mythical notions of monstrous births resulting from 'unnatural' conception – which in the case of cloning is not really conception at all. Paul Ramsey demanded to know whether 'in case a monstrosity – a subhuman or parahuman individual – results, shall the [cloning] experiment simply be stopped and this artfully created human life killed?' Notice that here the hypothetical developmental defect apparently renders the clone not deformed but in fact 'subhuman' (at least, until we consider 'killing' it, at which point its humanity returns). In assessing the safety of human cloning, we would therefore need to distinguish the real from the fanciful and not capitulate to arguments like those raised against IVF that we can 'never be sure' of the risks no matter what we do.

Some opponents of human cloning worry about its effects on society. Will cloning erode respect for the individual and for human life in general? To argue that cloning blurs the notion of an 'individual' is to misunderstand it according to a crude view of genetic determinism, as I discuss below – there is no reason to imagine that a human clone will be indistinguishable from his or her genetic parent. But this fear is not wholly groundless: the prospect of, say, fifty genetically identical human clones is genuinely disquieting. Why anyone would wish to make fifty clones is far from clear, however, and in any case this too would seem to be an argument for regulation, not an objection in principle. As for the alleged erosion of respect for human life, it isn't obvious why that should happen unless we were to deny people 'conceived' by cloning their humanity. The same objections were raised for IVF until it became apparent that the aim was to produce healthy, much-desired babies.

The bioethicist Alto Charo suggests a particular interpretation of this worry:

that human cloning would result in a person's worth or value seeming diminished because we would now see humans as able to be manufactured or 'handmade'. This demystification of the creation of human life would reduce our appreciation and awe of it and its natural creation.

Yet our response to the notion of 'manufacturing' humans must necessarily be personal, and all I suggest is that there is nothing inevitable or self-evident about it. The point I wish to make in this book is that our responses have generally been conditioned by a complex web of cultural associations and assumptions about the myths of anthropoeia, naturalness and the morality of manufacture. In other words, the feelings that Charo adduces have too long remained unexamined – in which case we will find it hard to be honest about them, with ourselves and with each other.

One unquestionably important ethical concern about making babies by cloning is that it might psychologically harm the child. This is an immensely difficult and necessarily speculative issue. One worry is that people produced by cloning of adult cells will be burdened with the knowledge of what lies in store for them – an equivocal supposition at best, which I will examine later. Other problems could originate not in the cloned individuals per se but in our perception and treatment of them: as philosopher Gregory Pence, an advocate of human cloning, puts it, 'Because we expect cloned children to be odd, we treat them as if they were odd, in effect making them odd.' Given current prejudices and assumptions, this seems distinctly possible, but an ethical argument that relies on (and even tries to engineer) a self-fulfilling prophecy does not seem particularly ethical to me. Besides, we have again the experience of IVF to draw upon – as we saw, opponents to this technique also argued that stigmatization, prurience and media attention would distort the lives of the children born by this means, but that fear has proved unfounded.

How will parents treat a child that is genetically identical to one of them and genetically unrelated to the other? Leon Kass's insistence that 'virtually no parent is going to be able to treat a clone of himself or herself as one treats a child generated by the lottery of sex' is an extraordinarily definitive judgement on the basis of no evidence whatsoever, seeming almost to constitute an instruction about how the genetic parent *should* treat the child. Neither Kass nor anyone else has the slightest idea if it is true, particularly when we recognize that the child will not necessarily be a physical replica of the adult (see below) and will certainly not be identical in their behaviour. But surely the important question is whether the parents will treat the child *well*. The question of asymmetry between the genetic and non-genetic parental relationship also cannot be answered with very much more than speculation, although it seems not so far removed from the situation of step-parenting, which we rightly do not prohibit or disapprove of because of alleged damage to the child's psyche.

The concern of both physical and psychological harm to cloned children faces a fundamental philosophical barrier beyond which it can be hard to advance. Specifically, the question is: harm in relation to what? How can we possibly assess the pros and cons of possible 'damage' to a person against the option of their *never having existed at all*? There are a few situations in which one might reasonably suggest that a child would better not be born at all than be born with the difficulties it would inevitably face: for example, if it will suffer a painful or otherwise deeply unpleasant medical condition for which there is no cure and for which the life expectancy is very short. Even then it is not clear quite what it can mean to say that the child is better off *not being* – rather, we have an intuition that the decision to prevent birth is a humane one. Yet there are many less extreme but highly undesirable situations into which a child might be born where that call is far harder to make. People survive, and even thrive, in the most adverse circumstances. How can one decide in advance that it would have been better if they had not been allowed to do so? Compared to some such scenarios that one might envisage, prohibiting a cloned child from being born into a loving, nurturing family because of vague and untestable fears about psychological impacts does not seem very reasonable.

The philosopher Derek Parfit has pointed out that the real basis for comparison here should not be against *non-existence* but against being born without having being cloned. That, of course, will work only when the option exists – which it might very well not. We might imagine, for example, a single person who wants a child to be cloned from their cells rather than being conceived via gamete donation or surrogacy. The ethical calculus even here is not obvious, but since alternatives exist, a decision to prohibit cloning could plausibly be defended. We might also think it unwise to allow someone to clone themselves simply because they are too egotistical to allow their genes to be 'diluted' with anyone else's. But we would need in these cases to examine the specifics of the situation, and insofar as psychological risks are concerned we should perhaps give more emphasis to the kind of environment the child would be born into than to any intrinsic harm associated with cloning per se. We might reasonably fear for the future of the (hypothetical) children allegedly cloned by the company Clonaid, operated by the quasi-religious Raelian cult who believe that humans are descended from aliens, or of those whose cloning is said to have been attempted by the publicity-seeking maverick and former physicist Richard Seed,* or by Iraqi scientists using the genes of Saddam Hussein. But we have waived responsibility if we allow our ethics to be shaped by the posturing of mavericks and fringe practitioners, billionaire fantasists and would-be Nobel laureates and pathological dictators. We must surely ask: *what are the motives* for wanting to bring a cloned child into the world?

Although we can do little more than speculate about the possible psychological harms that will be *intrinsically* inflicted on a cloned child, it isn't at all obvious that these could be so serious as to make this form of human procreation more damaging than others currently permitted. For example, there is no prohibition against a couple conceiving a child even if that child would certainly inherit a serious genetic defect. Indeed, cloning could offer a means of *avoiding* harmful genetic mutations that would otherwise certainly be inherited from one parent by sexual reproduction. And even if cloning were to create some confusion in a child about his or her genetic heritage, can we say with certainty that this burden would be any greater than it is for, say, anonymous gamete donation?

Opposition to human cloning, as with opposition to ectogenesis and other reproductive technologies both real and hypothetical, tends to be presented in terms of its purported failings compared with sexual reproduction by the prospective parents and gestation by the genetic mother. In most cases (although not when cloning is used to avoid serious genetic defects), the latter route to procreation is likely to be the least fraught with obvious potential problems. But of course these technologies are generally developed because that 'normal' option is not available. It is one thing to argue about which way of making people is 'best', another to decide which ways are permissible. The idealization of procreation within the nuclear family by conservative

---

* Seed has no scientific credentials that would make his claims remotely plausible. Gregory Pence cannot resist pointing out how 'Dick Seed', looking in a congressional hearing on cloning like a 'homeless man', seems to be 'an eccentric from central casting'.

bioethicists, even when a comparison of the alternatives against this 'gold standard' is spurious, is in the end an assertion that this is the only means of reproduction their belief systems will sanction.

And it is not to endorse total libertarianism to say that we should always think very carefully about interfering with people's reproductive rights. There are of course many precedents for doing so, but when we consider them – prohibition of contraception and artificial insemination, negative eugenics, blanket religious and social censure of sex outside marriage, caste systems – they do not tend to suggest that this is something to be encouraged. What would governments do, Gregory Pence asks, if cloning were as easy as insemination by turkey baster? How then would it be policed? Steed Willadsen asserts that 'in America, cloning [of humans] may be bad but telling people how they reproduce is worse'.

Fuzzy copies

Billionaires and tyrants dreaming of remaking themselves had better get one thing straight: a clone is not an identical copy. It is a (more or less) *genetically* identical copy. The distinction is profound, although in the earliest modern discussions of cloning scientists were apt to forget it. Biologist James Bonner offered the blithe assessment in 1968 that within fifteen years we would know how 'to order up carbon copies of people' in a process he called 'human mass production'. And Joshua Lederberg suggested that cloning might be a way of 'copying' talented people for the supposed good of humankind – a kind of naive positive eugenics. 'If a superior individual . . . is identified', he said,

why not copy it directly, rather than suffer all of the risks, including those of sex determination, involved in the disruptions of recombinations (sexual procreation)? Leave sexual reproduction for experimental purposes: when a suitable type is ascertained take care to maintain it by clonal propagation.

Early journalistic descriptions of human cloning encouraged this view that all our important human characteristics are written into our genome. David Rorvik's 1969 article for *Esquire* was illustrated with duplicates of Hitler along with (dating it nicely) Barbra Streisand and Mahalia Jackson. And in his 1969 book *The Second Genesis*, Albert

Rosenfeld fed the myth of the cloned 'replacement' child or relative with a grotesquely jejune remark:

In our current circumstances, the absence of a loved one saddens us, and death brings terrible grief. Think how easily the tears could be wiped away if there were no single 'loved one' to miss that much – or if that loved one were readily replaceable by any of several others.

That it borders on the comical to impute such indifference to the death of cloned individuals was acknowledged in a cartoon that appeared in the *Guardian* newspaper, in which a woman reassures a cab driver who has just run over and killed her husband. 'That's alright,' she says, 'I have another one upstairs.' The joke only works because of the assumption that clones are indistinguishable (not to mention that the husband here was presumably cloned at birth).

But cloning does not necessarily produce identical individuals at all, not even physically. The same genes can produce quite different phenotypes – different appearances and traits – depending on the environment in which they are expressed. This became apparent to Ian Wilmut and his colleagues at Roslin when they created Cedric, Cecil, Cyril and Tuppence: the four lambs, they said, were 'very different in size and temperament'. Neither was Dolly an identical replica of the ewe who provided the parent cell. In fact, the dissimilarities between clones and their genetic parents create doubts about how useful cloning will be in livestock breeding.

There are many possible reasons for these differences. Genes may become activated or silenced during embryonic development, and one can never be sure that this will happen in the same way in two separate organisms. Nutrition levels are a particularly important factor in determining physical characteristics. And all cells are susceptible to random 'noise': unpredictable variations in the amount of proteins produced from specific genes. Moreover, because genes typically interact with one another in complex networks, much as humans do in society, effect doesn't necessary follow in direct proportion to its cause: small random variations of this sort have the potential to create large knock-on effects. For these and other reasons that I mention in the next chapter, a genome does not wholly specify the organism that will spring from it. What's more, clones made by nuclear transfer also retain some genes from the enucleated egg cell. Some, present in the cytoplasm, play an important part in early development of the embryo; others are located in compartments outside the nucleus, called mitochondria (about three per cent of the DNA in our cells is mitochondrial, and the mitochondrial genes encode thirteen proteins).

And that's just the physical body. In psychological terms it would be absurd to imagine that any two human beings could be identical. That our psyche is shaped by experience is trivially obvious, but it is also worth remembering that there cannot even in principle be complete genetic predestination in the way the brain is wired, because it is simply too complicated. There are many billions of connections between synapses in the brain, yet we have only 21,000 or so genes in our entire genome.

The biologist Richard Lewontin has exposed the fallacy of genetic determinism thus:

It is a basic principle of developmental biology that organisms undergo a continuous development from conception to death, a development that is a unique consequence of the interaction of the genes in their cells, the temporal sequence of environments, through which the organisms pass, and random cellular processes . . . As a result, even the fingerprints of identical twins are not identical. Their temperaments, mental processes, abilities, life choices, disease histories, and deaths certainly differ, despite the determined efforts of many parents to enforce as great a similarity as possible.

So anyone who thinks that by cloning themselves they are going to produce an identical copy – a younger self – is deluded. Yet one can't help suspecting that some of the aspirations behind a desire to self-clone go beyond even this myth of rejuvenation: one senses a belief that somehow there will be continuity of consciousness, that this is a way of avoiding mortality.

The speciousness of genetic determinism undermines another objection to cloning: that, if clones are born at different times, the younger one will always be haunted by a vision of what he or she will become. This so-called 'Dorian Gray scenario' is articulated by bioethicist Arthur Caplan:

Twins that become twins separated by years or decades let us see things about our future that we don't want to. You may not want to know, at 40, what you will look like at 60. And parents should not be looking at a baby and seeing the infant 20 years later in an older sibling.

But while it's true that an older clone of yourself might give a pretty good indication of what lies in store for your physical appearance, there's no reason to think that this simulacrum should be very much more accurate a predictor than is a photo of your parents, especially if you and your clone do not share identical lifestyles. Besides, one could argue that there might equally be significant *benefits* of a glimpse into your genetic future: you might learn about genetic predispositions towards disease that could be avoided or minimized with appropriate lifestyle choices.

The supposedly identical clone also enters into the legendary terrain of the doppelgänger, the shadow self. The image of ranks of identical humans is unquestionably uncanny in the proper sense, for it renders the familiar (a 'normal' person) frightening. Karel Čapek understood this in *R.U.R.*, in which the robots are all given the same features. As the company's executives look out on the masses waiting patiently to exterminate them, they feel this unnerving effect:

Dr Gall: Listen Domin, we made a crucial mistake.

Domin: What mistake?

Dr Gall: We made the faces of the robots too much like one another. There are a hundred thousand faces staring up at us and they're all the same. A hundred thousand expressionless bubbles. This is like a bad dream.

Domin: If each of them were different . . .

Dr Gall: Then the sight of them wouldn't be so ghastly.

In his essay on the uncanny, Sigmund Freud considered that in mythical terms the 'double' served precisely the purpose that popular notions of self-cloning now encode: it was, he said, 'originally an insurance against the destruction of the ego, an energetic denial of the power of death . . . a preservation against extinction'. Freud points out that this is essentially an infantile impulse, so it shouldn't surprise us that the people who wish to clone themselves in the hope of attaining a kind of immortality are infantile people. And Freud notes that 'whatever reminds us of this inner "compulsion to repeat" is perceived as uncanny'.

Some popular representations of cloning exploit the frisson of the doppelgänger for unsettling or comic effect, for example in the movies *Multiplicity* (1996) or *The 6th Day* (2000). Since these depictions generally require the instant cloning of a 'normal' adult person, perhaps with an identical set of thoughts and memories, they are forced to be scientifically risible; but the key point is that cloning is being used here as an off-the-shelf, quasi-scientific justification for exploring a legendary trope.

A belief in 'identical copy' clones is not entirely attributable to a misunderstanding of the science by laypeople, for scientists have sometimes fostered the idea. While many biologists profess to disown strict genetic determinism, their pronouncements often seem to support it. 'We used to think our fate was in the stars', James Watson has proclaimed. 'Now we know, in large measure, our fate is in our genes.' And the developmental biologist Lewis Wolpert has said that 'it will one day be possible to predict from an embryo's genome how it will develop' – in all the fine details, including abnormalities. Wolpert suggests that the nematode worm, which has precisely 959 cells, should be the ideal system for making such predictions, but his message was more general and could easily be interpreted as applying to humans – which it patently cannot. These simplistic comments have been thoroughly assimilated by the media: when sociologists Dorothy Nelkin and M. Susan Lindee recently surveyed the cultural landscape of genetic determinism, they found references to genes for 'criminality, shyness, arson, directional ability, exhibitionism, tendencies to tease, social potency, sexual preferences, job success, divorce, religiosity, political leanings, traditionalism, zest for life, and preferred styles of dressing'.

Genetic determinism has motivated opposition to human reproductive cloning. A British lobby group called the All-Party Parliamentary Pro-Life Alliance stated in 1998 that the creation of a human clone would be unethical because complete specification of the clone's genetic constitution places its creators 'in an unacceptable position of control' – as though the genome determines all that will befall the cloned person. (Besides, sexual combination of gametes hardly offers a much wider genetic palette – don't all of us 'suffer' the tyranny of control by our biological parents' genes?) The Pro-Life Alliance went on to construct a *Brave New World* fantasy about a factory owner who creates his docile workforce by cloning workers with low intelligence and low expectations. Setting aside the fact that this industrialist clearly has unusually long-term vision (growing his workforce from babyhood), we are back in the realm of a fantasy society in which, frankly, the cloning of humans would seem the least of our problems. To put it another way, we are entitled to interpret such infantile arguments as transparent camouflage for more deeply rooted and less easily articulated fears about artificial procreation.

Such misconceptions about clones are alarmingly widespread even among those who lay claim to a professional platform for speaking about them. Bioethicist Laurie Zoloth bases her objections to cloning on the idea that it will generate a 'close as can be' replica, and that this would indeed be a clone's *raison d'être*. In a car crash of metaphors, she asserts that in child-rearing we must 'learn to have the stranger, not the copy, live by our side as though out of our side'. (Didn't Eve come, *asexually*, out of Adam's side?) Even in literal terms, it seems odd to imply that the cloned child of a couple would be, to the non-cloned parent, less of a 'stranger' than a child sharing half his or her genes. But Zoloth's real fear seems to be that, for reasons unspecified, the parents of a cloned child will (like Victor Frankenstein) fail to parent it as they would any other child:

The whole point of 'making babies' is not the production, it is the careful rearing of persons, the promise to have bonds of love that extend far beyond the initial ask and answer of the marketplace.

Taken literally, this damns IVF, and perhaps all other assisted conception, in the same breath as human cloning. And indeed Zoloth, like Leon Kass, considers all assisted reproduction technologies to embody a 'search for perfection', seeing them only in a mythical rather than a medical sense.

Making Hitler

Did I mention Hitler? One cannot speak about assisted conception (just as allegedly one cannot too long sustain an online debate) without his making an appearance. Through the contemporary landscape of anthropoeia he walks hand in hand with Frankenstein, a pasty-faced warning about 'where it all might lead'.

Leon Kass's demand in 1971 for international regulation of laboratory intervention into human procreation was at face value a very reasonable proposal, albeit one that has never proved possible. But his reasoning invoked the Hitler clause, given a topical flavour that was guaranteed to alarm US patriots:

The need for international agreements and supervision can readily be understood if we consider the likely American response to the successful asexual reproduction of 10,000 Mao Tse-tungs.

The image was repeated more vaguely by Kit Pedler, a medical scientist and science-fiction writer who acted as adviser to the BBC's dystopic *Doomwatch* drama series in the early 1970s. The work of Steptoe and Edwards, he said, meant that 'you have the means . . . of mass-producing people without the advent of a mother at all . . . A general might order 100,000 troops to be produced.' Even James Watson warned of 'the potentialities for misuse by an inhumane totalitarian government'.

Nightmare visions of mass-produced 'evil men' have haunted anthropoeia at least since Victor Frankenstein imagined his creatures mating and populating the world with fiends. But this is primarily a twentieth-century theme, in which we are invited to insert the fiend of our choice: in current times the Hitler role has been awarded to Saddam Hussein and Osama bin Laden. But the Führer of Nazi Germany remains a perennial favourite.

It will be clear from what I have said above that the idea of resurrecting Hitler by cloning is puerile nonsense. No matter; it was well entrenched even before human cloning was on the horizon of technological feasibility. At a conference in the late 1960s on 'The Manipulation of Man', an Italian geneticist asserted that 'manipulation means test-tube babies, Hitler and the Master Race', and we have seen how the Führer featured in the debate over embryo research. But it was Ira Levin's 1976 thriller *The Boys from Brazil*, and the corresponding movie made two years later, that crystallized these peculiar fears. The plot involves the cloning of Hitler from samples of his blood by Josef Mengele, hiding out in remote Brazil – an unconscious allusion to the way absolute evil is spread by contaminated blood in vampire legends. The comparison is apt, for to say that something in the nature of Hitler's genetic constitution brings about his inevitable evolution into an evil mastermind is to invoke an essentially magical idea. However, Levin makes a token effort to evade accusations of genetic determinism by having Mengele select parents for the clones who are a little like Hitler's: an older father and young mother. And the father must be killed, as Hitler's was, when the boys reach the age of fifteen – for why else did the Führer devise the Final Solution? The movie is memorable primarily for an exchange between Mengele and one of his young Hitler clones that should perhaps stand as an epigraph for the entire mythology of 'making Hitler':

Mengele: You are the living duplicate of the greatest man in history – Adolf Hitler!
Boy: Oh man, you're weird.

How about an entire army of Hitlers? There is some amusement to be found, perhaps, in the idea, each toothbrush-moustachioed private insisting that *he* should become the Führer. One has to ask, too, why anyone imagines that Hitler's undistinguished military career and rather feeble physique recommend him as a model soldier. And again we must be talking here of an evil government with remarkably long-term plans, prepared to raise all those baby Hitlers until they are ready for their adult roles as . . . as, well, I'm frankly unsure as to what. As massed evil incarnate, or something.

The image of a 'far-flung' country cloning armies of Hitlers or Pol Pots or Mao Tse-tungs, or of governments cloning Einsteins to do their thinking and Mozarts to entertain them, is risible in practical terms and involves a childish view not just of what cloning is but of what humanity is.* When people believe patently silly things, however,

---

* It is not just Hitler, Einstein and Mozart who are the fantasized targets of cloning. Some websites already purport to offer cloned copies of celebrities, including an 'instant Elvis'. Bioethicist Ronald Green thinks that, even if it seems a comical possibility, some doctors, dentists and hairdressers of celebrities may one day have to sign agreements not to use their genetic material in such a manner. Even if we'll never make an 'instant Elvis', we can be sure some people would be deluded enough to pay for one.

there is often a reason beyond sheer ignorance. It's very clear that fantasies about 'cloning Hitler' are in fact stories about reincarnation and metempsychosis – the transmigration of souls. Here, then, is another absurd delusion about artificial procreation drawing sustenance from the rich soil of myth.

Human or beast?

The production of human stem cells by SCNT – a form of cloning conducted for biomedical research – requires an oocyte to act as the host. These are not easy to obtain at present, since they have to be collected from donors who must endure hormone-regulating drugs and surgical intervention. It is possible that human eggs might one day not require human donors: they might be harvested from *in vitro* ovaries cultivated from stem cells, as has been done already with mice. But there is currently an inadequate supply of eggs for human stem-cell research, which is why some scientists are interested in using animal rather than human oocytes as the host. After nuclear transfer, the cells are genetically human – or almost so, since some genetic material from the animal remains in the cytoplasm and mitochondria.*

Embryos made by transferring human chromosomes into enucleated animal eggs are known as cytoplasmic hybrids or 'cybrids'. They are somewhat different from true hybrids, which are embryos made by the fusion of gametes of different (but closely related) species or of two subspecies or breeds. (A mule is a hybrid of this sort.) Both types of hybrid are genetically homogenous – every cell in the organism has essentially the same genetic composition – and are thus distinct from so-called chimeras, which are mosaics of cells of different species grown from an embryo to which cells from another embryo have been attached. Steen Willadsen reported a chimera of a goat and sheep, dubbed a 'geep', in 1984: it had a goat-like head but a woolly body like a sheep.

---

* Mitochondria, which have the function of producing the biochemical energy-storage molecule ATP, are rather similar in humans and other mammals, but not identical. It is not clear how much this matters, but it is one valid reason for concern about the viability of human–animal hybrids. Note, however, that in the form of SCNT used by Wilmut and colleagues, the entire cell containing the chromosomes is fused with the enucleated oocyte, and so it contributes its own cytoplasmic DNA and mitochondria too.

Chimeric embryo technology may make it possible for human organs to be grown in animals for transplantation.* American researchers have incorporated human stem cells into sheep fetuses, so that the lamb tissues such as blood, cartilage, muscle and heart contained up to forty per cent of human-like tissue, even though the animals look normal. And mice have been grown with tiny but functional kidneys that are essentially human. It is even conceivable that animals (for example, mice) might be created that can generate wholly human sperm and eggs from which a human embryo could be created. In 2003 a team in Shanghai fused human cells with rabbit eggs, making embryos that were allowed to develop for several days *in vitro* before being harvested for stem cells. Less plausibly, later that same year a maverick Greek Cypriot reproductive biologist named Panayiotis Zavos announced he had created 'human–cow' embryos that lived for around a fortnight and could theoretically have been implanted into a woman's womb.

Many countries, including Australia, Canada, France, Germany and Italy, have banned this kind of human–animal embryo research. In the UK, the production of human–animal hybrid embryos and stem cells has been regulated by the HFEA, which has so far granted licences to two groups to make human–animal cybrids.

There are proper concerns about the safety and health of embryos grown this way. But human–animal hybrids and chimeras provoke widespread disgust quite out of proportion to any such risks. It is obvious why this should be so: they invoke (even in the technical terminology) mythical monsters, part-human and part-animal. The mythical Chimera was a bizarre patchwork, possessing the body and head of a lioness with the head of a goat on its back and a snake's head at the end of its tail. Such visions are projected in the 2009 movie *Splice*, in which rogue genetic engineers mix the DNA of humans and other animals to produce a hybrid creature with a human upper body, bird-like legs, a prehensile tail, wings and a toxic sting. The flawed premise here is that body plans are determined simply by an assembly of genetic modules for 'arms', 'legs', 'wings' and so forth; but nonetheless *Splice* conjures a strong mythic resonance from its exploration of

---

* The converse procedure – animal organs being transplanted to humans in 'xenografts' – is well established. The animal organ must be specially prepared to minimize immune rejection by the recipient.

the themes of revulsion and transgression in cross-species sex (the composite, gender-shifting creature mates with both of its creators) and the consequent generation of monsters.

Although there seems no obvious reason why a cytoplasmic hybrid human embryo should grow into a creature that is less than wholly human in any meaningful sense (all the component molecules, apart from the handful of mitochondrial proteins, would be assembled under the guidance of human genes), we cannot shake the suspicion that the beast lurks somewhere within. One might consider it a trans-species version of the notorious 'one drop' view of racial bloodlines, in its insistence that any hint of non-humanness can never be eradi-cated. We cannot expunge images of the centaur, the minotaur, the harpy. This assumption is explored in H. G. Wells's *The Island of Dr Moreau*, in which the eponymous amoral doctor uses surgery to 'humanize' animals in order to explore 'the extreme limit of plasticity in a living shape'. Just as these 'animals half-wrought into the outward image of human souls' are doomed eventually to revert to their bestial nature, so we suspect that human beings bearing any fragment of animal tissue will be dominated by that taint.

This suspicion was apparent in Jeremy Rifkin's response to a report of embryos grown from human–bovine cybrid stem cells:

The developing embryo is a human embryo inside a cow egg. That means it is going to share with the cow cytoplasm as it develops. We don't know what kind of creature could develop from that. There is no precedent in history. It will be mostly human as it develops but it will share information and biological matter from the cow egg.

Needless to say, the cybrid is *not* a 'human embryo inside a cow egg', but a unified embryonic entity. What Rifkin conjures up is a micro-scopic analogue of the homunculus growing inside the womb of a cow: a legendary artificial being, now seeded not by sperm but by chromosomes.

During the UK debate over the regulation of human–animal chimeras and cybrids, the Scottish Council on Human Bioethics stated that 'in crossing the species barrier, the general understanding of what it means to be a human person would no longer be clear cut'. This is a curious remark, not least because it seems to challenge the objection to embryo research that the humanness and personhood of an embryo is defined by its genetic disposition. Besides, it is evident already that humanness cannot be construed in terms of a uniquely 'human' constitution of body tissues: xenotransplanted animal organs and our pervasive symbiotic microbiota supply just two of the considerations that make nonsense of this idea. And while it is possible to imagine a situation in which human and non-human genes are so interlarded in a genome that there is genuine ambiguity about species identity, no one is considering making such a genome, nor is there any reason why they should. Nor do we, in any event, have either the technical capability to construct it or the understanding needed to build a viable organism this way. While biologist Lee Silver has suggested that 'at some point in the future, science and society will move into the fuzzy area where we will be forced to draw an entirely arbitrary line between human beings and others', it's not obvious why or how this will happen, let alone that it somehow occurs the moment our cells contain any 'non-human' genes or components.

Clones are coming

Surveying the future of human cloning in 1970, Paul Ramsey wrote that

Aldous Huxley's fertilizing and decanting rooms in the Central London Hatchery (*Brave New World*) will become a possibility within the next fifteen to fifty years. I have no doubt they will become actualities – at least as a minority practice in our society.

Bioethicist Ronald Green, writing thirty years later, forecast the same time horizon for the first human clone, saying that within one or two decades, 'around the world a modest number of children (several hundred to several thousand) will be born each year as a result of somatic cell nuclear transfer cloning'.

But it is significant that Ramsey, who vehemently opposes human cloning (and most other reproductive technologies), and Green, who endorses it, provide very different contexts for this development. Ramsey can imagine cloning only in terms of Huxley's government-controlled eugenic population manufacture; Green feels that within this timescale 'cloning will have come to be looked on as just one more available technique of assisted reproduction among the many in use'. For example, human reproductive cloning would allow women with no ova or men with no sperm to have a biologically related offspring without recourse to artificial insemination or IVF using donor gametes.

My hunch is that Green is right, although it may take longer than another ten years before reproductive cloning is socially normalized. The motivations for it will surely be diverse. To avoid inherited genetic disease from one partner, for example, cloning may turn out to be unnecessary as pre-implantation diagnosis of IVF embryos becomes more routine, although it might still be preferred in the case where one partner carries many, perhaps unknown, mutations. Yet cloning might be used in conjunction with gene therapy (the replacement of harmful variants of genes): because replacement currently has a low success rate, it might be done for thousands of somatic cells, and one of the hits could then be used for cloning via SCNT. It is quite possible, particularly in countries that refuse to regulate, that cloning will occasionally be used to satisfy egotistical desires, or for the ethically dubious purpose of 'replacing' a lost child. Yet I cannot begin to understand why a government, however ruthless, would see any merit in Huxley's scheme.

Some say that reproductive human cloning has already been achieved. David Rorvik's sensationalist 1978 book *In His Image: The Cloning of a Man* was presented, and for a time received, as a piece of authentic reportage about the cloning of an eccentric millionaire named Max by a genetist dubbed 'Darwin', using SCNT. It was soon revealed as a hoax, which Rorvik defended on the grounds that he considered this a valid way to stimulate public debate about human cloning. Whatever the merits of that argument, it certainly proved an effective way to stimulate sales.

After Dolly, the boundaries of credibility are less easily discerned. In 2002 the Italian gynaecologist Severino Antinori, who had courted earlier controversy by conducting IVF on women over sixty, claimed to have cloned human embryos. Antinori said that he produced three pregnancies by cloning, but the details were not divulged and the assertion was never substantiated. Antinori believes that cloning is a valid form of assisted conception and, with the kind of theatricality that always seems to accompany claims of this nature, he faces down the

opprobrium of the Church in his native country by making comparisons with other alleged martyrs to science: 'The Church burnt
Giordano Bruno at the stake. They persecuted Galileo Galilei. And
now they're persecuting me.'

Antinori began his work on cloning in collaboration with Panayiotis
Zavos in Lexington, Kentucky, who announced in 2003 that he too
had created a human clone – a four-celled human embryo made by
SCNT – and transferred it to a woman's uterus, where it failed to
implant. And in 2009 Zavos purported to have cloned fourteen embryos
and transferred them to eleven women. 'There is absolutely no doubt
about it, and I may not be the one that does it, but the cloned child
is coming', he says. 'There is absolutely no way that it will not happen.'
Most other scientists in the field regard Zavos as a showman who
makes exaggerated and ill-supported claims: one bioethicist has accused
him of 'wallowing in a mix of publicity and fund-raising that rests on
a foundation of hype'. He shrugs off such accusations, revelling in his
outsider status: 'We are not regular scientists. We are the irregulars.
This country was built on cowboys. Cowboys are the guys that get
the job done.' And Zavos is perhaps the best placed of the several
claimants to achieve his goal, having previously collaborated with
cloning pioneer Karl Illmensee.

Like Antinori, Zavos believes that reproductive cloning is justified
when other methods of assisted conception have failed or are impossible. He is also prepared to accept requests to clone children who
have died: he alleged to have made an embryo from the frozen blood
of a ten-year-old girl killed in a car crash, using SCNT with an enucleated cow egg. Zavos announced, however, that he had no intention
of transferring this hybrid to the mother's womb, although the
mother insisted that she would countenance it if there was a chance
that the embryo could develop. Needless to say, Zavos has a long
list of potential patients, despite the estimated cost of between
$45,000 and $75,000 for a cycle of cloning. With his mansion and
fleet of cars, he fits the role of the alchemist drawing his fortune
from private investors who are persuaded that he knows the secret
of the philosopher's stone.

Then there is Clonaid, the 'first human cloning company in the
world', run from the Bahamas by a 'bishop' of the Raelian cult, which
insists that it has created the first cloned child, born on 26 December

2002. Some dismiss the company as nothing more than an unusually ambitious variant of the mail-order scam. Yet Clonaid says that it 'has received cloning requests from around the world', and that 'a surprisingly large number come from the Los Angeles/Hollywood area'. This much we can believe.

Mavericks and grandstanders aside, it seems highly likely that human clones will soon be among us. How will we receive them? We can barely even pose the question without making them sound alien, for it forges their identity from their origin, just as we continue to insist that some adults are 'test-tube babies'. It seems quite possible that the primary psychological challenges for people who come into being through cloning might have nothing to do with their genetic constitution and everything to do with the attitudes of their fellow citizens. Severino Antinori says of his alleged human clones that, unless attitudes change, they were likely to be perceived as 'monsters' in the (unspecified) country in which they would be born. Philosopher and bioethicist Gregory Pence even argues that 'clone' should itself be considered a prejudicial label to attach to a person, on a par with terms of racist and homophobic abuse that were once common parlance. There may be prescience in the German movie *Blueprint* (2003), one of the very few serious cinematic explorations of cloning, when the young clone of the central character (a concert pianist with multiple sclerosis) pins to her clothes a yellow Star of David with the word 'Clone' written on it.

The social reception of clones is thoughtfully explored in Pamela Sargent's 1976 book *Cloned Lives*, which, while pedestrian in literary terms, was unjustly marketed as a slice of pulp sci-fi. Sargent presents a scientifically accurate account of human cloning within a future scenario in which virtually all genetic engineering and biotechnology has been halted by a twenty-year moratorium imposed in 1980. Japanese geneticist Hidehiko Takamura seizes the chance in 2000, before any new moratorium can be imposed, to make six clones of astrophysicist Paul Swenson via ectogenesis. One of them dies *in vitro*, as it were, but the four boys and one girl* 'born' from this procedure struggle to find their places in society.

---

* Chromosomal manipulation is used to insert a second X chromosome in place of the male-specific Y in Paul's cells – which means that the girl, Kira, is not in fact a true clone.

In a discussion between Swenson, Takamura and a liberal Christian minister named Jon Aschenbach, Sargent rehearses – intelligently, if somewhat laboriously – the arguments for and against cloning. Jon says:

If it works, every narcissist alive will be trying to use it. You'll be interfering with the course of human evolution with no conception of what the results might be. What would happen in the long run if even a sizeable minority decided to reproduce in this way? You can't know.

Paul replies that 'We can't refuse to use something simply because it may be misused.'

Paul agrees to Takamura's plan for scientific rather than narcissistic reasons, although he is also partly motivated by the fact that his wife died before they could have children. And while Takamura is no saint – he advocates cloning of dead children by their heartbroken parents, and ends up in a relationship with the cloned girl Kira – neither is he an egotistical Frankenstein. The cloned children are well cared for (at least until Paul dies in 2016), but they struggle against the prejudices of society, whether expressed in playground jokes ('What comes in vanilla, vanilla, vanilla, vanilla and vanilla? Ice cream clones') or a relentlessly intrusive media. As undercover reporters vie for access to the clones' prospective foster-parents and throng to the ectogenetic births, Sargent accurately anticipates the antics of the press before the birth of Louise Brown:

The reporters crowded together on the other side of the glass, cameras aimed, tape machines busy, a multi-legged, many-eyed, curious being. *They're just babies*, Paul wanted to shout, *not monsters or genetic freaks, just babies. Make sure your cameras catch that.*

Rumours that these clones can communicate telepathically abound, a popular conviction again drawn from reality. In his 1968 book *The Biological Time Bomb*, British journalist Gordon Rattray Taylor wrote that

It is not mere sensationalism . . . to ask whether the members of human clones may feel particularly united, and be able to cooperate better, even if they are not in actual supersensory communication with one another.

And in a further blow to the simplistic picture of clear-sighted scientists trying to quell a tide of journalistic B-movie fantasy, this occult idea of what we might call 'telepathic DNA' was encouraged by none other than Joshua Lederberg, who speculated that a population of clones might enjoy more 'intimate communication' with one another. After all, he said (without offering any evidence), identical twins are 'notoriously sympathetic, easily able to interpret one another's minimal gestures and brief words'. While it is of course common for new scientific ideas to elicit speculations even from experts that can look naive in retrospect, it is more significant that Lederberg's remarks here are drawn from the same well of folk belief – the attribution of uncanny and quasi-supernatural powers to identical twins – that informs lay opinion. Lederberg's corollary – that cloning might furnish teams of identical, preternaturally sympathetic individuals who might work together effectively in challenging situations such as space travel – only compounds the solecism and suggests that, as the prospect of human cloning first hove into view, the scientists were as busy inventing the science fiction as they were combating it.

The greatest difficulties faced by Sargent's clones, however, stem from the psychological tension created by their multiplicity. In particular, the four cloned boys struggle each to establish an identity distinct from their brothers. In other words, the main psychological harm is wrought by Takamura's ill-considered and quite unnecessary plan to create so many of them. But Sargent makes it clear that the boys *do* establish their individuality, developing quite different personalities, lives and adult appearances.

The same is true of the clones in Caryl Churchill's 2002 play *A Number*, in which a man named Salter clones his young son only to find that the hospital doctors also clandestinely made 'a number' of other 'copies'. Perhaps unbalanced after his wife committed suicide, Salter badly mistreated the young boy until his behaviour became so disturbed that his father put him into care. But he first had the boy cloned so that Salter may have another chance to parent his 'perfect child' properly. So egregious are the sins of both Salter and the cloning doctors (the latter being explicitly just a plot device rather than a judgement of how scientists really behave) that the play cannot, and does not, pretend to be about the ethics of cloning per se. It is an exploration of how identity is formed by both nature and nurture, and in that respect the play challenges assumptions that clones will be identical: the three clones who feature all have very different personalities that reflect their developmental environments.*

Sargent and Churchill offer a sympathetic, measured analysis of what human reproductive cloning might mean. Yet if movies are any guide to social attitudes – and there is a great deal resting on that 'if' – it seems that human cloning is currently still viewed in the old Faustian terms. In *Godsend* (2004), a doctor helps a couple illegally clone their son, who was killed in an accident at the age of eight. But the doctor (a jobbing Robert De Niro) has secretly added some of his own dead son's DNA to the clone, and bad things begin to happen as the cloned boy approaches eight years of age. The film provoked Arthur Caplan to make a stinging indictment addressed directly to Hollywood:

Just as people were beginning to understand cloning, you have put greed before need and made a movie that risks keeping ordinary Americans afraid and patients paralyzed and immobile for many more years.

Caplan's complaint, however valid, is sadly futile, for in most respects cloning has simply become another tool in the cinematic evil scientist's arsenal, or a pseudo-justification for rerunning the mythical doppelgänger narrative. Sometimes there is lip service paid to the problem of giving these clones identical memories to their genetic parent so that this scenario can be played out – in *The 6th Day* memories are downloaded from a CD – but in comedies such as *Multiplicity* that derive their humour from the coexistence of identical selves, even this simple courtesy tends to be waived. The other inconvenient thing about clones as a narrative device is that the plot usually requires them to be of much the same age as the genetic parent – for example, in *The Island* (2005), where clones are grown by rich people as sources of spare parts – which is often glossed over by the miraculous production of fully grown adults.†

---

* In a TV version of the play filmed in 2008 with Rhys Ifans as the son, the final scene showed all twenty or so clones, each evidently with a different personality, proclivities and, to a considerable extent, appearance.
† The same problem, with the same under-the-carpet solution, was encountered in the 1979 movie *Parts: The Clonus Horror*, from which *The Island* itself seemed to be cloned – sufficiently so, indeed, to trigger a Tinseltown lawsuit.

Yet movie clones themselves are often sympathetic, or at least not monstrous; it is typically the scientists and corporations responsible for making them who are the evildoers. The archetype, therefore, is closer to Goethe's *Faust*, with its articulate and affable homunculus, than to *Frankenstein*. Conceivably, this says more about prevailing attitudes towards science and technology – particularly biotechnology – than it does about how we regard the clone: it is the depravity and corruption of the perpetrators that bring about calamity, not the intrinsic nature of the cloning process. That, however, may be too optimistic, for it is often implied that the degeneracy of the scientists or their masters is what makes them willing to conduct human cloning in the first place. Only evil people, in other words, would be prepared to transgress against nature to this degree.

A more weighty, and certainly more auspicious, response to cloning was offered by the writer and poet Naomi Mitchison in her 1975 novel *Solution Three*. It was auspicious because Mitchison was J. B. S. Haldane's sister and a close friend of Aldous Huxley, and because her children and grandchild Tim became respected biologists (Mitchison even makes an unidentified one of them a pioneer of cloning in *Solution Three*). Mitchison evidently shared little of her brother's fervent enthusiasm for human genetic modification – her stance on cloning is evident from the novel's outset, when she dedicates it to James Watson 'who first suggested this horrid idea'.

It is, however, an idea that Mitchison takes pains to make horrid, and the tainted Utopia of her rather reactionary story establishes something of a template for subsequent clone tales. It takes place in a future of unspecified date, probably not far removed from our own era. Following conflicts called the Aggressions and social unrest caused by population growth, world society has been reconstructed by a benign autocracy that looks like a well-meaning (and matriarchal) composite of Huxley's and Orwell's. Population is now held in check by the Code, otherwise known as Solution Three, according to which homosexuality is encouraged and reproduction is managed by cloning. Heterosexual relationships are not wholly forbidden, but they are considered deviant and widely deemed to be distasteful. Heterosexuality remains common, however, among the scientist class, called Professorials, in whom it is tolerated because their work is still needed to maintain the biotechnological population policy. The repeated insistence on how disgusting sex between a man and a woman is thought to be leaves us in no doubt that we are supposed to be witnessing a society that has become *contra naturam*, unwittingly horrible because its sexuality is so 'unnatural'.

All the Clones are genetic replicas of two revered figures known only as Him and Her, a black man and a white woman, and they are produced by SCNT and gestation in the uteruses of Clone Mums. No Clone is ever permitted an inner voice in the book – the story's chief protagonists are all 'normal' people – and indeed their flat speech and bland, placid and innocent manner, lack of fear and shallow aesthetic sensibility shows them to be unambiguously Other, a cipher for robots.* Like robots their knowledge and feelings are programmed into them, here by a tightly prescribed course of communal education. As the Professorial Miryam says of a 'Clone boy' who shows an unexpected aptitude for looking after her biological daughter Em, 'his empathy must be very well expanded and controlled'. The Clones, as the insistent capitalization implies, are evidently a race apart, virtually a different species. When Miryam's husband Carlo talks to a Clone assistant in the laboratory, he is surprised to find himself feeling that he is having a 'normal' conversation. Both the non-Clone characters and the novel's third-person narrative voice deny the Clones genuine personhood. A Clone named Bobbi, the lover of a historian named Ric, tells him that the love song Ric has composed 'showed your love for people', whereupon Ric replies, 'I wouldn't say it was people, Bobbi. Didn't you understand? – it was you.'

While children are, like Em, occasionally produced by heterosexual sex, it is clear that the growing population of Clones will eventually become the majority, and everyone accepts that at some point they will take over the world-governing Council: a gentle vision of the ascent of an unnatural super-race. Only at the end of the book does a crisis in plant breeding lead to the dawning realization in some members

---

* In popular culture, clones are yet again apt to be denied souls, either figuratively or literally. And indeed cloning poses a real dilemma for the Christian view of the soul: if ensoulment happens at conception then the clone *must* lack a soul, for there *is* no moment of conception. One can meanwhile argue on theological grounds that the absence of conception, more than the absence of a soul, absolves the cloned person from original sin.

of the Council that a genomic monoculture might not make for the most biologically robust of societies.

Mitchison lends support to the view that cloning will necessarily destroy the family. Clone Mums raise 'their' Clone babies until they reach an age, some time after puberty, at which they must undergo a 'strengthening', a kind of ritual simulation of the traumatic lives of Him and Her so that the Clones may become more like Them. After this, links between the Mums and their Clones are totally severed. And even while the Mums raise the babies, steps are taken to minimize environmental influences such as styles of mothering that might lead to differences between the children – the sameness of the Clones is *enforced*. In other words, Mitchison creates a society in which the fears evoked by artificial procreation in general, and by cloning in particular, are apparently validated but in fact are (yet again) imposed by authorial fiat.

Solution Three shows that capitulation to anthropoetic mythology need not be the result of political, moral or sexual conservatism, nor of inadequate scientific knowledge. What distinguishes the book, however, is its refusal to cast either scientists or social engineers and authorities as villains: as feminist literary critic Susan Squier remarks, the novel explores 'the feminist temptation to use reproductive technology to eradicate sexism, racism, and war, and to produce instead a peaceable, uniform, and predictable society'. The Council, far from seeking to outlaw or eradicate all dissent, is willing to accept and even learn from deviations and challenges to its Code. If this results in a plot that lacks much narrative tension, it does at least paint a picture of how a technocracy might avoid the worst evils of *Brave New World*. But it does clones no favours.

Regulating clones

While our governing institutions do not exactly demonize cloning, they are extremely wary of it, and seem keen to outlaw it. In March 2005, the United Nations General Assembly adopted the Declaration on Human Cloning, which opposes the reproductive and therapeutic cloning of human beings. The declaration (which is not legally binding) calls on states to 'prohibit all forms of human cloning inasmuch as they are incompatible with human dignity and the protection of human life'. It warns of 'the serious medical, physical, psychological and social dangers that human cloning may imply for the individuals involved', and of 'the potential dangers of human cloning to human dignity'.

In the United States, the National Bioethics Advisory Commission recommended in 1997 that human cloning by SCNT be banned for any purpose, whether reproductive or medical. And later that year 64,000 scientists signed a voluntary five-year moratorium on human cloning. Following suggestions that private human-cloning clinics might be established, the US Food and Drug Administration announced in 1998 that it had the right to prohibit this on safety grounds. Several federal bills have been drafted that propose to ban human cloning, but none has been passed by Congress. In 2001 the Human Cloning Prohibition Act was passed by the House of Representatives but not by the Senate. Similar bills have been considered in subsequent years, but none has become law; at the time of writing, the 2009 Act is still being debated by committee before going to a congressional vote. The 2001 Act proposed that:

It shall be unlawful for any person or entity, public or private, in or affecting interstate commerce, knowingly
   (1) to perform or attempt to perform human cloning;
   (2) to participate in an attempt to perform human cloning; or
   (3) to ship or receive for any purpose an embryo produced by human cloning or any product derived from such embryo.

The Act proposed that violations be punished by fines of at least $1 million, or by imprisonment.

In the UK, human reproductive cloning has been banned since 1990 by the Human Fertilisation and Embryology Act (although embryonic cloning for research is permitted under HFEA licence). And nineteen European countries signed a protocol in January 1998 prohibiting 'any intervention seeking to create a human being genetically identical to another human being, whether living or dead'.

As moratoria imposed to allow careful consideration of the issues, these positions seem to represent a reasonable compromise. But we can see that the 'serious dangers' of the UN declaration are by no means obvious, either medically or ethically. Will such questions be engaged in the future in a forthright and transparent manner, or will society prefer to use vague assertions about 'human dignity' and 'safety' as a veil in which controversy may be conveniently shrouded? The issue is not so much whether existing legislation and guidelines *should* be made more permissive, but whether a political space will ever be allowed to open up in which that possibility can be honestly debated. The current signs are not encouraging. But as there exists no global ban on human reproductive cloning, there is a strong chance that it will be carried out within a private institution somewhere in the world (if it hasn't been already), and will become a de facto reality without having been given the close consideration it is due. I suspect that we shall soon face with human cloning a situation similar to that which exists in the United States at present in IVF, whereby a free market may take advantage of the governmental refusal to regulate. If so, it should not be welcomed by either supporters or opponents of human cloning, and will certainly not be in the interests of potential children born from cloning. It will be an indictment of our squeamishness over the challenging questions of anthropoeia.

What is life? I am not going to answer that question. In fact, I
doubt if it will ever be possible to give a full answer . . . But it is
not a foolish question to ask.

   J. B. S. Haldane, *What Is Life?* (1949)

The human genome is literally a sort of digital text. It contains
some 30,000–100,000 separate stories, known as genes, which is
about as many as there are verses in the Bible.

   Matt Ridley (2000)

Reproductive technologies and cloning are the anthropoetic techniques of the age of biotechnology: manipulations of human cells
and genetic material that can bring about the conception and
growth of new people. They imply that we already have substantially
new ways of making people, and that the role of artifice in these
'artificial' forms of procreation is likely to increase year by year. That
is the complicated, some will say unpalatable, reality we must
confront.

But it does not quite acknowledge the full extent of the capabilities
that biotechnology seems now to possess. In the most rigorously
materialistic picture, our burgeoning understanding of life's molecular
constitution in its full atomistic detail presents the kind of prospect
hinted at by Diderot's d'Alembert and by the attempts at a chemical
synthesis of protoplasm in the nineteenth century: to make a person
from nothing but raw, insensate ingredients, from the atoms up. This
is the vision that sits unspoken at the heart of the genomic era, in

which the perennial question 'what is life?' is answered with reference to the 'language' of genes and DNA.

Consider, for example, how shortly before the chemical structure of the entire human genome was unveiled by the international Human Genome Project (HGP), President Bill Clinton called this sequence of molecular building blocks 'the very blueprint of life'. Or how John Sulston, director of the Wellcome Trust's Sanger Centre and one of the HGP's principal architects, called the genome 'the set of instructions to make a human being'.

A blueprint is a plan for building something – it explains how to *construct* the object in question – while the purpose of an instruction manual is to tell you how the parts fit together. These metaphors, in the view of social scientists Brigitte Nerlich and Robert Dingwall, implied that 'the information provided by the human genome gives scientists the power to build human beings from scratch'. If so, the sequencing of the human genome, which was largely completed by 2001, should have been a landmark in anthropoeia.

Yet it was nothing of the sort. And the reason why is that these metaphors are inappropriate. If the human genome sequence* is an instruction book, it is one for which we can't understand most of the instructions, and in which many may be meaningless, misspelled, garbled or inter-mixed. Moreover, we don't know in what order to read the instructions, many of them require further annotation or alteration before they acquire their proper meaning, and much of the essential information has been omitted because it is taken as read.

In their enthusiasm for the extraordinary technology that enables us to decode entire genomes composed of billions of molecular units, genetic scientists have added fresh impetus to the myth of genetic determinism that haunts our view of the clone. On the other hand, these prodigious efforts have in the end only highlighted the inadequacies of the popular picture of life as a string of genes. Without quite intending to, the Human Genome Project and its related genomic enterprises have tested this conventional picture of 'how humans

---

* You might reasonably ask: *whose* genome sequence? That deduced by the international HGP consortium was a representative collage of several individual genomes, each different in many fine details from yours or mine.

work' to breaking point, and have consequently shown just how remote the idea of making people from scratch still remains.

Life as information

Beliefs about the feasibility of making artificial people have always been conditioned by the prevailing view of what life is. When the universe was seen as an innately fecund matrix, permitting bees and vermin to emerge from putrescence and even quickening metals with a kind of vegetative life, it was easy to imagine that sentient beings might body forth from insensate matter. The mechanical model of mankind fostered the notion that a 'spark of life' – after the discovery of electricity, literally that – might animate a suitably arranged assembly of organic parts. The golden age of chemistry in the second half of the nineteenth century led to a conviction that the right blend of chemical ingredients alone will suffice: nature's diverse grandeur grew from primordial colloidal jelly.

But by the 1930s, the goal of understanding – and as a consequence, that of making – life was starting to look discouragingly difficult. Scientists working with the latest tools for visualizing the sub-cellular world, such as X-ray crystallography and the electron microscope, began to appreciate how intricately structured living organisms are at the finest scales. As H. G. Wells and his son George, along with Julian Huxley, explained in their trilogy *The Science of Life* (1929–30), 'To be impatient with the biochemists because they are not producing artificial microbes is to reveal no small ignorance of the problems involved.'

But the discovery of the chemical structure of DNA, the molecule responsible for genetic inheritance, by Francis Crick and James Watson in 1953 revitalized the belief that life could be made from scratch. When Crick and Watson, along with their collaborator Maurice Wilkins of King's College London, were awarded the Nobel Prize in 1962, the *Sunday Times* said that 'their work heralds the day when man may be able to create simple forms of life himself at will, in a test-tube'.*

---

* The 'test tube' had by then become the symbolic emblem of all chemical synthesis. Anyone who has worked in a chemistry laboratory knows that it is absurd to imply that all the processes enacted therein are conducted in a test tube. This container takes the iconic place of the alchemical alembic and crucible, the locus where things are transformed. It goes without saying that this symbolism was transferred wholesale to the 'test-tube baby'.

We have now nearly scaled that peak. 'Artificial' microbes – bacteria with bespoke, scratch-built DNA – were created in 2010 by biotechnology entrepreneur J. Craig Venter and his illustrious team of co-workers. They used chemical methods to build a synthetic genome, based on that of a naturally occurring bacterium but with some components omitted and others added, and inserted it into a bacterial cell, in effect reprogramming the cell to act like a different and part-designed organism.

However impressive Venter's achievements, and those of other so-called synthetic biologists who seek to engineer life, they do not imply that we have conquered life's mysteries. In some respects, the successes make life look even more impenetrable, since they remind us how dependent we are on the basic designs that evolution has produced. Venter's 'synthetic cell' is in fact more like a motor car constructed by copying an existing vehicle one part at a time (with occasional, modest modifications to some parts), without fully understanding how internal-combustion engines and transmissions and gears work. 'Frankly,' said bioengineer Jim Collins of Boston University in the wake of Venter's announcement, 'scientists do not know enough about biology to create life.'

That is a salutary message for a generation of biologists inculcated into a triumphalist vision of their discipline. For Crick and Watson's decoding of the structure of DNA rapidly bred false confidence about our grasp of how life operates. This misplaced optimism has been actively encouraged by many scientists working in the field of genomics (the elucidation of the chemical structure of DNA-inscribed genomes), who would have us believe that by sequencing* the entire human genome we have decoded the 'book of life'. Now, the story goes, anyone can literally read for themselves what it takes to make a human being, for this sequence is printed in a multi-volume tome that sits on the shelves of the Wellcome Institute in London. This misleading picture of life as genome sequence not only damages the public's perception of biology – and threatens to distort the agenda of biological research – but also leads directly to new variants of the old myths

---

* The 'sequence' of a gene refers to the arrangement of the four types of chemical building block (nucleotides) that it contains. These four units, denoted A, C, G and T, are arrayed in a linear fashion along the double helix. A gene sequence consists of a string of nucleotides – for example, TGACTCACCC . . .

about making people. All our nightmares about creating artificial beings have been projected onto the genome. The homunculus is no longer a little man, but a double-helical molecule. The misshapen monster is no longer a shuffling hulk created by cross-species intercourse or by tampering with fetal growth, but lurks in mutant genes and hybrid genomes, which come to be seen as malformed or alien material woven into our vital fabric. The parts that we assemble to make Frankenstein's creature are not limbs and heads and brains, but genes and other fragments of DNA. To copy human beings, we no longer need Aldous Huxley's 'rows upon rows of gravid bottles' – we need the DNA-manipulating technology of cloning. At the same time, genes supply a new mechanistic, Cartesian model of humankind – we are said to be 'machines controlled by genes'. It is neither coincidental nor irrelevant that we use the old metaphors of mutants and machines to describe this new science. And when, in suggesting that making the human form entails the readout of a digital code, we employ language borrowed from our latest, most advanced technology (computer science), we are revealing more about our changing cultural view of anthropoeia than we are providing a prescription for actually doing it.

Crick and Watson's discovery did change the game in a fundamental way. The complex biochemistry of proteins had previously testified that life's secrets reside not in the issue of elemental composition that dominated discussions of *sarcode* and protoplasm, but in the three-dimensional structures of molecules. And by the 1940s scientists also knew that inheritance was controlled by genes, which were recognized as molecular units located on the chromosomes. They understood that Darwinian evolution happened by random mutation and natural selection among these genes, those that convey an adaptive advantage being preferentially spread through a population. They knew that chromosomes contain DNA, but also that they contain proteins, and it wasn't entirely clear until Crick and Watson's work which of the two substances was the gene-carrier. But the real revelation of their discovery was that it showed how chemical structure could encode genetic instructions: life became a question of *information*.

This was not wholly unanticipated. In 1944, the physicist Erwin Schrödinger had written how a 'tiny speck of material, the nucleus of the fertilised egg, could contain an elaborate code-script involving all the future development of the organism'. Moreover, he explained how this code might be imprinted, somewhat like Morse, in 'a well-ordered association of atoms'. But not until Crick and Watson published their diagram of a sequence of nucleotides strung like beads along the elegant entwined helices of DNA did it become obvious how this might happen.

The new picture looked alluringly simple. The information in a gene specifies which amino acids should be conjoined in the corresponding protein. As Crick and others went on to show, each amino acid in the backbone of a protein is represented in DNA by a so-called codon, a permutation of three of the four nucleotides. The gene that encodes a particular protein structure is first *transcribed* from DNA to an intermediary molecule called messenger RNA (mRNA), and the string of amino acids that make up the protein is assembled piece by piece on the mRNA template in a process called *translation*. While the relationship of gene sequence to protein structure is now known to have considerably more subtlety than this, the basic notion of a genetic code that mediates the translation of DNA sequence into protein sequence (which in turn determines the protein's structure and function) still prevails.

Armed with the code that converts chemical structure into biological information – that relates, as we would now say, genetic sequence (genotype) to physical traits (phenotype) – scientists in the nascent field of molecular biology began the 1960s with immense optimism. The former director of the US National Institutes of Health, William Sebrell, claimed that 'we're going to see within the next generation the artificial creation of living things starting with the virus – and that's almost here'.* The president of the American Chemical Society, Charles Price of the University of Pennsylvania, declared in 1965 that 'the synthesis of life is now within reach', and argued that it should be made a national goal, which, with funding

---

* Sebrell was a little premature. Nobel laureate Arthur Kornberg did indeed make the genome of a virus by chemical synthesis in 1967, but the first active virus made by DNA synthesis solely from a record of the sequence was reported in 2002, when Eckard Wimmer and colleagues at the State University of New York in Stony Brook made the polio virus from strands supplied by mail order from commercial companies. If this sounds dangerous, it was meant to: Wimmer and colleagues wanted to show how easy it had become to produce pathogens from their sequence alone.

on a par with the space missions, could be realized within twenty years. And in 1963 an article in the *New York Times* claimed that 'Ultimately, we may be able to fashion living species to order, to "manufacture" living organisms with specific properties just as we now produce machines or instruments.' Those powers, the article declared, would be 'God-like'.

For a heady couple of decades there seemed indeed to be little that genetic science could not do to reconfigure life. Natural enzyme molecules were commandeered to selectively excise genes or fragments of DNA and to splice new ones into the helix; these techniques supply the basis of genetic engineering. Since our physical and, quite possibly, our mental characteristics were deemed to be under close genetic control, it looked plausible that humankind might be modified in all manner of ways: the dream of making people gave way to the imminent prospect of transforming people.

In the early years many experts naively assumed that the human form was almost infinitely plastic and that just about all traits were determined in a transparent way by the corresponding genes. At a symposium on 'The Future of Man' in London in 1960, J. B. S. Haldane declared that it would soon be possible to tailor human characteristics to social needs. By 'gene grafting', he said, one might make a human with a prehensile tail (it wasn't clear which social need this would satisfy). We might make astronauts free from the unnecessary encumbrance of legs – in zero gravity these limbs merely get in the way, unnecessarily increasing one's oxygen requirement. We could engineer colonists for Jupiter: short and squat, perhaps even quadripedal. We might make humans who could, like plants, photosynthesize their sugars out of thin air, or who have eyes in new, convenient places. Joshua Lederberg forecast 'man–animal chimeras' with 'varying proportions of human, subhuman and hybrid tissue', while the Rand Corporation, a US military think tank, predicted that 'para-humans' – human-like animals – would be created for low-grade labour.

It is peculiar how little mention was made in these speculations of the potential of genetic engineering for the amelioration of disease, which presumably seemed mundane by comparison. Instead, they took their lead from mythical sources: their context is the fabulous beast, the headless and dog-headed men of John Mandeville. We must see them not as serious statements about the world as it is or might be, but as expressions of the mythic unconscious. At least, that is the generous view.

Bad books

The gene-centred view of life culminated in the Human Genome Project, the international effort between 1993 and 2001* to identify each of the three billion or so nucleotides of the human genome. The primary stated motive of this project was to create a database of human genes that might be later used in biomedicine: to develop an understanding of gene-based diseases and to guide research towards cures. According to the project's director Francis Collins, the HGP would 'revolutionize medicine' by making it possible to tailor drug-based therapies to a patient's personal genomic constitution – for example, maximizing the efficiency of the drugs and minimizing side effects that might be manifest only in people carrying specific gene variants. Although, ten years after the HGP released its first draft of the genome, this revolution in genetic medicine has yet to arrive, and although it is no easy matter to progress from knowledge of a disease-linked gene sequence to the development of drug therapies, the project stimulated advances in sequencing technologies that allowed it to be completed sooner than expected and which now promise to make personalized gene sequencing fast and affordable. The technical developments were accelerated by competition between the international, publicly funded consortium and a parallel effort privately run by Craig Venter's biotechnology company Celera Genomics Inc. – a race to obtain the complete sequence that ended in an uneasy agreement to call it a draw.

All this was laudable, even if the real (or at least the immediate) medical potential of the HGP was frequently exaggerated. But, just as with the Apollo moon missions, stimulating massive investment in speculative new technologies tends to require more than vague promises of practical innovations and future social benefits. It demands grand

---

* That is the formal duration of the project, but the sequencing efforts began a little before 1993, while the project had only quasi-official status, and they did not end in 2001. The draft of the genome announced in June 2000 was preliminary, with several gaps. A 'complete' genome was announced in 2003, but even this had holes that are still being filled at the time of writing.

rhetoric. Advocates of the HGP therefore deemed it necessary to argue that decoding the human genome was equivalent to 'reading the book of life'. According to Bill Clinton it would in fact constitute the 'revelation' of nothing less than 'the language in which God created life'.

This was not, for once, a politician's grandiose, over-extrapolated or simplistic interpretation; Clinton was doing no more than smarten up the apparel in which genome scientists themselves had draped their endeavour. The use of metaphors of codes and information in genetics stems from the earliest days of molecular biology in the 1960s, probably under the influence of cybernetics and information theory, which flourished during the Cold War. But the imagery of nature as a book goes back much further, being prominent in the natural philosophy of the Renaissance and the early modern period. 'Philosophy', wrote Galileo, 'is written in a great book which is always open before our eyes, but we cannot understand it without first applying ourselves to understanding the language and learning the characters used for writing it.'

DNA seemed so ideally suited to this long-standing trope that the metaphor was taken literally – the nucleotides are not only denoted symbolically as letters but are spoken of as such, with the corollary that all the other attributes of written language must carry over to the genome. Ever since Watson and Crick's seminal paper, the 'book of life' has been the dominant conceit of genetics. 'What more powerful form of study of mankind could there be', asked Francis Collins, 'than to read our own instruction book?' Clinton merely embellished the religious connotations of this imagery (implicit in its Renaissance context), in which the genome sequence becomes a holy book of revelation.

Caught in the web

The full complexity of the molecular-scale view of life is slowly becoming clear, and the orderly picture in which genes act as templates for task-specific proteins is only one aspect of it. For one thing, genes do not generally act on their own but interact with one another, switching each other on and off or subtly attenuating and modifying one another's activity. Individual genes can have several biological roles, and some generate a number of different proteins, perhaps in a tissue-specific manner, rather than possessing a unique one-to-one relationship to a particular protein. Researchers are starting to think of the genome not as a linear list of instructions but as a dense web of interactions, containing feedback loops and other control elements that make it hard, perhaps impossible, to understand the function of many genes in isolation.

But it's worse than that. Most DNA does not encode proteins at all. It used to be thought that this 'non-coding' DNA was just useless junk accumulated over three billion years of evolution, but we now know that most of it appears to have a biological function. A great deal of non-coding DNA is transcribed by the cell into RNA, but there is no translation of this RNA to protein: the RNA is the end in itself. Transcribing DNA to RNA costs energy, and it is very unlikely that cells would bother to do so unless there was a good reason for it. In some cases the RNA molecules regulate the activity of coding genes; in others, we haven't a clue what they are up to.

Moreover, an organism's phenotype – the traits it exhibits – can be altered without changing its genome. This is called an epigenetic effect. Epigenetic influences on the development of organisms have been known for a long time. As I mentioned in the previous chapter, identical twins with the same genomes don't necessarily have precisely the same physical characteristics or disposition to genetic disease. The effect of genes may be altered by their environment: differences in the activity of genes in identical twins become more pronounced with age, for example, as the messages in the genome get progressively more modified. If we want to persist in considering genomes as 'instruction books' at all, they are thus being constantly revised and edited.

A common type of epigenetic change involves chemical modification of DNA, such as the attachment of tags that 'silence' the genes. These modifications can be strongly influenced by environmental factors such as diet. In effect, epigenetic alterations constitute a second kind of code that overwrites the primary genetic instructions. An individual's genetic character is therefore defined not just by his or her genome but also by the way it is epigenetically annotated. So rapid genetic screening will provide only part of the picture for the kind of personalized medicine that the HGP was said to promise: merely possessing a gene doesn't mean that it is 'used'.

Even the basic idea of a gene as a self-contained unit of biological information is now being thrown into question. The molecular machinery that transcribes DNA, creating the intermediary mRNA molecule that acts as a template for a protein, doesn't just start at the beginning of a gene and stop at the end: it seems regularly to overrun into another gene, or into regions that don't seem to encode proteins at all. Thus, many RNA molecules aren't neat copies of genetic 'words', but contain fragments of others and appear to disregard the distinctions between them. If this looks like sloppy work, that may be because we simply don't understand how the processes of transcription to mRNA and translation to proteins really work, and have developed an oversimplistic way of describing them.

This is backed up by observations that some RNA transcripts are composites of 'words' from completely different parts of the genome, as though the copyist began writing down one word, then turned several pages and continued with another word entirely. If that's so, one has to wonder whether the notions of copying, translation and books really have much value at all in describing the way genetics works. Worse, they could be misleading, persuading scientists that they understand more than they really do.

If the metaphor of a book is really so bad, why was it so popular? Scientists are constantly trying to visualize and express what they discover using concepts that are already familiar – not just when they communicate outside their field but also when they talk among themselves. This is natural, and probably essential in order to gain a foothold on the slopes of new knowledge. After Watson and Crick, genetics – and by extension, life itself – looked about to become comprehensible at last, because we thought we knew how to talk about information: it consists of discrete, self-contained and stable packages of meaning that may be kept in databanks, copied, translated. Meaning is constructed by assembling those entities into linear strings organized by grammatical rules. The delight that accompanied the discovery of the structure of DNA half a century ago was that nature seemed to use this model too.

The notion of information stored in the genome, passed on by copying, and read out as proteins, still seems basically sound. But it is looking increasingly doubtful that nature acts like a librarian or a computer programmer. Perhaps genetic information is parcelled and manipulated in ways that have no direct analogue in our own storage and retrieval systems. This means that there is no 'book of life', no blueprint for a human being, in any meaningful sense. And it suggests that a wholly bottom-up approach to anthropoeia, beginning with DNA alone, is likely to be fruitless.

When life loses its meaning

If a definition of life is not to be found in the genome, then where else might we look? Modern scientific attempts to define 'life' have generally eschewed explicit reference to genes and focused on more abstract concepts. Many have followed Erwin Schrödinger in recognizing that a key characteristic of living systems is their ability to create order and organization from the raw materials and energy sources in their environment – a view that harks back to Aristotle. Because the second law of thermodynamics seems to insist that everything that happens spontaneously in the universe leads to greater *disorder* – to an increase in entropy – it might seem that life somehow miraculously evades this law. But that is an illusion: living systems maintain their organization at the expense of increasing the disorder around them, most notably by generating heat. The net effect of life is an increase in entropy, but this takes the form of localized ordering compensated by more dispersed entropy production. Or as Schrödinger put it, 'What an organism feeds upon is negative entropy'.

That, however, is not an exclusive property of life: order and organization may appear spontaneously in non-living systems too, such as the growth of snowflakes and the patterns that bloom in flowing fluids. So we seem unlikely to find down this path what it is that makes life unique.

I will not describe all the other attempts that have been made to locate a definition of life; it suffices to say that none so far has succeeded in finding a description that includes all systems we recognize as living while excluding all those we do not. Not the least of the difficulties of doing so is that we cannot even be sure where that boundary lies, which is to say, which phenomena we are seeking to include and exclude. Simple synthetic molecules, for example, have been devised that can replicate when supplied with their component parts: replication is not in itself a defining characteristic of life. Some definitions merely build up a tautological list of the common hallmarks, such as metabolism, replication and the ability to undergo natural selection. In the end it seems that 'life' is a little like Justice Potter Stewart's famous definition of hard-core pornography – it is hard to say precisely what it is, but we know it when we see it.

This is another way of saying that 'life', like personhood, is not a scientific concept at all, but a colloquial word that, while having some value in scientific discourse, should not be thereby mistaken as a technical term. Scientific attempts to draw up criteria for what constitutes 'life' only bolster the misguided popular notion of a sharp demarcation between animate and inanimate matter – a belief more appropriate to eighteenth-century vitalism than to twenty-first-century chemical biology. Of course, in one sense 'death' is a threshold that the human being may cross all too abruptly. But this does not imply that the human body is turned to inanimate matter at this stage. Tissues from dead bodies can be cultured to sustain their life, and gametes from recently deceased people continue to be viable. The 'life' that is lost in the death of the organism is clearly not the life of its constituent cells. It is meaningful, perhaps even scientifically meaningful, to speak of a person's 'life' in this context, but we are not obviously talking about the same thing when we speak of the 'life' of cells, or of 'making life' in the laboratory.

That's a crucial distinction from the anthropoetic perspective. It is one thing to seek to understand what motivates the living cell: what determines its actions and functions. We can appreciate that this sort of 'livingness' is an attribute that gets passed from one organism to its progeny, and which sustains each of its tissues. But humankind is not an undifferentiated mass of 'livingness' – it is made up of discrete lives, individual autonomous bundles of life that function as separate organisms. This perception is so fundamental to human experience that we can't easily imagine how things might be otherwise. Either there is an 'I' or there is not. Even organ transplants do not confuse the distinction between us and somebody else. Yet we have seen that the contemporary anthropoeia of assisted conception, embryo research, stem-cell science and cloning undermines this picture. It is as simple as that, and all the furious arguments that these developments provoke must begin there.

Epilogue
Unnatural Creations

It is advisable to sift the merits of knowledge, and clear it of the disgrace brought upon it by ignorance, whether disguised (1) in the zeal of divines, (2) the arrogance of politicians, or (3) the errors of men of letters.

Francis Bacon, *Advancement of Learning* (1605)

Whether mythic or scientific, the view of the world that man builds is always largely a product of his imagination.

François Jacob, 'Myth and Science' (1982)

The role of human agency in the shaping of 'nature' is already one of the major narratives of the twenty-first century. We are now re-engineering the planet on a global scale: altering its climate, its biosphere, its topography, and the composition and circulation patterns of its oceans. We reverse the flow of rivers, hollow out mountains, create artificial islands, reshape the surface of the world. Most people's experience of what we call nature now takes place in a highly constructed and managed environment very different from the indigenous landscapes of the early Holocene. We have created countless new varieties of plant and animal, and are producing hybrids assembled from genomic fragments of wildly different organisms. While he would have applauded such things, Francis Bacon inadvertently gave us the formula for deploring them when he spoke enthusiastically of the 'domination of nature'.

And we are beginning to blur the very notion of who 'we' are. Genomes can in principle be designed and assembled to order. Tissue engineering replaces medical prosthesis: living flesh supplants plastic and steel. Machines are controlled directly by electrical activity in the brain; memories can be selectively excised by drugs.

In its capacity to invert the whole concept of subject and object, the creation of artificial humans is the ultimate 'unnatural' act. Like the theory of Darwinian evolution, it seems to erase centuries of conviction that there is something special, something favoured, about human existence. Despite, or perhaps because of, our obsession with the question 'why are we here?', the prospect of finding a proximal, concrete, mechanical answer is widely felt to be uncanny, possibly even hideous. And what might be most shocking of all is the discovery that 'artificial people' will be in the end indistinguishable from us – that they are every bit as human as we are. Every myth of this outrageous act has insisted otherwise.

I have attempted to argue in this book that our sense of what is 'unnatural' has nothing to do with the physical world – with any relation to Aristotle's *physis*. It is a metaphysical and moral category, born out of a perpetually uneasy relationship with *techne* and imbued with judgements of value and propriety. This does not make it mere superstition, for societies have always needed to delineate moral categories and codes of behaviour. The 'unnatural' is the true Pandora's box that issued from the transgressive creativity of Prometheus: a repository not of real evils but imagined ones, which are often worse. But like Pandora, we cannot resist opening the lid.

We have also seen that the perception of transgression in creating humans by artificial means does not arise from the creation of life per se. The idea that making life is perverted, hubristic, blasphemous or indeed 'unnatural' is essentially a modern construct – it is the consequence of a reified view of nature. Until the Enlightenment, it was commonly assumed that anyone could make lowly forms of life, for example in the ancient bee-making practice of *bougonia*, and that spontaneous generation happened all the time in nature. Arguments about these practices pertained to the quality and character of 'artificial life' – was it inferior to, equivalent to, or better than 'natural' life? – and not the morality of the act. It was only after a common origin of all species was asserted in the nineteenth century that there was any real frisson associated with artificial life in general.

Yet making people was different – because people were favoured by God, and distinguished by their rational souls. Anthropoeia was therefore potentially hubristic and impious for several reasons. First, because it was not deemed to occur of its own accord in nature, raising suspicions of demonic intervention – a criticism, you might say, of the means rather than the end. Second, because human life, uniquely, was a divine mystery, a gift of God or gods. And third, because it raised awkward questions about the soul. The first matter, concerning *technique*, tests the problematic relationship between *physis* and *techne*, between ordained nature and contingent art. One way of reading the latter two issues within a secular context is to say that they symbolize the confusions that anthropoeia creates about our own nature. If we can be raised from inanimate or dead matter, where does our life reside, and from which source does it spring? These issues compel us to question our privileged status in God's creation.

And so we surround the 'artificial person' with prejudices and archetypes filtered through the scientific and cultural lens of our age. These images then *reach well beyond artificial people in themselves*. That is clear in the way, during the twentieth century, 'artificial beings' made on an industrial scale came to represent a fear of the faceless (or equivalently, the same-faced) masses, while artificial procreation was seen paradoxically as the way to tame those hordes by eugenic selection.

The far-reaching symbolism of the artificial person is apparent also in the way genetically modified organisms entering the human food chain quickly become dubbed 'Frankenfood'. Here the synthetic creature made by Victor Frankenstein does not simply exemplify tampering with life but serves as the paradigm for it. Craig Venter's 'synthetic microbe' has been dubbed 'Frankenbug', and even biofuels made by genetic modification of plants have been called 'Frankenstein fuels'. The 'Franken' label both imprints the stigma of unholy meddling and assures us that Faustian retribution will follow. It is no longer deemed necessary to establish the direct relevance of this branding: a survey of public attitudes to stem-cell research in 2007, for example, was titled 'Creating Frankenstein' despite the absence of any subsequent reference to Mary Shelley's story. Allusions to Frankenstein supply a ready-made moral context that is considered to need no further elaboration, as when a *Washington Times* columnist reported on the creation of 'embryo-free' stem cells under the title 'Dr Frankenstein plods on':

The temptation to create human life artificially goes back to Mary Shelley's 'Frankenstein' . . . Man's first temptation in the Garden comes when the serpent assures Adam and Eve that they need only eat the forbidden fruit, and 'ye shall be as gods'. In this case, by creating and destroying human life in a laboratory – for a good purpose, of course. There's always a good purpose. For there is no evil man cannot justify, especially if we were set on it all along.

If this sounds as though it was written a century ago, we can now see that, in effect, it was.

It is my contention here that all of the current debates about human embryo research, stem cells, cloning, genetic modification, and bioethics and biotechnology generally, regardless of whether they have any direct link to the creation of artificial humans, cannot be interpreted without understanding the cultural history of that idea and its relation to themes of 'naturalness'. Only by examining the old myths, legends and stories and the ways that they have been modified and mutated by the ages can we grasp the fears and preconceptions that teem beneath the surface of these discussions.

A bioethics of now

For some people, the questions raised by modern research in reproductive biology involve a titanic, Manichaean struggle between good and evil. Mythical imagery comes naturally to this sort of person. At the 2002 meeting of George W. Bush's Council on Bioethics, Leon Kass said:

The burgeoning technological powers to intervene in the human body and mind justly celebrated for their contributions to human welfare are also available for uses that could slide us down the dehumanizing path toward a brave new world or what C. S. Lewis called in a powerful little book by that name 'the abolition of man'. Thus just as we must do battle with the antimodern fanaticism and barbaric disregard for human life, so we must avoid runaway scientist [sic] and the utopian project to remake humankind in our own image . . . Safeguarding the human future rests on our ability to steer a prudent course avoiding the inhuman Osama bin Ladens on the one side and the post-human Mustapha Mond, Aldous Huxley's spokesman for the brave new world, on the other.

You might be wondering what Osama bin Laden and international terrorism have to do with stem-cell and embryo research. You might, indeed, wonder why the chair of a committee convened to provide the US president with advice on new biomedical technologies has his head filled with chiliastic terror, churning with visions of fictional totalitarian states, Islamist terrorism, 'inhuman' demons and the end of humankind. But such was the tone of that meeting: with the attacks on the World Trade Center still a fresh horror, the issues all too readily assumed apocalyptic proportions. Charles Krauthammer seized on the old notion that 'some people' would 'manufacture extremely intelligent, extremely powerful, extremely resistant people' – nothing less than a 'super-race'. Piling myth upon myth, Robert George spoke for the rights of embryos by invoking the US Declaration of Independence (although we might suspect that Jefferson and Franklin did not have embryos in mind) and, as if it were synonymous, the abolition of slavery.

Such delirium aside, how are we to conduct an ethical census in the face of so many different opinions? Kass and others who take a conservative stance on new biomedical technologies have often asserted that cool, logical reasoning cannot be the sole arbiter of ethical decisions. Intuitions, they say, encode valid moral judgement even if it cannot be easily translated into philosophical argument. Sometimes the gut knows best – this is Kass's 'wisdom of repugnance'.

Take the case carefully adduced by biologist Lee Silver for why it could be considered acceptable to clone embryos genetically engineered to lack heads and thus the capacity to feel sensation, to a stage where they have primitive but distinct organs that might be harvested later for implantation in a person grown from the same fertilized egg. Although he acknowledged that most people would find this repugnant, he admitted that he could see nothing wrong with it 'philosophically or rationally'. This led Charles Krauthammer to write in *Time* how amoral, indeed immoral, some embryo research might become. Kass has commented that 'Repugnance is the emotional expression of deep wisdom, beyond reason's power fully to articulate it . . . Shallow are the souls that have forgotten how to shudder.' But for Silver, emotions 'beyond reason' are not necessarily indicative of 'deep wisdom' at all, but may be responses 'irrelevant or entirely inappropriate to the lives we live today'.

The problem is that both are probably right. Kass is quite wrong to idealize repugnance in general, for we have seen how it can arise out of cultural myths and fears whose origins are irrelevant to the issue at hand. But personally I suspect that our repugnance at the creation of headless embryos grown for spare parts is worth listening to.* And I suspect our instinctive unease is also a good reason (if not perhaps the main one) to exercise extreme caution in the arena of making babies by human cloning. Sometimes there may be wisdom in heeding feelings that we cannot defend with airtight logic. How, though, can we know when that is so, and when not?

I believe that a first step in making our ethical deliberations humane is to recognize that when new moral questions come to light, new moral structures and institutions are needed to resolve them: one cannot simply look up the answer in old books and old codes of conduct. The practice of casuistry – a reconsideration of moral standpoints when new, extraordinary evidence arises – has a long tradition both in law and in religion. That is surely what is needed now to negotiate the daunting terrain of modern anthropoeia.

And we need to seek answers to specific and probable scenarios, not to shore up overarching generalities with the formidable weight of myth. Sociologist Michael Mulkay suggests, for example, that

we should abandon any attempt to try to settle, once and for all, the ontological standing of the human embryo . . . we should concentrate instead on the practical task of specifying the degree to which and the circumstances under which new human entities can be manipulated and, indeed, destroyed in order to bring about beneficial outcomes.

Jean Rostand advocated the same pragmatic point of view in 1959:

It would be vain to give a single overall solution to so many problems. Our task will be to improvise the solution for each one of them when the moment comes, taking account of the state of people's minds, of the collective

---

* Whether Silver agrees with that or not, he is a realist: 'I doubt that any open society anywhere will accept the laboratory creation of headless but otherwise whole functioning human bodies, even if such creatures become feasible. The sense of repugnance is too great, too instinctive, too deep-seated and too close to being universally held.'

mentality, of the social and moral situation. In this sort of thing there is no absolute rule, no dogmatic norm, no 'moral recipe' . . . As each decision arises, we shall have to face the risk of miscalculation and mistaken action.

One might object that an empirical, ad hoc approach to ethical decisions risks making them contingent on the arbitrary here and now and denying that there are any moral absolutes. Do we really think we are wiser than St Augustine, Thomas Aquinas, Immanuel Kant, or the experience accumulated in holy books? But we do not assert such superiority of judgement by merely allowing that we know some of the relevant facts that were hidden from our forebears. We do them a disservice to imagine that, given the choice, they would have excluded this information from their deliberations. What wisdom we can find is increased, not undermined, by recognizing the contingency of our position.

And too rarely do we consider the ethical implications of *in*action or prohibition, which moral absolutes tend to recommend. 'Do nothing' is not a recipe for doing no harm in the future. If the science in this arena is sometimes guilty of hype, perhaps that is partly because we put a greater burden of proof on the positive than on the negative. Besides, with scientific advances so often dependent on serendipity, should we rush to foreclose the possibility of unforeseen benefits through fear of vague or imaginary dangers?

Far too often, when we try to imagine how technologies will play out, we do so by drawing on old myths and taboos. Myths and taboos exist for a reason, but that reason is not about predicting the future. They are apt to lead us down blind alleys, to seduce us with grand narratives when the really important issues are rooted in the particularities of our times and cultures. For example, IVF, invented as a cure for infertility, now confronts us with difficult questions about the socio-economic imbalances of surrogate births. This is a situation created by the specific social context in which IVF has evolved, whereas early (and now evidently spurious) fears that it would lead to monsters or super-races very obviously have a different pedigree. Mulkay states this most appositely: 'If we wish to understand, and to control, the impact of new technologies, we need to think of them as emergent forms of social action embedded within larger and diverse social settings.'

The roads of charity

We cannot be confident, when contemplating the ethics of, say, assisted reproductive technologies, that we can anticipate all the complexities and dangers or that we can resolve all the social and moral dilemmas. But we need to recognize when we have reverted to mythological thinking, particularly the myths of anthropoeia. Myths are protean in themselves – we can assign them interpretations to suit our purposes, and in doing so, draw spuriously on their authority.

Here, for example, is what Leon Kass said in 1979, in the wake of the birth of Louise Brown:

What does it mean to hold the beginning of human life before your eyes, in your hands? Perhaps the meaning is contained in the following story:

Long ago there was a man of great intellect and great courage. He was a remarkable man, a giant, able to answer questions that no other human being could answer, willing boldly to face any challenge or problem. He was a confident man, a masterful man. He saved his city from disaster and ruled it as a father rules his children, revered by all. But something was wrong with his city. A plague had fallen on generation; infertility afflicted plants, animals, and human beings. The man confidently promised to uncover the cause of the plague and to cure the infertility. Resolutely, dauntlessly, he put his sharp mind to work to solve the problem, to bring the dark things to light. No secrets, no reticences, a full public inquiry. He raged against the representatives of caution, moderation, prudence, and piety, who urged him to curtail his inquiry; he accused them of trying to usurp his rightfully earned power, of trying to replace human and masterful control with submissive reverence. The story ends in tragedy: He solved the problem but, in making visible and public the dark and intimate details of his origins, he ruined his life, and that of his family. In the end, too late, he learns about the price of presumption, or overconfidence, of the overweening desire to master and control one's fate. In symbolic rejection of his desire to look into everything, he punishes his eyes with self-inflicted blindness.

Sophocles seems to suggest that such a man is always in principle – albeit unwittingly – a patricide, a regicide, and a practitioner of incest. We men of modern science may have something to learn from our forebear, Oedipus.

But is that what Sophocles wanted to suggest? That a resolute determination to cure disease leads to patricide, regicide, incest? Might one point out that the cause of the plague, as Sophocles tells it, is the abomination that Thebes is harbouring, namely Oedipus himself, who has already unknowingly killed the former king and committed incest before he even begins his search for a cure? And that these sins were thrust upon him through no fault of his own, but because humankind's impotence before divine fate is the wellspring of Greek tragedy?

It is not to deny the wisdom and profundity of myths to point out that their purpose is not so much to tell us what to expect in life as to serve as a receptacle for our dreams and nightmares. The legend of Oedipus does not just encode the challenges that a child faces as his or her sexuality emerges, but expresses the parental fear of acknowledging it, and being able to cope with it appropriately. By the same token, Gregory Pence suggests that the fears we adduce about the awful consequences of cloning (and by extension, anthropoeia in general) are projections of those we feel already. The concern that a man who has a daughter cloned from his wife will want to make love to her is at root the worry that he might harbour incestuous thoughts about a 'normal' biologically related daughter (who might, after all, equally resemble his younger wife). The idea of a cloned daughter merely gives us licence to project that fear elsewhere. It is the same, Pence argues, with our worries about 'designer children', or clones who are expected to possess the cherished attributes of the parent or of a dead child, or of monstrous births, or of the absence of family bonds in children born through assisted conception: we already fear what we might want, and what we might get, in our children, and these images simply provide a locus for those apprehensions. 'If we can't accept this,' says Pence, 'debates about cloning will never be honest.'

Even if we accept (as I think we should) that myth may have something to teach us, let's not forget that it is a form of storytelling, not a kind of allegorical historical determinism. In real human history we have abundant instances of people who have set out resolutely to find a cure for disease and have done so, and saved lives, and somehow avoided killing their fathers and marrying their mothers in the course of their research. Personally, I would rather that decisions on biomedical ethics be made by informed and thoughtful living individuals, and not by Sophocles, Goethe, Nathaniel Hawthorne, Mary Shelley or Aldous Huxley.

As Kass unwittingly reveals here, perhaps the biggest hazard for contemporary bioethics is grandiosity. Grand narratives have a lapidary quality that seems inevitably to fall back on legendary sources: the dangers they foresee bulk large but have only the vaguest of outlines. They tend to speak about humanity while neglecting actual people and thereby neglecting humane compassion. They may have more appeal and glamour than the kind of painstaking, modest casuistry exemplified, however imperfectly, by the Warnock committee or the UK's Human Fertilisation and Embryology Authority; but I know which of these authorities I would trust. When Leon Kass asserts that technologies for intervening in conception threaten to initiate 'the erosion, perhaps the final erosion, of the idea of man as something splendid or divine, and its replacement with a view that sees man, no less than nature, as simply more raw material for manipulation and homogenization', he insists on only the mythic choices: man as miracle, man as mud. We need instead to see humans, however they come into the world and whatever their genetic inheritance, as people, and moreover as people in the society of other people. It seems to me that until we can acknowledge the births of Louise Brown and of millions of other IVF babies as something splendid (and, if you wish, divine), we are never going to find the wisdom and humanity to navigate the difficult paths that modern medicine and biology have opened up.

What is splendid about IVF – when it succeeds – is not so much that a human life is created, for that happens somewhere in the world every few seconds, under slightly different conditions but in much the same biological manner. It is that human *art* is thereby turned to sympathetic ends. Reasons for creating or not creating humans by 'artificial' means should focus not on those means, nor on abstract (and probably irresolvable) moral considerations about, say, the 'sanctity of life', but on the intentions and outcomes: the effect on actual human lives and on society. As Bonnie Steinbock says, 'respect for human life is hardly expressed when embryos or presentient fetuses are protected at the expense of the "vital human interests" of already existing persons'. The difficulty is in distinguishing vital human interests from whims or selfish, egotistical, irresponsible desires. There is no easy way of doing that, no ready-made formula or doctrine. We have to find our path, cautiously, humbly, humanely, while – this is really all I ask – being as honest with ourselves as we can be about why we believe and presume the things that we do.

We should no longer be seduced by Prometheus – either by the noble Prometheus of the English Romantics who lionized his defiance of the king of Olympia, or the Faustian modern Prometheus of Mary Shelley, who found knowledge but not wisdom. There is no shame in grand ambitions, but no point in them either if in the end they do not, in however small a way, serve humanity here and now. While neither Francis Bacon nor St Paul can be accused of being overburdened with compassion, they together tell us something worth heeding in Bacon's *Advancement of Learning* (1605): 'It is merely from its quality when taken without the true corrective, that knowledge has somewhat of venom or malignity. The corrective which renders it sovereign is charity, for, according to St Paul, "Knowledge puffeth up, but charity buildeth."' Some may agree with Leon Kass in his cold assessment of the perils of IVF: 'The road to Brave New World is paved with sentimentality – yes, even with love and charity.' But as St Paul recognized, roads built with charity need not, and most probably will not, end in mythical dystopias.

Acknowledgements

I am fortunate to be part of such a fine team. Just as previously, my agent Clare Alexander and all at Bodley Head – Will Sulkin, Jörg Hensgen, Kay Peddle and Hannah Ross – have been supportive, efficient, thoughtful and generous. David Milner was again a conscientious and discerning copy-editor. While I was writing this book, my home team increased in number by one, and for that and for everything else I thank my wife Julia.

As ever, I am the greedy beneficiary of the scholarship of others, and in particular I have derived great inspiration (and simply stolen much else) from William Newman, Lorraine Daston, Katharine Park, Susan Merrill Squier, Debora Spar and Jon Turney. Several researchers have kindly provided information of various sorts; among them, Brigitte Nerlich generously read through the entire manuscript, and Kate Roach made perceptive comments on Chapter 4. It is only in retrospect that I realise that most of the interactions I have had about the material and ideas in this book have been with women. I remark on it because, to the (admittedly debatable) extent that my topic is based in science, that is unusual. I suppose it is because women are motivated and perhaps compelled to do most of the hard thinking about how and why we make people, and it is they who most often identify, examine and bravely challenge the myths.

Philip Ball
London, October 2010

Notes on Sources

(Biblical quotes are from the New International Version.)

Introduction: It's Not Natural

Page 2 'Whatever today's embryologists may do': Edwards (1989), p. 70.
• 'The necessity or otherwise for experiments on human embryos': ibid. • 'The trouble really started way back in the 1930s': ibid. • **Page 3** 'Nearly every literate person perceives *natural*': Silver (2006), p. xiii. • 'Un-natural is not simply non-natural': H. Puff, in Daston & Vidal (eds) (2004), p. 245. • 'The ultimate goal may be to produce a child': T. Skeet, *Parliamentary Debates*, Hansard (Commons), 4 February 1988, col. 1225. • **Page 4** 'a technology that wants to substitute true paternity and maternity': Pope John Paul II (2004), 'Address of John Paul II to the Members ofthe Pontifical Academy for Life, 21 February 2004'. Available at http://www.vatican.va/holy_father/john_paul_ii/speeches/2004/february/documents/hf_jp-ii_spe_20040221_plenary-acad-life_en.html.

Chapter 1: Art Versus Nature?

Page 5 'So, over that art': W. Shakespeare, *The Winter's Tale*, Act IV, scene iii. • 'All Nature is but Art, unknown to thee': A. Pope, quoted in Daston & Vidal (eds) (2004), p. 123. • 'In our dealings with nature': Haldane (1949), p. 260. • 'Would not the laboratory production of human beings': Kass (1971a), here p. 784. • **Page 9** 'Is *techne* a continuation of nature's activity': Bensaude-Vincent & Newman (2007), p. 3. • 'the sun and the heavens, and the earth and yourself': Plato, *Republic*, transl. B. Jowett, Book X. • **Page 10** 'A painter will paint a cobbler, carpenter': ibid. • 'Had he in his lifetime friends who loved': ibid. • 'the very term *mechanomai*, the verbal form': Newman (2004), p. 22. • **Page 11** 'nature is a cause, a cause that operates for a purpose': Aristotle, *Physics*, Book II, Part 8. • **Page 12** 'This is most obvious in the animals other than man': ibid. • **Page 13** 'This, then, is what is called concoction by boiling': Aristotle, *Meteorology*, Book IV, Part 3, 381a9–12. Available at http://classics.mit.edu/Aristotle/meteorology.4.iv.html. • **Page 14** 'We make, by art . . . trees and flowers to come': F. Bacon (1627), *The New Atlantis*. See, e.g., F. Bacon (2005), *The Great Instauration and New Atlantis*, ed. J. Weinberger, p. 72. Harlan Davidson, Arlington Heights, IL. Text available at http://oregonstate.edu/instruct/phl302/texts/bacon/atlantis.html. • 'to do that which without art would not be done': F. Bacon, *Descriptio globi intellectualis*, in J. Spedding, R. L. Ellis & D. D. Heath (eds) (1857–74), *The Works of Francis Bacon*, Vol. 5, p. 506. Longman, London. • 'the fashion to talk as if art were something different': ibid. • **Page 15** 'So unaccurate is [Art] in all its productions': Hooke (1665/2007), p. 47. • 'Nature always takes the shortest way': Ibn Khaldun (1958), *The Muqaddimah*, transl. F. Rosenthal, Vol. 3, p. 280. Routledge & Kegan Paul, London. • **Page 16** 'Nor does art make all these things, rather it helps nature': W. R. Newman (1991), *The Summa perfectionis of Pseudo-Geber*, p. 12. Brill, Leiden. • **Page 17** 'All other forms of life the earth brought forth': Ovid, *Metamorphoses*, Book I, 416–21, quoted in Newman (2004), p. 166. • **Page 18** 'They find a two-year calf with sprouting horns': Virgil (1956), *Virgil's Georgics: A Modern English Verse Translation*, transl. S. P. Bovie, 299–312. University of Chicago Press, Chicago. • **Page 19** 'Art is weaker than nature and does not overtake it': Newman (1991), p. 49. • 'hold prisoner in a bottle that God': V. Biringuccio (1942), *Pirotechnia*, transl. C. S. Smith & M. Gnudi, p. 40. MIT Press, Cambridge, MA. • **Page 20** 'deceive the ignorant populace as to the alchemical fire': quoted in E. J. Holmyard (1990), *Alchemy*, p. 149. Dover, Mineola, NY. • 'not in the nature of things': quoted in Newman (2004), p. 83. The text of the decretal is translated into French in R. Halleux (1979), *Les textes alchimiques*, pp. 124–6. Brepols, Turnhout. • 'In one of these isles be folk of great stature': quoted in Denny & Filmer-Sankey (1973), p. 51. • **Page 21** 'A Spaniard came to Florence, who had with him': L. Landucci (1927), *A Florentine Diary from 1450 to 1516 by Luca Landucci*, transl. A. de Rosen Jervis, p. 272. E. P. Dutton, New York. • 'Now mistakes come to pass even in the operations': Aristotle, *Physics*, Book II, Part 8. • **Page 22** 'predict future things': Isidore of Seville (1911), *Etymologiarum sive originum libri*, Book XX, ed. W. M. Lindsay, Vol. 2, 11.3.1. Clarendon Press, Oxford. • 'come by the divine will': Isidore of Seville, *Etymologiae*, Book XI, Chapter 3, quoted in Asma (2009), p. 75. • 'The order imposed on things by God': T. Aquinas (1975), *Summa contra gentiles* 3.99.9, transl. V. J. Bourke, Vol. 3, Part 2, p. 78. University of Notre Dame Press, Notre Dame, IN. • **Page 23** 'monsters are things that appear outside the course of Nature': A. Paré (1983), *On Monsters and Marvels*, transl. J. L. Pallister, preface. University of Chicago Press, Chicago. • 'Each monster is the result of iniquity': Jacob (1989), p. 27. • 'in all respects hateful': F. Bacon (1607), 'Of atheism'. Available at http://www.authorama.com/

essays-of-francis-bacon-17.html. • **Page 24** 'If philosophers deprived nature of skill and autonomy': Daston & Park (1998), p. 301. • 'all things proceed, according to the artificer's first design': R. Boyle, 'Free inquiry', in T. Birch (ed.) (1772), *The Works of the Honourable Robert Boyle*, Vol. 5, p. 163. London. Reprinted with an introduction by D. McKie (1965–6). Georg Olms, Hildesheim. • **Page 26** 'The first rule of reason is the law of nature': T. Aquinas (1964–73), *Summa theologiae*, transl. T. Gilby et al., Vol. 28, p. 104. Eyre and Spottiswoode, London. • **Page 27** 'It may be as much a mistake to claim': L. Kass, in Kristol & Cohen (eds) (2002), p. 67.

Chapter 2: Work of the Gods

Page 28 'A Greek it was who first opposing dared': Lucretius (1921), *On the Nature of Things*, transl. W. E. Leonard, Book I, p. 5. J. M. Dent, London. • 'A mighty lesson we inherit': Byron (1816), 'Prometheus'. Available at http://englishhistory.net/byron/poems/prometheus.html. • 'Promethean ambitions': Pope John Paul II (2002), 'Address of John Paul II to the Members of the Pontifical Council for Health Pastoral Care, 2 May 2002', para. 4. In *L'Osservatore Romano* (Engl. edn), 22 May, p. 9. Available at http://www.vatican.va/holy_father/john_paul_ii/speeches/2002/may/documents/hf_jp-ii_spe_20020502_pc-hlthwork_en.html. • **Page 29** 'The chemical or physical inventor is always a Prometheus': Haldane (1924), in Dronamraju (ed.) (1995), p. 36. • **Page 30** 'He found that the other animals were suitably furnished': Plato (2005). • **Page 32** 'all creatures that come into being are earth and water': quoted in Hall (1969), Vol. I, p. 39. • **Page 35** 'There is something which unites magic and applied science': Lewis (1943/1973), p. 88. • **Page 36** 'the first modern man': Haldane (1924), in Dronamraju (ed.) (1995), p. 36.

Chapter 3: Recipes for a Little Man

Page 37 'This hour will crown a wondrous undertaking': Goethe (1949/1959), Part II, p. 99. • 'I do not at all believe that this man': in University of Manchester MS John Rylands lat. 65, 205v. Quoted in Newman (2004), p. 189. • 'tinkering with natural human generation was a widespread': Newman (2004), p. 165. • **Page 39** 'When the flesh had become firm': quoted in Scholem (1965), pp. 172–3. • **Page 42** 'repressing what you know, which is that in this life': M. J. Harrison, *Guardian* Review, 26 February 2005. • '*Never Let Me Go* makes you want to have sex': ibid. • **Page 43** 'a true and living infant, having all the members': (Pseudo?-)Paracelsus (1537), *De rerum natura*, I, xi, 316–17. Quoted in N. Goodrick-Clarke (1999), *Paracelsus: Essential Readings*, p. 175.

North Atlantic, Berkeley, CA. • 'Bodys Appeare & are often thought Absolutely destroy'd': R. Boyle (*c*.1655), 'Essay of the holy Scriptures'. In M. Hunter & E. B. Davis (eds) (2000), *The Works of Robert Boyle*, Vol. 13, p. 204. Pickering & Chatto, London. • **Page 44** 'a hundred snakes where each is as big': (Pseudo?-)Paracelsus (1537) *De rerum natura*, in K. Sudhoff (ed.) (1922–33), *Theophrastus von Hohenheim genannt Paracelsus, Sämtliche Werke*, Vol. II, p. 346. Oldenbourg, Munich. Translation by W. Newman. • 'ashamed of the grosse sense of it': H. More (1656), *Enthusiasmus Triumphatus*, p. 46. J. Flesher, London. • **Page 46** 'Whoever therefore believes any created thing': quoted in Newman (2004), p. 56. • **Page 48** 'O what a world of profit and delight': C. Marlowe (1604/1994), *Doctor Faustus*, p. 5. Dover, New York. • 'Faustus is gone; regard his hellish fall': ibid., p. 56. • **Page 49** 'In madness, half suspected on his part': Goethe (1949/1959), Part I, p. 41. • 'That old style we declare': ibid., Part II, p. 100. • **Page 50** 'Now hope may be fulfilled': ibid. • **Page 51** 'Most strangely made, as I have heard him say': ibid., p. 148. • 'To hold the lordly ocean from the shore': ibid., p. 221. • 'Faust is saved by his will to strive': P. Wayne, introduction to Goethe (1949/1959), Part I, p. 25.

Chapter 4: It Lives!

Page 53 'I ought to be thy Adam': Shelley (1818/1831/1994), p. 96. (Subsequent quotations from *Frankenstein* are referred to this volume.) • **Page 54** 'the idea of an entirely man-made monster': Mellor (1988), p. 38. • 'switched-on magic, souped-up alchemy': J. Rieger, in Shelley (1818/1831/1974), p. xxvii. • 'it is precisely the electrification of Paracelsus': Turney (1998), p. 21. • **Page 55** 'the wildest extravagances of human fancy': Godwin (1834/2006), p. 6. • **Page 56** 'Man is a creature of boundless ambition': ibid., p. 16. • 'I am convinced I could represent myse[lf] to you': P. B. Shelley (1812), letter to William Godwin, 3 January. In F. L. Jones (ed.) (1964), *The Letters of Percy Bysshe Shelley*, Vol. 1, p. 21. Clarendon Press, Oxford. • 'various philosophical doctrines . . . and among others': Shelley, p. 8. • **Page 57** 'Whence did the principle of life proceed?': ibid. p. 49. • 'a new species would bless me as its creator': ibid., p. 51. • **Page 58** 'turns away from alchemy and the past': B. Aldiss (1975), *Billion Year Spree*, p. 27. Corgi, London. • 'skips the science': J. Rieger, in Shelley (1818/1831/1974), p. xxvii. • 'waste your time upon this; it is sad trash': Shelley, p. 37. • 'When I returned home my first care': ibid., p. 38. • 'It may appear strange that such should arise': ibid. • **Page 59** 'an uncouth man, but deeply imbued in the secrets': ibid., p. 44. • 'The professor stared. "Have you", he said': ibid. • 'I had a contempt for the uses of modern natural philosophy': ibid., p. 45. • 'The modern masters promise very little': ibid., p. 46. • **Page 60** 'What would the ancient philosophers': quoted in Schaffer (1983), here p. 8. • 'that branch of natural philosophy in which the greatest improvements': Shelley, p. 47. • 'days and nights in vaults and charnel-houses': ibid., p. 50. • **Page 61** 'The astonishment which I had at first experienced': ibid., pp. 50–1. • 'I began the creation of a human being': ibid., p. 51. • **Page 62** 'With an anxiety that almost amounted to agony': ibid, p. 55. • **Page 63** 'Instead of slowly endeavouring to lift up the veil': H. Davy (1802), *A Discourse, Introductory to a Course of Lectures on Chemistry*, p. 9. John Johnson, London. • '"I shall make metals", he cried': H. Balzac (1834), *La recherche de l'absolu*, Chapter 6. Vve Béchet, Paris. Quoted in Schummer (2006), p. 117. • **Page 64** 'that fiendish employment of decomposing all things': quoted in ibid., p. 113. • **Page 65** 'I have never beheld anything so utterly destroyed': Shelley, p. 39. • 'a flash of lightning illuminated the object': ibid., p. 73. • 'Perhaps a corpse would be reanimated': ibid., p. 8. • 'we are come at last to touch the celestial fire': quoted in Schaffer (1983), p. 9. • 'The fable of Prometheus is verify'd': ibid. • **Page 66** 'The jaw began to quiver, the adjoining muscles': J. Aldini (1803), *An Account of the Late Improvements in Galvanism*, p. 193. Cuthell & Martin/J. Murray, London. • **Page 67** 'A she-ass received the wourali poison': C. Waterton (1839), *Wanderings in South America*, 4th edn, p. 75. London. • 'Its power of exciting muscular motion in apparently dead animals': A. Walker (1799), *A System of Familiar Philosophy*, p. 391. London. • 'the contrast between the animal functions': quoted in Hall (1969), Vol. II, p. 227. • 'ether, electricity, air, or magnetism': James Graham (1784), *A Lecture on Love: or, Private advice to Married Ladies and Gentlemen*, privately printed, London. Quoted in M. Marsh & W. Ronner (1996), *The Empty Cradle*, p. 21. Johns Hopkins University Press, Baltimore. • **Page 68** 'On the 26th day each figure assumed the form': Crosse (1838). • 'With regard to Mr Crosse's insects etc.': M. Faraday, letter to C. F. Schönbein, 21 September 1837, in G. W. A. Kahlbaum & F. V. Darbishire (eds) (1899), *The Letters of Faraday and Schoenbein 1836–1862*, p. 33. Williams & Norgate, London. • **Page 69** 'a very dark business, and such as no Christian man': quoted in J. A. Secord, in Gooding, Pinch & Schaffer (eds) (1989), p. 359. • 'I have met with so much virulence and abuse': Crosse (1838). • **Page 70** 'To examine the causes of life': Shelley, p. 49. • 'I became acquainted with the science of anatomy': ibid. • 'His limbs were in proportion, and I had selected': ibid., p. 55. • **Page 71** 'How dare you sport thus with life?': ibid., pp. 95–6. • 'Treat a person ill, and he becomes wicked': P. B. Shelley, quoted in St Clair (2004), p. 358. • 'Learn from me, if not by my precepts': M. Shelley (1818), *Frankenstein: The 1818 Text*, ed. & intro. M. Butler (1993), p. 34. Oxford University Press, Oxford. • **Page 72** 'Seek happiness in tranquillity': Shelley, p. 210. • 'our culture's most penetrating literary analysis': Mellor (1988), p. 38. • 'suggests both psychological and historical determinism': Schummer (2006), p. 121. • 'contrasted what she considered to be "good" science': Mellor (1988), p. 89. • 'that scientific research which attempts to describe': ibid., p. 90. • **Page 73** 'violated the rhythms of nature': ibid., p. 102.

• 'to manipulate life-forms in ways previously reserved': ibid., p. 114. • **Page 74** 'The Idea that possessed his life had operated': N. Hawthorne (1850), 'Ethan Brand'. Available at http://www.ibiblio.org/eldritch/nh/eb.html. • **Page 75** 'Why was I? What was I?': Shelley, p. 124. • 'Did I request thee, Maker, from my clay': J. Milton, *Paradise Lost*, Book X, 743–5. Available at http://www.dartmouth.edu/~milton/reading_room/pl/book_10/index.shtml. • 'It moved every feeling of wonder and awe': Shelley, p. 125. • **Page 76** 'Fiend, I defy thee! with a calm, fixed mind': P. B. Shelley (1820), *Prometheus Unbound*, Act I, 262–5. Available at http://andromeda.rutgers.edu/ ~jlynch/Texts/ prometheus.html#1. • **Page 77** 'In the endurance, and repulse': Byron (1816), 'Prometheus'. • 'By stealing the female's control over reproduction': Mellor (1988), p. 115. • 'She might become ten thousand times more malignant': Shelley, p. 160. • **Page 78** 'One of the first results of those sympathies': ibid., pp. 160–1. • **Page 79** '*Frankenstein* is a book about what happens when a man': Mellor (1988), p. 40. • **Page 80** 'For by Art is created that great Leviathan, called a Common-wealth': T. Hobbes (1651/1985), *Leviathan*, ed. & intro. C. B. Macpherson, p. 81. Penguin, London. • 'an ingenious mechanism . . . the piecing together': F. Schiller (1967), *On the Aesthetic Education of Man*, ed. & transl. E. M. Wilkinson & L. A. Willoughby, p. 35. Oxford University Press, Oxford. • '[We] should approach to the faults of the state': E. Burke (1968), *Reflections on the Revolution in France, and on the Proceedings in Certain Societies in London Relative to that Event*, ed. C. C. O'Brien, p. 194. Penguin, Harmondsworth. • 'Out of the tomb of the murdered monarchy in France': E. Burke (1907), *The Works of the Rt Hon. Edmund Burke*, Vol. 6, p. 88. Oxford University Press, Oxford. • 'France is as a monstrous Galvanic Mass': T. Carlyle (1896–9), *The Works of Thomas Carlyle*, Vol. 3, p. 113. Chapman & Hall, London. • 'another unnatural disgusting fiction': T. Carlyle (1818/1970), in *The Collected Letters of Thomas and Jane Welsh Carlyle*, i, 124, 30 April 1818. Duke University Press, Durham, NC. • **Page 81** 'The actions of the uneducated seem to me typified': E. Gaskell (1848/1970), *Mary Barton: A Tale of Manchester Life*, ed. S. Gill, p. 219. Penguin, Harmondsworth. • 'an unfortunate but morally irresponsible creature': Baldick (1987), p. 87. • 'a tissue of horrible and disgusting absurdity': Anon. (1818), *Quarterly Review*, **18**, 385. • **Page 82** 'bordering too closely on impiety': Anon. (1818), *Edinburgh Magazine and Literary Miscellany* **2**, 249–53. • 'These volumes have neither principle': quoted in St Clair (2004), p. 360. • 'Do not go to the Opera House to see the Monstrous Drama': quoted in Holmes (2008), p. 334. • 'But Mr Devildum – have not I heard': R. B. Peake (1823), *Another Piece of Presumption*. Quoted in Forry (1990), p. 163. • **Page 83** 'The striking moral exhibited in this story': ibid., p. 5. • 'It lives! It lives!': Peake (1823), quoted in ibid., p. 143. • 'like Dr Faustus, my master is raising the Devil': ibid., p. 138. • 'The story is not well managed . . . I was much amused': quoted in ibid., p. 4. • **Page 84** 'the

Creature's deformed body mirrors an evil nature': ibid., p. 22. • 'Instead of the fresh colour of humanity': quoted in ibid., p. 194. • **Page 85** 'I really do think, at least it seems so to me': H. Milner (1826), *Frankenstein: or, The Man and the Monster!*, quoted in ibid., p. 193. • 'pale student of unhallowed arts . . . mock[s] the stupendous': Shelley, p. 9. • 'silent workings of immutable laws': ibid., p. 92. • 'chance – or rather the evil influence': ibid., p. 44. • 'words of fate, enounced to destroy me': ibid., p. 241. • 'as if my soul were grappling with a palpable enemy': ibid., p. 46. • **Page 86** 'a large intricate machine – like a galvanic battery': P. Webling (1927), *Frankenstein: An Adventure in the Macabre*, quoted in Forry (1990), p. 95. • **Page 87** 'the self can only engender the self in a parthenogenetic': ibid., p. 100. • **Page 88** 'The truth of a myth . . . is not to be established': Baldick (1987), p. 4. • **Page 89** 'After *Frankenstein*, the figure of the scientist in fiction': ibid., p. 142. • 'Devoted to the pursuit of knowledge at the expense of humane values': Tudor (1989b), p. 589. • 'Is one who has solved the secret of life to be considered mad?': ibid., p. 590. • **Page 90** 'Had I approached my discovery in a more noble spirit': Stevenson (1979), p. 85. • 'The human mind will only stand so much': quoted in Tudor (1989b), p. 589. • **Page 91** 'I reflected how fearful the combat which I momentarily expected': Shelley, p. 188. • 'most ardently desires his bride when he knows': Mellor (1988), p. 121. • 'Frankenstein dedicates himself to his scientific experiment with a passion': ibid., pp. 121–2. • 'feel uncomfortable with their emotions and sexuality': quoted in ibid., p. 108. See E. F. Keller (1978), 'Gender and science', *Psychoanalysis and Contemporary Thought* I, 409–33. • 'The magicians knew, at least imaginatively': D. H. Lawrence (1918/1962), in *The Symbolic Meaning: The Uncollected Versions of 'Studies in Classic American Literature'*, ed. A. Arnold, p. 36. Centaur Press, Arundel. • **Page 92** 'given him life but stripped him of a soul': Willis (2006), p. 80. • 'In leaving the creation of the monster equivocal': ibid., p. 84. • 'Nature prevents Frankenstein from constructing a normal human being': Mellor (1988), p. 123. • 'envisions nature as a sacred life-force': ibid., p. 124.

Chapter 5: Descartes' Daughter

Page 94 'This is no Workmanship of Humane Skill': J. Edwards (1696), *A Demonstration of the Existence and the Providence of God, from the Contemplation of the Visible Structure of the Greater and the Lesser World*, Part 2, p. 124. Jonathan Robinson, London. • 'This has been the fallacy of our age': D. H. Lawrence (1918/1962), in *The Symbolic Meaning: The Uncollected Versions of 'Studies in Classic American Literature'*, ed. A. Arnold, p. 37. • 'There is my library': quoted in Huxley (1874). • 'a physiologist of the first rank': Huxley (1874). • **Page 95** 'I do not recognize any difference between the machines': R. Descartes (1647/1897/1910), *Principes de la philosophie*, in *Oeuvres de Descartes*, ed.

C. Adam & P. Tannery, Vol. 9, pp. 321–2. Léopold Cerf, Paris. • **Page 96** 'be lodged in the human body exactly like a pilot': Descartes (1637), Part V. • 'After the error of those who deny the existence of God': ibid. • **Page 97** 'I wouldn't want to publish a discourse': quoted in Wood (2002), p. 8. • **Page 98** 'an empty word to which no idea corresponds': J. O. de La Mettrie (1747), *L'homme machine*, p. 59. E. Luzac, Leiden. See J. O. de La Mettrie (1912), *Man a Machine*. Open Court Publishing Co., Chicago. • 'resides in . . . the very substance of the parts': La Mettrie (1912), p. 131. • 'each part contains in itself springs whose forces are proportioned to its needs': ibid. • 'In the eighteenth century, to deny the soul's immateriality': Vartanian (1960), p. 24. • 'mainspring of the watch': Hall (1969), Vol. II, p. 52. • **Page 99** 'a trick of skill, a ruse of style': J. O. de La Mettrie, 'Histoire naturelle de l'âme' (1745), in *Oeuvres philosophiques* (1753). Amsterdam. Published with *Man a Machine* (1912/1960), transl. G. C. Bussey et al., p. 143. University of Chicago Press, Chicago. • 'that these proud and vain beings, more distinguished by their pride': ibid. • 'knows nothing, but nothing of what he has done': D. Diderot (1875), 'Eléments', in *Oeuvres de Diderot*, Vol. 9, p. 272. Paris. • **Page 100** 'That, sir, is my own handiwork. He can sing and he can act': Needham & Wang (1956), p. 53. • 'liver, gall, heart, lungs, spleen, kidneys, stomach and intestines': ibid. • **Page 101** 'the diabolical art . . . a device under the ground': Odoric of Pordenone, *Relatio*, published in *Sinica franciscana*, 1 (1929), p. 473. Quoted in Daston & Park (1998), p. 93. • 'performed and entertained and danced and capered and gambolled': quoted in ibid., p. 90. B. de Sainte Maure (1904–12), *Le Roman de Troie*, Vol. 2, 14657–918. Translated in Sullivan (1985), p. 14. • 'only of pieces of wood and hinges': quoted in Scholem (1965), p. 199. • 'which appeared so wonderful to the ignorant multitude': D. Brewster (ed.) (1830), *Edinburgh Encyclopedia* Vol. 1, p. 62. Blackwoods, Edinburgh. • **Page 102** 'dared to investigate the secrets of creation': quoted in C. Bailly (1947), *Automata: The Golden Age, 1848–1914*, p. 14. Sotheby's, London. • **Page 105** 'gives a better and more intelligible Theory of Wind-Musick': in J. de Vaucanson (1742), *An Account of the Mechanism of an Automaton or Image Playing on the German-Flute*, transl. J. T. Desaguliers, p. 21. London. • 'made manifest, not a reduction of animals to machinery': J. Rifkin, in Bensaude-Vincent & Newman (eds) (2007), p. 262. • **Page 107** 'Nor will this appear at all strange to those': Descartes (1637), Part V. • 'will have respiration, circulation, quasi-digestion, secretions, heart': Le Cornier de Cideville-Fontenelle, 15 December 1744, in Abbé Tougard (1912), *Documents concernants l'histoire littéraire du XVIIIe siècle*, Vol. 1, p. 53. • **Page 108** 'Man is . . . to the ape . . . what Huyghen's': quoted in Toulmin & Goodfield (1962/1982), p. 315. • **Page 109** 'were there such machines exactly resembling in organs': Descartes (1637), Part V. • **Page 110** 'It appears to me to be a very remarkable circumstance': quoted in Huxley (1874). • **Page 111** 'the soul is a mere spectator of the movements': C. Bonnet (1755), *Essai de Physiologie*, Chapter xxvii. Quoted in Huxley (1874). • 'Give the automaton a soul which contemplates its movements': ibid. • 'It is not to be denied that Supreme Power could': ibid. • 'What proof is there that brutes are other than': Huxley (1874). • **Page 112** 'in the condition of one of Vaucanson's automata': ibid. • 'eminently virtuous and respectable . . . guilty of the most criminal acts': ibid. • **Page 114** 'There is really no end to the march of invention': in Poe (ed. Harrison) (1965), p. 263. • **Page 116** 'If Doctor Johann Faust had been living in the age of Goethe': Villiers (2001), p. 3. • **Page 117** 'What a pity you hadn't invented it before': quoted in Wood (2002), p. 123. • **Page 118** 'exact imitations of our nerves, arteries and veins': Villiers (2001), p. 130. • 'Since our gods and our hopes are no longer anything but *scientific*': ibid., p. 164. • 'Dear friend, don't you recognize me?': ibid., p. 192. • **Page 119** 'It seems to me that I've come into the world of Flamel': ibid., p. 62. • 'Every man bears the name of Prometheus without knowing it': ibid., p. 67. • '"Tis he! . . . Ah! said I, opening my eyes wide in the dark': ibid., p. 8. • **Page 120** 'There exist true Vaucansons in this province': I. Bloch (1908), *The Sexual Life of Our Time in its Relations to Modern Civilization*, transl. M. E. Paul, pp. 648–9. Rebman, London. • 'the love heroine of which is such an artificial doll': ibid., p. 649. • 'Picture to yourselves the most exquisite figure': Hoffmann (1814). • **Page 121** 'Vaucanson . . . and all that crowd were barely competent makers of scarecrows': Villiers (2001), p. 61. • **Page 122** 'video recordings generated this way gave a so realistic reproduction': N. Negroponte, quoted in Negrotti (2002), p. 7. • 'is that class of the frightening which leads back': Freud (1919/2003). • 'doubts whether an apparently animate being is really alive': E. Jentsch, quoted in ibid. • 'The child twists and turns his toy, he scratches it': C. Baudelaire (1853/1994), 'The philosophy of toys', in *Essays on Dolls*, p. 24. Penguin, London.

Chapter 6: Protoplasmic Beings

Page 123 'We must not read into [living organisms] either a chemical retort': quoted in Rostand (1959), pp. 29–30. • 'You would seek me with less fervour': ibid., p. 27. • 'the event on which this fiction is founded has been supposed': in Shelley (1818/1831/1994), p. 11. • **Page 125** 'The life of the whole (animal or vegetable) would seem to be': G. L. Leclerc, Comte de Buffon (1829–33), in *Oeuvres complètes de Buffon*, Vol. II, p. 220. Paris. • 'Nothing at first, then a living point': Diderot (1769). • **Page 126** 'Man splitting himself up into an infinity of atomized men': ibid. • 'A warm room, lined with small container cups': Diderot (1769). • 'If there's a place where the human being divides': ibid. • **Page 127** 'This *power to live* belongs not to the constituent parts': J. J. Berzelius (1812), 'Öfversigt af djurkemiens framsteg och närvarande tillstånd', published in English as 'A view of the progress and present state of animal chemistry',

London, 1818. In Teich (ed.) (1992), p. 445. • 'We look for the cause of animal phenomena in a transcendental substrate': J. C. Reil (1796), 'Von der Lebenskraft', *Archiv für die Physiologie* 1, 8. In Teich (ed.) (1992), p. 439. • **Page 128** 'pulpy, homogeneous, gelatinous substance': Dujardin, quoted in Hall (1969), Vol. 2, p. 177. • 'self-moving wheel': F. Unger (1852), *Botanische Briefe*, p. 151. Vienna. Parts transl. by B. Paul (1854) as 'Botanical letters to a friend', *Quarterly Journal of Microscopical Science* 2, 123–6. • 'one kind of matter which is common to all living things': Huxley (1869). • 'is the clay of the potter: which, bake it and paint it': ibid. • 'I can find no intelligible ground for refusing to say that the properties': ibid. • **Page 129** 'if the cornea of the eye and the enamel of the teeth': J. H. Stirling (1869), *As Regards Protoplasm in Relation to Professor Huxley's Essay on the Physical Basis of Life*. Edinburgh. Quoted in Hall (1969), Vol. 2, p. 310. • 'Probably the most popular theory about the nature of life': quoted in G. Geison (1969), 'The protoplasmic theory of life and the vitalist-mechanist debate', *Isis* 60, 273–92, here p. 273. • **Page 130** 'It can be said that a chemist in his laboratory': C. Bernard (1878/1974), *Phenomena of Life Common to Animals and Vegetables*, transl. R. P. Cook & M. A. Cook, p. 96. Dundee. • **Page 131** 'Determinism is the only scientific philosophy possible': ibid., p. 174. • 'a collective term to designate the substances that constitute': Wilson (1923), p. 2. • 'the more critically we study the question': ibid, p. 6. • 'we are forever conjuring with the word': ibid., p. 30. • **Page 132** 'If on the one hand the morphologist strives to elucidate': F. Hofmeister (1901), 'Die chemische Organisation der Zelle', *Naturwissenschaftliche Rundschau* 16, 581, here p. 614. Transl. in Teich (ed.) (1992) pp. 504–6. • **Page 133** 'It is clear that a special feature of the living cell': F. G. Hopkins (1914), 'The dynamic side of biochemistry', *Report on the 83rd Meeting of the British Association* 652, here p. 664. • **Page 134** 'I am the last person to defend the view that these drops': O. Bütschli, in E. R. Lankester (1890), 'Professor Bütschli's experimental imitation of protoplasmic movement', *Quarterly Journal of Microscopical Science*, s2–31, 99–104, here p. 103. Available at http://jcs.biologists.org/cgi/reprint/s2-31/121/99.pdf. • **Page 135** 'when the Earth was still wholly or partially': E. Pflüger (1875), 'Beiträge zur Lehre von der Respiration. I. Ueber die physiologie Verbrennung in den lebendigen Organismen', *Pflügers Archive* 10, 251, here p. 340. Transl. in Teich (ed.) (1992). • 'It seems well within the range of scientific explanation': H. Williams (1900), 'Some unsolved scientific problems', *Harper's Magazine* 100, 774–83, here p. 774. • 'Now, when ultra-violet light acts on a mixture of water, carbon dioxide, and ammonia': J. B. S. Haldane (1929), 'The origin of life', *Rationalist Annual* 1929, 3, here p. 7. • **Page 136** 'material which had formerly been structureless': A. I. Oparin (1924), *The Origin of Life*, p. 229. Moscow, reprinted in English in 1938, Macmillan, New York. • **Page 137** 'Take a drop of sea water . . . that contains some nitrogenous material': L. Pasteur (1864/192–39), *Oeuvre* *de Pasteur*, Vol. II, 328–46. Masson et Cie, Paris. Transl. G. Geison (1995), *The Private Science of Louis Pasteur*, p. 111. Princeton University Press, Princeton. • 'I do not think I shall behold the synthesis of anything so nearly alive': Haldane (1929), p. 10. • 'Creation of Life. Startling Discovery of Prof. Loeb.': Anon. (1899), *Boston Herald*, 26 November, p. 17. • **Page 138** 'The development of the unfertilized egg, that is an assured fact': quoted in Pauly (1987), p. 101. • 'The invention of artificial parthenogenesis represented an attack': ibid., p. 94. • 'chemical citizens, the son of Madame Sea-Urchin': quoted in Rostand (1959), p. 11. • **Page 139** 'alas, the professor, so learned in biology and chemistry': W. A. Evans (1912), 'Prof Loeb has fatherless frog', *Chicago Daily Tribune*, 25 September, p. 2. • 'the shameful bondage of needing a man to become a mother': quoted in Pauly (1987), p. 101. • **Page 140** 'We must either succeed in producing living matter artificially': Loeb (1912), pp. 5–6. • 'The idea is now hovering before me that man himself': Loeb (1890), letter to E. Mach, quoted in Pauly (1987), pp. 4, 51. • 'We eat, drink and reproduce not because mankind': Loeb (1912), p. 31. • **Page 141** 'I wanted to take life in my hands and play with it': quoted in Pauly (1987), p. 102. • 'premised the necessity of a soul where she should have predicated': ibid., p. 103. • 'the tanks full of prisoners kept in running water': G. P. Serviss (1905), 'Artificial creation of life', *Cosmopolitan* 39, 459. • **Page 142** 'The combination of these elements into a colloidal compound': E. A. Schäfer (1912), *Nature* 90, 7–19, here p. 10. • 'the most that we can look for is no more than to create': Anon. (1912), *Lancet*, 7 September, p. 703. • 'it is some consolation to think that the life-producing scientist': quoted in Anon. (1912), *Public Opinion* 102, 241. • 'we should be able to turn out living germs of our own creation': Anon. (1912), 'Origin of life', *Manchester Guardian*, 11 September. • **Page 143** 'In time, this *thing* might, by industrious coaxing, be induced to become an *animal*': Anon. (1912), 'More life', *Daily Mirror*, 6 September. • 'This idea of artificially creating life . . . may lead to the manufacture': Anon. (1912), 'Life as per recipe', *Daily Herald*, 6 September. • 'The life creators will join with the eugenicists': ibid. • 'The Mexican consul in Trieste reports that Prof. Herrera': Anon. (1910), *New York Times*, 4 October. Quoted in Turney (1998), p. 69. • **Page 144** 'The general public is prone to believe that to have caused': Rostand (1959), p. 24. • 'We merely plagiarize nature, and our plagiarism has not the perfection': quoted in ibid., p. 24. • 'a privileged value to the living thing': ibid., p. 27. • 'As the result of the alteration in physical conceptions': Whyte (1927), p. 83. • 'Could an infinitely wise physicist order the necessary chemicals': ibid., p. 47. • **Page 145** 'it is *shortage of time* that our ambitious scientist is up against': ibid., p. 51. • 'Only an International Institute of Evolutionary Research': ibid, p. 52. • 'Life is a chemical reaction; death is the cessation of that reaction': F. Stockridge (1912), 'Creating life in the laboratory', *Cosmopolitan* 52, 774–81.

Chapter 7: Animating Clay

Page 146 'Living beings are always imbued with magic': Jacob (1989), p. 36.
• 'Making artificial people is an industrial secret': Čapek (1921). • **Page 147** 'armchair inventor': Daston & Park (1998), p. 95. • **Page 148** 'When the sages say, whoever has no sons is like a dead man': quoted in Scholem (1965), p. 194. • 'are forms and seals [that] collect the supernal and spiritual': quoted in Idel (1990), p. 169. • 'After saying certain prayers and holding certain fast days': quoted in Scholem (1965), p. 200. • **Page 149** 'we may describe the Golem practices as an attempt of man': Idel (1990), p. xxvii. • 'We have lived too long under the terror of the matchless perfection': B. Schulz (1934/1977), *The Street of Crocodiles*, transl. C. Wieniewska, pp. 60–1. Penguin, New York. • 'The Golem did not disappear and even in the time': quoted in Idel (1990), p. 256. • **Page 150** 'When this is done the golem collapses and dissolves': quoted in Scholem (1965), pp. 200–1. • 'to give him a real soul, *neshamah*, is not in the power of man': quoted in ibid., p. 196. • **Page 151** 'to give birth to a man without [using] the male semen': quoted in Idel (1990), p. 171. • 'If it were not rash to draw a modern comparison': Newman (2004), p. 187. • **Page 153** 'As little did his scheme partake of the enthusiasm of some natural philosophers': Melville (1856/1986); see http://www.melville.org/bell-towr.htm. • 'With him, common sense was theurgy; machinery, miracle': ibid. • 'supplying nothing less than a supplement to the Six Days' Work': ibid. • 'So the creator was killed by his creature': ibid. • **Page 155** 'Well, this is what happened. It was in 1920': this and all subsequent quotes from *R.U.R.* are taken from http://ebooks.adelaide.edu.au/c/capek/karel/rur/. • **Page 163** 'I wished to write a comedy, partly of science, partly of truth': quoted in Klíma (2002) p. 78. • 'I ask whether it is not possible to see in the present societal conflict': quoted in ibid., p. 79.

Chapter 8: The Bottled Babies

Page 165 'Developments in this direction are tending to bring mankind': Haldane (1923), in Dronamraju (ed.) (1995) p. 29. • 'A squat grey building of only thirty-four storeys': Huxley (1983), p. 1. • **Page 166** 'is perfectly right, and Mr Huxley has included nothing in this book': J. Needham, review of *Brave New World* in *Scrutiny*, May 1932, I, pp. 76–9, here p. 77. • **Page 167** 'An impersonal generation will take the place of Nature's hideous system': Huxley (1921), p. 27. • **Page 168** 'Dear Grandpater have you seen a Water-baby?': J. Huxley (1970), *Memories*, p. 18. Penguin, London. • 'I never could make sure about that Water Baby': ibid., p. 20. • **Page 170** 'We can take an ovary from a woman': Haldane (1924), p. 64. • **Page 171** 'leaven the population with fifty thousand irresponsible': Smith (1930), p. 15. • 'the exact human counterpart of the worker bee': ibid., p. 16. • **Page 172** 'Consciousness itself might end or vanish in a humanity': Bernal (1929), p. 47. • 'distaste and hatred which mechanization has already': ibid., p. 58. • 'Scientific corporations might well become almost independent states': ibid., p. 78. • 'a human zoo, a zoo so intelligently managed': ibid., p. 79. • 'One would not dream of buying cattle without thoroughly examining them': quoted in Rostand (1959), p. 88. • **Page 173** 'If we want to prevent the human race from degenerating': quoted in ibid. • 'at the yearly output by unfit parents of weakly children': F. Galton (1869/1892), *Hereditary Genius*, p. xx. Macmillan, London. • 'the present miserably low standard of the human race to one in which': ibid., p. xxvii. • 'If a twentieth part of the cost and pains were spent in measures': ibid., quoted in Gosden (1999), p. 61. See also F. Galton (1865), 'Hereditary talent and character', *Macmillan's Magazine*, June, pp. 165–6. • **Page 174** 'the hindrance of marriages and the production of offspring': ibid. • 'criminals, idiots and imbeciles': quoted in J. S. Smith (1914), 'Marriage, sterilization and commitment laws aimed at decreasing mental deficiency', *Journal of the American Institute of Criminal Law and Criminology* 5 (September), 367. • 'Had it not been for ectogenesis there can be little doubt': Haldane (1923), in Dronamraju (ed.) (1995), p. 42. • 'only form of salvation which science and ingenuity can suggest': quoted in Ferreira (2009), p. 45. • **Page 175** 'a hopelessly perverted movement': Muller (1935), pp. 10–11. • 'The Nazi racial theory is a mere rationalization': Huxley (1942), p. 50. • 'Until we equalize nutrition, or at least nutritional opportunity': ibid., p. 47. • 'Once the full implications of evolutionary biology are grasped': ibid., p. 34. • 'The upper economic classes are presumably slightly better endowed': ibid., p. 66. • **Page 176** 'control of the human germ-plasm': ibid., p. 69. • 'the prevailing individualist attitude to marriage': ibid., p. 77. • 'the implications of the recent advances in science and technique': ibid. • 'females possessing characters particularly excellent': quoted in Ferreira (2009), p. 35. • **Page 178** 'There is no sexual reformer but must wish that woman': Paul (1930), p. 35. • 'The man of creative genius is more valuable to society': Smith (1930), p. 164. • 'if [this] mode of creation were to be extrapolated to humans': N. L. Jones (2004), 'Three mothers make a baby: is that sex? Yes, or maybe?', paper for the Center for Bioethics and Human Dignity, Deerfield, Illinois. Available at http://www.cbhd.org/content/three-mothers-make-a-baby-sex-yes-or-maybe. • 'their vanity in every kind of sterile pursuit that gave them': Ludovici (1924), p. 81. • 'When once artificial impregnation is an every-day occurrence': ibid., p. 90. • **Page 179** 'the fertilized ovum [to] be matured outside the female body': ibid., p. 91. • 'enlisting the cow or the ass into our service': ibid., p. 92. • 'will mature very much as chickens now do in incubators': ibid., p. 93. • 'intensified bovinity or asisinity': ibid., p. 92. • 'public centres will be provided where the Borough Council': ibid. • 'and in a very short while it will be a mere matter of routine':

ibid., p. 95. • 'The popular press will say that men are redundant': quoted in H. Pearson (2004), 'Mouse created without father', Nature Science Update, 21 April. Available at http://cmbi.bjmu.edu.cn/news/0404/82.htm. • 'The end of men?': E. Cook (2009), 'The end of men? Scientists create sperm in the lab out of stem cells', *Daily Mirror*, 8 July. Available at http://www.mirror.co.uk/news/top-stories/2009/07/08/the-end-of-men-115875-21503346/. • 'on the brink of a society without any need for men': M. Hanlon (2009), 'Are we on the brink of a society without any need for men?', *Daily Mail*, 8 July. Available at http://www.dailymail.co.uk/debate/article-1198202/MICHAEL-HANLON-Are-brink-society-need-men.html. • **Page 180** 'I submit, then, that the first demand for any alternative system': Firestone (1970), p. 233. • 'shitting a pumpkin': ibid., p. 189. • **Page 181** 'He lifted one of the black velvet curtains': S. W. Ellis (1930). • **Page 182** 'was used only in cases of grave need': Sargent (1976), p. 16. • **Page 183** 'If I have the time and money for experiments': quoted in Rosen (2003). • **Page 184** 'Something much more sophisticated than tubes and pumps will be needed': Gosden (1999), p. 202. • 'That's my final goal': quoted in Rosen (2003). • 'A strong desire to have a child of their own genes and hereditary features': quoted in ibid. • 'the introduction of ectogenesis might be the decisive event': Ferreira (2009), p. 51. • 'artificial wombs could lead to a commodification of the whole process': quoted in Rosen (2003). • **Page 185** 'at stake in this debate is the very meaning of human pregnancy': Rosen (2003). • 'It seems apparent that many women . . . would avail themselves': Rorvik (1971), p. 84. • 'What kind of child will we produce from a liquid medium inside a plastic box?': J. Rifkin (2002), 'The end of pregnancy', *Guardian*, 17 January. Available at http://www.common-dreams.org/views02/0117-05.htm. • 'How will the elimination of pregnancy affect the concept': ibid. • **Page 186** 'Other feminists view the artificial womb': ibid. • 'We may find ourselves without a product of any kind': quoted in Rosen (2003). • **Page 187** 'Why do we want the embryo to be housed in its mother?': C. Krauthammer, in President's Council on Bioethics, October 2003. Quoted in Rosen (2003). • **Page 188** 'has slowed to a crawl': Gosden (1999), p. 202. • **Page 189** 'enhance rather than diminish the value we place on the fetus': ibid., p. 200. • 'Even the phrase "artificial womb" appears at odds with itself': Rosen (2003). • 'There is something about being born of a human being': ibid.

Chapter 9: Opening the Bottle

Page 190 'Now, it appears to many, the sexual act . . . has been dragged out': G. Leach (1970), 'The test-tube fantasy', *Observer*, 1 March. • 'We are entering an age of human manufacture': C. Krauthammer, in President's Council on Bioethics transcript, 2 January 2002. • 'mass-producing people without the advent of a mother at all': quoted in P. Wright (1970), 'Progress towards "test-tube" baby', *The Times*, 24 February, p. 1. • 'We have the awful knowledge': W. Gaylin (1972), 'The Frankenstein myth becomes reality', *New York Times Magazine*, 5 March. • 'these are not perverted men in white coats doing nasty experiments': Anon. (1969), 'What comes after fertilization?', *Nature* 221, 613. • **Page 191** 'Huxley's prophetic vision had become reality': F. Golden (1978), 'The first test-tube baby', *Time*, 31 July, 46–52. • 'assisted reproduction continues to provoke anxiety': K. Valerius (1999), in Kaplan & Squier (eds) (1999), pp. 172–3. • **Page 192** 'destabilizes any simple equation of the biological': ibid., p. 183. • 'a kind of metonym, seed crystal or icon for configurations of person': D. J. Haraway (1997), 'The virtual speculum in the new world order', *Feminist Review* 55, 22–72. • 'It is not our custom to give the younger daughter in marriage': Genesis 29:26. • 'Give me children, or I'll die!': ibid., 30:1. • **Page 193** 'God has taken away my disgrace': ibid., 30:23. • 'Sterility happens like-wise, from the Womans Disgust': quoted in J. McMath (1694), *The Expert Midwife*, p. 3. Mosman, Edinburgh. • **Page 194** 'why does only one sperma-tozoon enter an egg?': Edwards & Steptoe (1980), p. 8. • 'ninety-six human beings grow where only one grew before': Anon. (1936), 'Brave New World', *New York Times*, 28 March, p. 14. • **Page 195** 'No father to guide them': J. D. Ratcliff (1937), 'No father to guide them', *Collier's Magazine*, 20 March, p. 137. • 'Man's value would shrink, the mythical land of the Amazons': ibid. • 'Not since Professor Jacques Loeb hatched fatherless sea-urchin larvae': Anon. (1936), 'Brave New World', *New York Times*, 28 March, p. 14. • **Page 196** 'seemed to have almost an obsession with human eggs': R. M. Henig, in PBS docu-mentary 'Test tube babies', broadcast on 'American Experience'. Transcript available at http://www.pbs.org/wgbh/americanexperience/features/transcript/babies-transcript/. • 'His manner was strange because he didn't look people in the eye': ibid. • 'I don't want any human eggs fertilized here': Edwards & Steptoe (1980), p. 45. • **Page 197** 'Sometimes, at night, I could hear below': ibid., p. 50. • 'births may be by proxy': quoted in ibid., p. 55. • **Page 198** 'Life is created in test tube': quoted in ibid., p. 88. • 'the scientists are now in a position not only to replace the fertilized egg': quoted in Anon. (1969), 'New hope for the childless', *Guardian*, 14 February, p. 1. • 'Life, or at least the potentiality of life, resides in the oocyte': A. Tucker (1969), 'Conception in the lab', *Guardian*, 15 February. • **Page 200** 'Orwell's baby farm': quoted in Gosden (1999), p. 26. • 'When new methods of human reproduction become available': *Life* (cover), 13 June 1969. • **Page 201** 'Test tube babies have no souls': quoted in A. Mitchison (2010). 'Getting the baby you want', *Guardian* Weekend, 3 April. • 'Radical new techniques in biology promise (or threaten) a near-future world': *Life*, 13 June 1969, p. 37. • **Page 202** 'breeding a race of intellectual giants': Anon. (1969), 'Life in the test-tube', *The Times*, 15 February, p. 9. • 'Ultimately we could have the know-how to breed': quoted in Edwards & Steptoe (1980), p. 89. • 'predestinators of Aldous Huxley's prophetic fiction': Rorvik (1971), p. 73. • **Page 203** 'When man assumes – some might say *presumes*': ibid. • 'Hafez . . . foresees the day, perhaps only ten or fifteen years hence': ibid., p. 32. • **Page 204** 'Ultimately, Dr Hafez believes, entire colonies of men': ibid., p. 33. • 'It is a technique that will be used not only to improve': ibid., p. 4. • 'mass-producing test-tube babies': quoted in Anon. (1969), 'Test-tube fertility hope for women', *The Times*, 15 February, p. 1. • 'because their imaginations have already been dramatically doom-lit': Edwards & Steptoe (1980), p. 13. • **Page 205** 'Ban the test tube baby': quoted in ibid., p. 108. • 'unethical medical experimentation on possible future human beings': P. Ramsey (1972), 'Shall we "reproduce"?', *Journal of the American Medical Association* 220, 1346–50 and 1480–5, here p. 1346. • 'The spirit of Frankenstein did not die with the Third Reich': quoted in Rorvik (1971), pp. 80–1. • **Page 206** 'Some people oppose this type of operation because they feel': quoted in Singer & Wells (1984), p. 31. • **Page 208** 'She was born at 11.47 pm with a lusty yell': P. Gwynne (1978), 'All about that baby', *Newsweek*, 7 August, p. 44. • 'How would you like to be the world's first test tube baby?': quoted in PBS documentary 'Test tube babies', broadcast on 'American Experience'. • 'When they say "test tube baby", everybody had the impression': quoted in ibid. • 'completely out of order': quoted in 'Vatican official criticises Nobel win for IVF pioneer', BBC news, 4 October 2010. Available at http://www.bbc.co.uk/news/health-11472753. • 'Techniques involving only the married couple': Catechism of the Catholic Church, Part 3, Section 2, Chapter 2, Article 6, 2377. Available at http://www.vatican.va/archive/catechism/p3s2c2a6.htm#III. • **Page 211** 'To reduce the shared life of a married couple': quoted in Singer & Wells (1984), p. 52. • **Page 212** 'In no other case does the Catholic Church espouse': ibid., p. 50. • **Page 213** 'To say that a child ought to spring from the sexual union': G. C. Meilaender (1997), 'Hello Dolly', *Dialog* 36(2), Summer. Quoted in Peters (2003), p. 169. • 'They could become complete egoists, knowing that': quoted in Singer & Wells (1984), p. 49. • 'Any attempt to set up an antinomy between natural and biologic': quoted in ibid., p. 41. • **Page 214** 'Those who welcome and those who fear the advent of "human engineering" ground their hopes': Kass (1971a), p. 779. • 'The complete depersonalization of procreation shall be, in itself': ibid., p. 784. • 'fosters and supports in parents an adequate concern for': ibid. • **Page 215** 'transfer of procreation to the laboratory will no doubt weaken': ibid. • 'without safeguards and serious study of safeguards': Hansard (Lords), Vol. 432, 9 July 1982, col. 1001. • **Page 216** 'it will be difficult to forestall dangerous present and future applications': L. Kass (1979), 'Ethical issues in human in vitro fertilization, embryo culture and research, and embryo transfer', testimony to the US Ethics Advisory Board, p. 32. In *HEW Support of Research Involving Human* In Vitro *Fertilization and Embryo Transfer*, 4 May.

Ethics Advisory Board, Department of Health, Education, and Welfare. • 'Once the genies let the babies into the bottle': ibid., p. 21. • 'this blind assertion of will against our bodily nature': Kass (1985), p. 114. • 'Is it a right to have your own biological child?': L. Kass, 'Making babies: the new biology and the "old" morality', in Kristol & Cohen (eds) (2002), pp. 55–6. Originally published in *The Public Interest*, Winter 1972. • **Page 217** 'Those who seek to submerge the distinction between *natural* and *unnatural* means': ibid., p. 56. • 'You can only go ahead with your work if you accept': quoted in Edwards & Steptoe (1980), pp. 124–5. • **Page 218** 'Even if only a single abnormal baby is born': quoted in ibid., p. 129. • 'It doesn't matter how many times the baby is tested': Kass (1971a). • **Page 219** 'Many, if they consider the matter at all, thought such a procedure': Edwards & Steptoe (1980), p. 5. • 'half-hoped that the first child born through such a method': ibid. • **Page 220** 'scientist-controlled, technological reproduction . . . necessitates interference from the state': P. Spallone (1989), *Beyond Conception: The New Politics of Reproduction*, p. 183. Bergin, Granby, MA. • 'the hunger of the infertile is ravenous, desperate': L. Zoloth, in Lauritzen (ed.) (2001), p. 139. • **Page 221** 'It is only when we recognize our anxieties': M. Ryan, in ibid., pp. 55–6. • 'the integrity of sexuality, reproduction, and parenthood': ibid., p. 66. • **Page 222** 'the moment when men will play God and be the sole procreator': R. D. Klein (1984), 'Taking the egg from the one and the uterus from the other', *Seeds of Change* 4, 92–7, here p. 94. • 'a constant source of experimental subjects for research': Klein (1990) [footnote 3]. • **Page 223** 'is not that they are defying nature, but that they are': van Dyck (1995), pp. 189–90. • 'carry it to a place where they may hatch it out': quoted in H. M. Pachter (1951), *Paracelsus: Magic Into Science*, p. 74. Henry Schuman, New York.

Chapter 10: Anthropoeia in the Marketplace

Page 224 'An embryo is not a little baby in the freezer': R. Deech, quoted in A. Ferriman (1996), '2,500 "orphan embryos" to be destroyed within days', *Independent on Sunday*, 7 July, p. 1. • 'As long as [assisted] conception remains a furtive trade': Spar (2006), p. 67. • 'infertility therapy has become big business': J. Palca (1988), 'US *in vitro* fertilization in limbo according to OTA', *Nature* 333, 388. • **Page 226** 'There were walls daubed with whitewash': quoted in Challoner (1999), p. 61. • 'None of the traditions involved in the debate could simply reapply': A. R. Chapman (1999), *Unprecedented Choices: Religious Ethics at the Frontiers of Genetic Science* pp. 123–4. Fortress, Minneapolis. • **Page 227** 'We have a situation where society as a whole': Lauritzen (ed.) (2001), p. 15. • 'It may seem strange that a federal government': C. Tauer, in ibid., p. 148. • **Page 228** 'a baby, guaranteed': quoted in ibid, p. 134. • **Page 229** 'one of the greatest causes of public disquiet': *Human Fertilization and Embryology:*

A Framework for Legislation, p. 7, note 56. HMSO, London, 1987. • 'Scientists are to be banned from creating superbeings': Anon. (1987), *Today*, 27 November. • 'the further we get away from nature': Hansard (Lords), 31 October 1984, col. 577. • 'I once watched a science-fiction film about a man': ibid., col. 576. • 'If we take the rectification of genetic defects to its logical conclusion': Hansard (Commons), 15 February 1985, col. 659. • **Page 230** 'debasement of human generation': Hansard (Lords), 31 October 1984, col. 566. • 'thirst for knowledge': Hansard (Commons), 4 February 1988, col. 1227. • 'I do not know when the soul enters the body': Hansard (Lords), 31 October 1984, col. 563. • **Page 231** 'such research can make an important contribution to preventing': ibid., 7 December 1989, col. 1057. • 'Wife who found joy backs embryo research': quoted in Mulkay (1997), p. 70. • 'Science-fiction imagery was almost entirely absent': ibid, p. 122 • **Page 232** 'any embryo researcher has tried to produce a monster': Anon. (1989), 'Embryonic journey', *New Scientist*, 2 December, p. 24. • 'overweening arrogance of our perverted science': *Daily Telegraph*, 26 April 1990. Quoted in Mulkay (1997), p. 77. • 'There is only one minor distinction between the life of a human being': *Evening Standard*, 9 February 1990. Quoted in ibid., p. 80. • 'No person shall bring about the creation of an embryo': Human Fertilisation and Embryology Act 1990, 3:1. Available at http://www.opsi.gov.uk/acts/acts1990/ukpga_19900037_en_1. • **Page 234** 'Throughout Christianity's two-thousand-year history': Pope John Paul II (1995), *Evangelium Vitae*, 25 March, para. 61. Available at http://www.vatican.va/holy_father/john_paul_ii/encyclicals/documents/hf_jp-ii_enc_25031995_evangelium-vitae_en.html. • 'Christians have a rather difficult time because it is no good': Hansard (Lords), 8 March 1989, col. 1553. • 'Did you not pour me out like milk': Job, 10:10–11. • **Page 235** 'it is not the fact that when an animal is formed': Aristotle, *On the Generation of Animals*, transl. A. L. Peck (1942, reprinted 1979), 2.3, 736b, 1–5. Harvard University Press, Cambridge, MA. • **Page 236** 'unformed like water': quoted in Jones (2004), p. 67. • 'if the fetus is already formed and animated': quoted in G. R. Dunstan & M. Sellars (eds) (1988), *The Status of the Human Embryo: Perspectives from Moral Tradition*, p. 47. King Edward's Hospital Fund, London. • 'all living things begin as an egg': quoted in Jones (2004), p. 161. • **Page 237** 'Christian compassion must embrace *both* mother and child': ibid., p. 211. • 'Although the questions of when life or personhood begin': Warnock et al. (1984), para. 11.9. • **Page 238** 'Unless we can give a convincing account of "special respect"': in Lauritzen (ed.) (2001), p. 28. • 'serious moral consideration': National Institutes of Health (1994), *Report of the Human Embryo Research Panel*. Bethesda, MD. Available at http://www.bioethics.gov/commissions/. Quoted in Laurtizen (ed.) (2001), p. 35. • 'a moral calculus [was] to be constructed to do the necessary': D. Callahan (1995), 'The puzzle of profound respect', *Hastings Center Report* 25, No. 1, p. 40. • 'thinking, intelligent being, that has reason and reflection': J. Locke (1690), *An Essay Concerning Human Understanding*, Book II, Chapter 27, para. 9. Available at http://oregonstate.edu/instruct/phl302/ texts/locke/locke1/Book2c.html#Chapter%20XXVI. • **Page 239** 'An embryo is not deprived of its life by being killed': in Lauritzen (ed.) (2001), p. 24. • **Page 240** 'A fertilized egg may try to develop in the [Fallopian] tubes': Hansard (Commons), 23 April 1990, col. 92. • **Page 241** 'Biology is itself too surprising to be really amusing material': C. Haldane (1932), '*Brave New World*: a novel', *Nature* **129**, 597–8. • **Page 243** 'while the egg and sperm are alive as cells': quoted in Silver (1997), p. 49. • 'Recent findings of human biological science . . . recognize that in the zygote': quoted in ibid. • 'modern genetic science has demonstrated that, from the first instant': ibid. • **Page 244** 'a thing either is or is not a whole human being': 'Statement of Professor George (joined by Dr Gomez-Lobo)', in L. Kass (ed.) (2002), *Human Cloning and Human Dignity: The Report of the President's Council on Bioethics*, pp. 294–306. PublicAffairs, New York. • 'awaken[ing] a revulsion that is not naturally present': Cohen (2008), p. 74. • 'life and soul are irreducibly myste-rious': Kass (2002), *Life, Liberty, and the Defense of Dignity: The Challenge for Bioethics*, p. 296. Encounter, San Francisco. • **Page 245** 'The scientific quest to understand the inner workings of life': C. Campbell, in Lauritzen (ed.) (2001), p. 45. • 'the fundamental wonder of life itself': ibid., p. 46. • 'the soul has been understood first and foremost': Jones (2004), p. 248. • **Page 246** 'One ounce of fact more or less will have more weight': D. MacDougall (1907), 'Hypothesis concerning soul substance together with experimental evidence of the existence of such substance', *Journal of the American Society for Psychical Research* 1(5), 237–75. Available at http://www.scribd.com/doc/20281719/ 21-Grams-Hypothesis-Concerning-Soul-Substance-Together-with-Experimental-Evidence-of-The-Existence-of-Such-Substance. • 'at the first breath taken after birth': quoted in Silver (2006), p. 9. • 'is not introduced into the body after birth': Lactantius, *On the Workmanship of God, or the Formation of Man*, 17, transl. W. Fletcher, in A. Roberts & J. Donaldson (eds) (1885), *The Fathers of the Third and Fourth Centuries*, Vol. VII. Eerdmans, Grand Rapids, MI. • **Page 247** 'when the little body is prepared': M. Luther, *Sermon on the Day of the Conception of the Mother*. Quoted in Jones (2004), p. 143 • 'there is really nothing *possible* for man': K. Rahner (1968), 'Experiment: Man', *Theology Digest* 16 (February), 60. • 'an element within the totality of man which can be encountered': K. Rahner (1978), *Foundations of Christian Faith: An Introduction to the Idea of Christianity*, transl. W. V. Dych, p. 30. Seabury Press, New York. • **Page 248** 'Theologians need only be able to affirm that human consciousness': K. Rahner (1988), 'Natural science and reasonable faith', in K. Rahner, *Theological Investigations*, Vol. 21, transl. H. M. Riley, p. 43. Crossroad, New York. • 'the human person in the world': quoted in Klein (2008), p. 6. • 'completely rejects the notion of a disembodied, unworlded soul': Klein (2008), p. 7. • 'As the new body is generated, God gives it a new soul': Jones (2004), p. 227. • **Page 249** 'The problem with twinning seems less our inability to tell': ibid. • **Page 250** 'This is not an easy question': ibid., p. 228. • 'a *human* life, preparing and calling for a soul': quoted in Boyd et al. (1986), p. 94. • **Page 252** 'Because no one wants to define baby-making as a business': Spar (2006), p. xi. • 'We need to view reproductive medicine as an industry': ibid., p. xv. • 'We need to give serious thought to whether': L. Andrews (1998), 'Human cloning: assessing the ethical and legal quandaries', *Chronicle of Higher Education* 44(23) February, p. B4–5. • 'It's no use being coy about the baby market': Spar (2006), p. 233. • **Page 253** 'which is ideological rather than technical': van Dyck (1995), p. 11. • 'The creation of need is an intricate process of image-building': ibid. • 'the use of reproductive technologies reinforces the ideal': ibid., p. 25. • **Page 254** 'becomes incorporated in the myth of natural reproduction': ibid., p. 135. • 'The beauty and power of IVF are that it allows doctors': ibid. • 'a dark and dangerous place, a hazardous environment': J. Fletcher (1974), *The Ethics of Genetic Control: Ending Reproduction Roulette*, p. 103. Anchor Books, Garden City, NY. • **Page 255** 'The ethics boat the president has launched is stacked': A. L. Caplan (2002), 'A council of clones'. Available at http://bioethics.net/articles.php?view Cat=2&articleId=35. • 'sickness in theocon bioethics goes beyond imposing a Catholic agenda': S. Pinker (2008), 'The stupidity of dignity', *New Republic*, 28 May. Available at http://www.tnr.com/article/thestupidity-dignity. • **Page 256** 'Men ought not to play God before they learn': Ramsey (1970), p. 138. • 'The phrase "playing God" has very little cognitive value': Peters (2003), p. 2. • **Page 257** 'although the phrase "playing God" is foreign to theologians': ibid., p. 13. • 'with certain important driving forces behind the growth': L. Kass, in President's Council on Bioethics transcript, 2 January 2002. • 'I do not think the sign of the birthmark is superficial': ibid. • **Page 259** 'atemporal index of human repulsion to the hubris': Newman (2004), p. 3.

Chapter 11: Make Me Another

Page 260 'In the not-too-distant future, it will be looked at': G. Stock (2000), *New York Times*, 1 January. Quoted in Lauritzen (ed.) (2001), p. 132. • 'After fielding numerous inquiries about the possible cloning': available at http://www.clonaid.com/page.php?18 • 'You said thing, these things': C. Churchill (2002), *A Number*, p. 11. Nick Hern, London. • **Page 262** 'lately research worker at Middlesex Hospital': J. Huxley, 'The Tissue-Culture King', *Yale Review*, April 1926. Available at http://www.revolutionsf.com/fiction/tissue/index.html. • 'It is the merest cant and twaddle to go on asserting': ibid. • **Page 263** 'even in mammals': J. Huxley (1922), 'Searching for the elixir of life', *Century Illustrated Monthly Magazine*, February, p. 625. • 'he could by taking various parts': quoted in Turney (1998), p. 76. • 'incorporated into a myth of a collective scientific endeavour': Turney (1998), p. 77. • 'Could you love a chemical baby?': Strangeways Archives, Spears Archive, Box 5, Wellcome Trust. Quoted in Klotzko (ed.) (2001), p. 85. • **Page 264** 'Living cells have a unique capacity to self-assemble': V. Mironov (2007), 'Tissue engineering and bioprinting', available at http://visualcultureandbioscience. blogspot.com/2007/03/mironov-tissue-engineering-and-bio.html. • 'if we are able to bioprint all human organs': Mironov (2006), 'On art and science: bioprinting and Pygmalion's dream?', available at http://www.musc.edu/bio printing/html/bioprinting_art.html. • **Page 265** 'technology is probably not the biggest problem': ibid. • **Page 266** 'If a biologist takes a tiny fragment of tissue': Rostand (1959), pp. 13–14. • **Page 270** 'every speck of flesh [is] charged with seminal virtue': ibid., p. 13. • **Page 271** 'A clone again, naturally': quoted in Wilmut, Campbell & Tudge (2000), p. 229. • **Page 272** 'is a matter of morality': W. J. Clinton, White House press conference, 4 March 1997. Quoted in, e.g., http://edition.cnn.com/TECH/9703/04/clinton.cloning/. • 'a serious evil': L. Kass (2001), 'Preventing a Brave New World: why we should ban human cloning now', *New Republic*, 17 May, pp. 30–9. • 'One doesn't expect Dr Frankenstein to show up': quoted in Wilmut, Campbell & Tudge (2000), p. 246. • 'Science on the way to cloned people': quoted in ibid. • 'Dolly the cloned sheep kills a lamb': quoted in ibid. • 'In the popular media, nightmare scenarios of laboratory mistakes': D. Brock, in Lauritzen (ed.) (2001), p. 94. • 'quickly came to dread the pleas from bereaved families': Wilmut, Campbell & Tudge (2000), p. 16. • **Page 274** 'Frankensteinian hubris': Kass (1997). • **Page 275** 'fetuses produced by nuclear transfer are ten times more likely': Wilmut, Campbell & Tudge (2000), p. 326. • **Page 276** 'in case a monstrosity – a subhuman or parahuman individual': Ramsey (1970), p. 78. • 'that human cloning would result in a person's worth or value': A. Charo, in Lauritzen (ed.) (2001), p. 110. • **Page 277** 'Because we expect cloned children to be odd': Pence (2004), p. 51. • 'virtually no parent is going to be able to treat a clone': quoted in ibid., p. 47. • **Page 279** 'an eccentric from central casting': ibid., p. xi. • **Page 280** 'in America, cloning may be bad but telling people': quoted in Lauritzen (ed.) (2001), p. 193. • 'to order up carbon copies of people': J. F. Bonner (1968), *Evening Bulletin* (Philadelphia), 23 February. Quoted in Ramsey (1970), p. 65. • 'If a superior individual . . . is identified': J. Lederberg (1996), 'Experimental genetics and human evolution', *Bulletin of the Atomic Scientists*, October, p. 8. • **Page 281** 'In our current circumstances, the absence of a loved one': A. Rosenfeld (1969), *The Second Genesis*, pp. 185– 6. Prentice-Hall, Englewood Cliffs, NJ. • 'That's alright, I have another one upstairs': quoted in Klotzko (ed.) (2001), p. 86. • 'very different in size and temperament': Wilmut, Campbell & Tudge (2000), p. 17. • **Page 282** 'It is a basic principle of developmental biology': R. Lewontin (1997), 'The confusion over cloning', *New York Review of Books*, October, pp. 18–23. • **Page 283** 'Twins that become twins separated by years or decades': A. L. Caplan, quoted in G. Kolata (1993), 'The hot debate about cloning human embryos', *New York Times*, 26 October. Available at http://www.nytimes.com/1993/ 10/26/health/the-hot-debate-about-cloning-human-embryo.html. • 'Listen Domin, we made a crucial mistake': Čapek (1921). • 'originally an insurance against the destruction of the ego': Freud (1919/2003). • **Page 284** 'whatever reminds us of this inner "compulsion to repeat" is perceived': ibid. • 'We used to think our fate was in the stars': J. D. Watson, quoted in L. Jaroff, J. M. Nash & D. Thompson (1989), 'The gene hunt', *Time*, 20 March. Available at http://www.time.com/time/magazine/article/0,9171,957263-1,00.html. • 'it will one day be possible to predict from an embryo's genome': L. Wolpert (2009), in 'What can DNA tell us? Place your bets now', *New Scientist*, 8 July. Available at http://www.newscientist.com/article/mg20327161.100-is-dna-predictive-you-can-bet-on-it.html. • 'criminality, shyness, arson, directional ability, exhibitionism': D. Nelkin & M. S. Lindee, in Klotzko (ed.) (2001), p. 84. • 'in an unacceptable position of control': quoted in Challoner (1999), p. 148. • **Page 285** 'learn to have the stranger, not the copy': L. Zoloth, in Lauritzen (ed.) (2001), p. 140. • 'The whole point of "making babies" is not the production': ibid., p. 140. • **Page 286** 'The need for international agreements and supervision': Kass (1971a), p. 787. • 'you have the means . . . of mass producing people': quoted in P. Wright (1970), 'Progress towards "test-tube" baby', *The Times*, 24 February, p. 1. • 'the potentialities for misuse by an inhumane totalitarian government': J. D. Watson, in Kristol & Cohen (eds) (2002), p. 48. • 'manipulation means test-tube babies': quoted in M. Bateman, 'East, West, which manipulators work best?', *Sunday Times*, 1 March 1969. • **Page 290** 'the extreme limit of plasticity in a living shape': Wells (1995), p. 133. • 'animals half-wrought into the outward image of human souls': ibid., p. 177. • 'The developing embryo is a human embryo inside a cow egg': J. Rifkin, quoted in M. Fox (1998), 'Cloned cells – Frankenstein or savior of humanity?', Reuters news report, 12 November. • 'in crossing the species barrier, the general understanding': Science and Technology Committee Fifth Report (2007), 3:32. Available at http://www.publications. parliament.uk/pa/cm200607/cmselect/cmsctech/ 272/27206.htm#note80. • **Page 291** 'at some point in the future, science and society will move': Silver (2006), p. 187. • 'Aldous Huxley's fertilizing and decanting rooms': Ramsey (1970), p. 104. • 'around the world a modest number of children': R. Green, in Lauritzen (ed.) (2001), p. 114. • 'cloning will have come to be looked on as just one': ibid. • **Page 293** 'The Church burnt Giordano Bruno at the stake': quoted in F. Rocco (1993), 'But how old, how safe, and how much?', *Independent*, 25 July. Available at http://www.independent.co.uk/ news/uk/but-how-old-how-safe-and-how-much-fiammetta-rocco-talks-to-the-italian-obstetrician-dr-severino-antinori-1487011.html. • 'There is absolutely no doubt about it': quoted in A. Zitner & S. Simon (2001), 'Fringe cloning venture raises troubling issues', *Los Angeles Times*, 22 April. Available at http://www.newsmakingnews.com/contents4,21,01.htm#Fringe%20Cloning%20Venture%20Raises%20Troubling%20Issues. • 'wallowing in a mix of publicity': quoted in ibid. • 'We are not regular scientists': quoted in ibid. • **Page 294** 'has received cloning requests from around the world': Clonaid press release, 8 July 2009. Available at http://www.clonaid.com/page.php?18. • **Page 295** 'If it works, every narcissist alive will be trying to use it': Sargent (1976/1981), p. 29. • 'The reporters crowded together on the other side of the glass': ibid., p. 73. • 'It is not mere sensationalism': quoted in ibid., p. 75. • **Page 296** 'notoriously sympathetic, easily able to interpret one another's minimal gestures': quoted in Ramsey (1970), p. 71. • **Page 297** 'Just as people were beginning to understand cloning': A. L. Caplan (2004), '*Godsend* no blessing for cloning research', on MSNBC.com, 30 March. Available at http://www.msnbc.msn.com/id/4865240/. • **Page 298** 'who first suggested this horrid idea': Mitchison (1975/1995), p. 4. • **Page 299** 'his empathy must be very well expanded and controlled': ibid., p. 158. • 'showed your love for people': ibid., p. 74. • **Page 300** 'the feminist temptation to use reproductive technology': S. M. Squier, in ibid., p. 162. • 'prohibit all forms of human cloning' (and following): United Nations Declaration on Human Cloning, available at http://www.un.org/law/cloning/. • **Page 301** 'It shall be unlawful for any person or entity, public or private': Human Cloning Prihibition Act, 2001, Sec. 302. Available at http://usgovinfo.about.com/library/bills/blhr2505.htm. • 'any intervention seeking to create a human being': Council of Europe 'Additional Protocol' to the Convention for the Protection of Human Rights and Dignity of the Human Being with regard to the Application of Biology and Medicine, on the Prohibition of Cloning Human Beings, ETS No. 168, 12 January 1998. Available at http://conventions.coe.int/Treaty/EN/Treaties/Html/168.htm.

Chapter 12: Assembly Instructions Not Included

Page 303 'What is life? I am not going to answer that question': Haldane (1949), p. 58. • 'The human genome is literally a sort of digital text': M. Ridley (2000), 'Life, the universe, and the little things', *Daily Telegraph*, 27 June, p. 26. • **Page 304** 'the very blueprint of life': W. J. Clinton, State of the Union Address, 27 January 2000. Available at http://www.insidepolitics.org/speeches/clinton00.htm. • 'the set of instructions to make a human being': J. Sulston, interview with the BBC, 26 June 2000, quoted on http://news.bbc.co.uk/1/hi/sci/tech/805803.stm. • 'the information provided by the human genome': Nerlich & Dingwall (2003), p. 15. • **Page 305** 'To be impatient with the biochemists': H. G. Wells, G. Wells & J. Huxley (1930), *The Science of Life*, Vol. 2, pp. 434–5. Amalgamated Press. • 'their work heralds the day when man': T. Margerison (1962), 'Architects of life', *Sunday Times* Magazine, 9 December. • **Page 306** 'Frankly, scientists do not know enough about biology': J. J. Collins (2010), 'Life after the synthetic cell', *Nature* **465**, 422–4, here p. 424. • **Page 307** 'tiny speck of material, the nucleus of the fertilised egg': Schrödinger (1944/2000), p. 74. • **Page 308** 'a well-ordered association of atoms': ibid. • 'we're going to see within the next generation': quoted in L. Galton (1962), 'Science stands at awesome thresholds', *New York Times*, 2 December, p. vi. • 'the synthesis of life is now within reach': quoted in Rorvik (1971), p. 89. • **Page 309** 'Ultimately, we may be able to fashion living species to order': Anon. (1963), 'Probing heredity's secrets', *New York Times*, 12 September, p. 36. • **Page 311** 'the language in which God created life': W. J. Clinton, White House press conference on the publication of the first draft of the human genome, 26 June 2000. Available at http://www.ornl. gov/sci/techresources/Human_Genome/project/clinton2.shtml. • 'Philosophy is written in a great book which is always open': Galileo (1623/1890–1909), *IL Saggiatore*, in *Opere*, Vol. IV, p. 232. Florence. Quoted in Jacob (1989), p. 29. • 'What more powerful form of study of mankind could there be': F. Collins, at White House press conference, 26 June 2000. Available at http://www.ornl.gov/sci/techresources/Human_Genome/project/clinton2. shtml. • **Page 314** 'What an organism feeds upon is negative entropy': Schrödinger (1944/2000), p. 88.

Epilogue: Unnatural Creations

Page 316 'It is advisable to sift the merits of knowledge': F. Bacon (1605/1620/1944), p. 3. • 'Whether mythic or scientific, the view of the world': F. Jacob (1982), 'Myth and Science', in Jacob (1989), p. 362. • **Page 318** 'Frankenstein fuels': M. Lynas (2006), 'Frankenstein fuels', *New Statesman*, 7 August. • **Page 319** 'The temptation to create human life artificially goes back': P. Greenberg (2008), 'Dr Frankenstein plods on', *Jewish World Review*, 23 January. Reprinted in *Washington Times*, 25 January 2008. Available at http://www.jewishworldreview.com/cols/greenberg012308.php3. • 'The burgeoning technological powers to intervene': L. Kass, in President's Council on Bioethics transcript, 2 January 2002. • **Page 320** 'some people . . . manufacture extremely intelligent, extremely powerful, extremely resistant people': C. Krauthammer, in ibid. • 'Repugnance is the emotional expression of deep wisdom': L. Kass (1997), 'The Wisdom of Repugnance', *New Republic* 216, 17–26, 2 June. • 'irrelevant or entirely inappropriate to the lives': Silver (2006), p. 169. • **Page 321** 'I doubt that any open society anywhere will accept': ibid., p. 171. • 'we should abandon any attempt to try to settle, once and for all':

Mulkay (1997), pp. 161–2. • 'It would be vain to give a single overall solution': Rostand (1959), p. 68. • **Page 322** 'If we wish to understand, and to control, the impact': Mulkay (1997), p. 163. • **Page 323** 'What does it mean to hold the beginning of human life': L. Kass, in Kristol & Cohen (eds) (2002), p. 69. • **Page 324** 'If we can't accept this, debates about cloning': Pence (2004), p. 15. • **Page 325** 'the erosion, perhaps the final erosion': Kass (1985), p. 37. • 'respect for human life is hardly expressed when embryos or presentient fetuses': B. Steinbock, in Lauritzen (ed.) (2001), p. 54. • **Page 326** 'It is merely from its quality when taken without the true corrective': Bacon (1605), p. 4. • 'The road to Brave New World is paved with sentimentality': L. Kass, in Kristol & Cohen (eds) (2002), p. 60.

Bibliography

A. E. Adams (1994). *Reproducing the Womb*. Cornell University Press, Ithaca.

B. Aldiss (2000). *Frankenstein Unbound*. House of Stratus, Kelly Bray, Cornwall.

I. Aristarkhova (2005). 'Ectogenesis and mother as machine'. *Body and Society* 11(3), 43–59.

Aristotle (*c.*350 BC). *Physics*, transl. R. P. Hardie & R. K. Gaye. Available at http://classics.mit.edu/Aristotle/physics.html.

I. Asimov (1983). *The Complete Robot*. Voyager, London.

S. T. Asma (2009). *On Monsters*. Oxford University Press, New York.

M. Atwood (1985). *The Handmaid's Tale*. Fawcett Crest, New York.

F. Bacon (1605/1620/1944). *Advancement of Learning and Novum Organum*. Willey, New York.

C. Baldick (1987). *In Frankenstein's Shadow*. Clarendon Press, Oxford.

P. Ball (2006). *The Devil's Doctor: Paracelsus and the World of Renaissance Magic and Science*. Heinemann, London.

S. Bann (ed.) (1994). *Frankenstein, Creation and Monstrosity*. Reaktion Books, London.

M. Bedau et al. (2010). 'Life after the synthetic cell'. *Nature* 465, 422–4.

S. Bedini (1964). 'The role of automata in the history of technology'. *Technology and Culture* 5(1).

B. Bensaude-Vincent & W. R. Newman (2007). *The Artificial and the Natural*. MIT Press, Cambridge, MA.

J. Bering (2006). 'The folk psychology of souls'. *Behavioural and Brain Sciences* 29, 453–62.

M. Berman (1988). *All That is Solid Melts Into Air*. Penguin, London.

J. D. Bernal (1929). *The World, the Flesh and the Devil: An Enquiry into the Future of the Three Enemies of the Rational Soul*. Routledge & Kegan Paul, London.

J. D. Biggers (2010). Editorial, *Human Reproduction* 25, 2156.

B. Bildhauer & R. Mills (eds) (2003). *The Monstrous Middle Ages*. University of Wales Press, Cardiff.

S. Blackburn (1999). *Think*. Oxford University Press, Oxford.

K. Boyd, B. Callaghan & E. Shotter (1986). *Life Before Birth*. SPCK, London.

V. Brittain (1929). *Halcyon, or the Future of Monogamy*. Kegan Paul, Trench, Trubner & Co., London.

A. Brown (2005). *J. D. Bernal: The Sage of Science*. Oxford University Press, Oxford.

S. Butler (1872/1985). *Erewhon*. Penguin, London.

K. Čapek (1921). *R.U.R.*, transl. D. Wyllie. eBooks@Adelaide, available at http://ebooks.adelaide.edu.au/c/capek/karel/rur/.

J. Challoner (1999). *The Baby Makers*. Channel 4 Books, London.

C. Churchill (2002). *A Number*. Nick Hern, London.

R. A. Clack (2000). *The Marriage of Heaven and Earth*. Greenwood, Westport.

E. Cohen (2008). *In the Shadow of Progress*. Encounter Books, New York.

J. J. Cohen (1999). *Of Giants: Sex, Monsters, and the Middle Ages*. University of Minnesota Press, Minneapolis.

J. J. Cohen (ed.) (1996). *Monster Theory: Reading Culture*. University of Minnesota Press, Minneapolis.

A. Crosse (1839). 'Description of some Experiments made with the Voltaic Battery', *American Journal of Science and Arts* 35, 125–37.

A. Crosse (1838). 'Note on a kind of Acarus', *Annals of Electricity, Magnetism & Chemistry* 2, 246–57.

L. Daston & K. Park (1998). *Wonders and the Order of Nature*. Zone, New York.

L. Daston & F. Vidal (eds) (2004). *The Moral Authority of Nature*. University of Chicago Press, Chicago.

S. de Beauvoir (1949/2009). *The Second Sex*. Jonathan Cape, London.

N. Denny & J. Filmer-Sankey (1973). *The Travels of Sir John Mandeville*. Collins, London.

R. Descartes (1637). *Discourse on the Method of Rightly Conducting the Reason, and Seeking Truth in the Sciences*. Available at http://www.literature.org/authors/descartes-rene/reason-discourse/.

L. de Rooy & H. van den Bogaard (2009). *Forces of Form*. Hans/AMC-Vossiuspers UvA, Amsterdam.

D. J. de Solla Price (1964). 'Automata and the origins of mechanism and mechanistic philosophy'. *Technology and Culture* 5, 9–23.

D. Diderot (1769). *D'Alembert's Dream*, transl. I. Johnson. Available at http://records.viu.ca/~johnstoi/diderot/dalembertsdream.htm.

K. R. Dronamraju (ed.) (1995). *Haldane's Daedalus Revisited*. Oxford University Press, Oxford.

Editorial (1987), 'IVF remains in legal limbo'. *Nature* 327, 87.

R. G. Edwards (1981). 'Test-tube babies, 1981'. *Nature* 293, 253–6.

R. G. Edwards (1989). *Life Before Birth: Reflections on the Embryo Debate.* Hutchinson, London.

R. G. Edwards & P. C. Steptoe (1980). *A Matter of Life.* Hutchinson, London.

R. G. Edwards, B. D. Bavister & P. C. Steptoe (1969). 'Early stages of fertilization in vitro of human oocytes matured in vitro'. *Nature* 221, 632–5.

R. G. Edwards, P. C. Steptoe & J. M. Purdy (1970). 'Fertilization and cleavage in vitro of preovulator human oocytes'. *Nature* 227, 1307–9.

S. W. Ellis (1930). 'Creatures of the Light'. *Astounding Stories of Super-Science,* February. Available at http://www.web-books.com/Classics/ON/B1/B1511/04MB1511.html.

D. Evans (ed.) (1996). *Creating the Child.* Kluwer Law International, The Hague.

D. Evans & N. Pickering (eds) (1996). *Conceiving the Embryo.* Martinus Nijhoff, The Hague.

A. Ferreira (2009). 'The sexual politics of ectogenesis in the To-day and To-morrow series'. *Interdisciplinary Science Reviews* 34, 32–55.

S. Firestone (2003). *The Dialectic of Sex.* Farrar, Straus & Giroux, New York.

S. Fishel & E. M. Symonds (eds) (1986). *In Vitro Fertilization: Past, Present, Future.* IRL Press, Oxford.

S. E. Forry (1990). *Hideous Progenies.* University of Pennsylvania Press, Philadelphia.

S. Freud (1919/2003). *The Uncanny.* Penguin, London. Available at http://people.emich.edu/acoykenda/uncanny1.htm.

J. B. Friedman (2000). *The Monstrous Races of Medieval Art and Thought.* Syracuse University Press, Syracuse.

D. M. Fryer & J. C. Marshall (1979). 'The motives of Jacques de Vaucanson'. *Technology and Culture* 20, 257–69.

D. G. Gibson et al. (2010). 'Creation of a bacterial cell controlled by a chemically synthesized genome'. *Science* 329, 52–6.

W. Godwin (1834/2006). *Lives of the Necromancers.* Echo Library, Teddington.

J. W. von Goethe (1949/1959). *Faust,* Parts I and II, transl. P. Wayne. Penguin, London.

D. Gooding, T. Pinch & S. Schaffer (eds), (1989). *The Uses of Experiment.* Cambridge University Press, Cambridge.

R. Gosden (1999). *Designing Babies: The Brave New World of Reproductive Technology.* W. H. Freeman, New York.

S. J. Gould (1977). *Ontogeny and Phylogeny.* Harvard University Press, Cambridge, MA.

I. Hacking (1991). 'Artificial phenomena'. *British Journal for the History of Science* 23, 235–41.

S. Hadlington (1987). 'British government hedges bets on embryo research'. *Nature* 330, 409.

N. Haire (1927). *Hymen, or the Future of Marriage*. Kegan Paul, Trench, Trubner & Co., London.

J. B. S. Haldane (1923). 'Daedalus, or science and the future', a paper read to the Heretics, Cambridge on 4 February. Reprinted in Dronamraju (ed.) (1995), pp. 23–50.

J. B. S. Haldane (1924). *Daedalus, or Science and the Future*. Kegan Paul, Trench, Trubner & Co., London.

J. B. S. Haldane (1963). 'Biological possibilities for the human species in the next ten thousand years'. In *Man and His Future*, ed. G. E. W. Wolstenholme. Little, Brown, Boston.

J. B. S. Haldane (1949). *What Is Life?* Lindsay Drummond, London.

T. S. Hall (1969). *Ideas of Life and Matter* (2 vols). University of Chicago Press, Chicago.

L. Harris (1969). 'The *Life* Poll: Brave New World – with reservations'. *Life* 13 June, 52–5.

N. Hawthorne (1843/1987). 'The Birth-mark'. In *Selected Tales and Sketches*. Penguin, London. Available at http://etext.virginia.edu/toc/modeng/public/HawBirt.html.

R. Haynes (1994). *From Faust to Strangelove: Representations of the Scientist in Western Literature*. Johns Hopkins University Press, Baltimore.

R. M. Henig (2004). *Pandora's Baby: How the First Test Tube Babies Sparked the Reproductive Revolution*. Houghton Mifflin, New York.

E. T. A. Hoffmann (1814). *Automata*. Available at http://www.munseys.com/book/11232/Automata.

E. T. A. Hoffmann (1816/1982). 'The Sandman'. In *Tales of Hoffmann*, transl. R. J. Hollingdale. Penguin, London.

R. Hoffmann (1995). *The Same and Not the Same*. Columbia University Press, New York.

R. Holmes (2008). *The Age of Wonder*. Harper Press, London.

R. Hooke (1665/2007). *Micrographia*. BiblioBazaar, Charleston, SC.

M. Hunter (2009). *Boyle: Between God and Science*. Yale University Press, New Haven.

A. Huxley (1932/1998). *Brave New World*. Harper Perennial, New York.

A. Huxley (1921). *Crome Yellow*. Chatto & Windus, London.

J. Huxley (1942). *The Uniqueness of Man*. The Scientific Book Club, London.

T. H. Huxley (1869). 'On the physical basis of life'. *Fortnightly Review* 5, 129. Available at http://alepho.clarku.edu/huxley/CE1/PhysB.html.

T. H. Huxley (1874). 'On the hypothesis that animals are automata, and its history', *Fortnightly Review* 95, 555–80. Available at http://alepho.clarku.edu/huxley/CE1/AnAuto.html.

M. Idel (1990). *Golem: Jewish Magical and Mystical Traditions on the Artificial Anthropoid*. State University of New York Press, Albany.

F. Jacob (1989). *The Logic of Life and The Possible and the Actual.* Penguin, London.

H. S. Jennings (1925). *Prometheus, or Biology and the Advancement of Man.* Kegan Paul, Trench, Trubner & Co., London.

M. H. Johnson, S. B. Franklin, M. Cottingham & N. Hopwood (2010). 'Why the Medical Research Council refused Robert Edwards and Patrick Steptoe support for research on human conception in 1971', *Human Reproduction* 25, 2157–74.

D. A. Jones (2004). *The Soul of the Embryo.* Continuum, London.

H. F. Judson (1979). *The Eighth Day of Creation.* Cold Spring Harbor Laboratory Press, New York.

E. A. Kaplan & S. M. Squier (eds) (1999). *Playing Dolly: Technocultural Formations, Fantasies and Fictions of Assisted Reproduction.* Rutgers University Press, Piscataway, NJ.

L. R. Kass (1971a). 'The new biology: what price relieving man's estate?'. *Science* 174, 779–88.

L. R. Kass (1971b). 'Babies by means of in vitro fertilization: unethical experiments on the unborn?'. *New England Journal of Medicine* 285, 1174–9.

L. R. Kass (1985). *Toward a More Natural Science: Biology and Human Affairs.* Free Press, New York.

L. R. Kass (1997). 'The wisdom of repugnance: why we should ban cloning of humans'. *The New Republic*, 2 June, pp. 17–26. Available at http://www.catholiceducation.org/articles/medical_ethics/me0006.html.

A. Kimbrell (1993). *The Human Body Shop.* HarperCollins, New York.

J. Kitzinger & C. Williams (2003). 'Forecasting the future: legitimising hope and calming fears in the embryo stem cell debate'. *Social Science and Medicine* 61, 731–40.

R. D. Klein (1990). 'IVF research: a question of feminist ethics'. *Reproduction and Genetic Engineering: Journal of International Feminist Analysis* 3(3). Available at www.finrrage.org/pdf_files/IVF/IVF_Research_A_Question_of_Feminist_Ethics_1991.pdf.

T. Klein (2008). 'Karl Rahner on the soul'. *St Anselm Journal* 6.1, Fall. Available at http://www.anselm.edu/Institutes-Centers-and-the-Arts/Institute-for-Saint-Anselm-Studies/Saint-Anselm-Journal/Archives/Vol-6-No-1-fall-2008.htm.

I. Klíma (2002). *Karel Čapek: Life and Work.* Catbird, North Haven, CT.

A. J. Klotzko (ed.) (2001). *The Cloning Sourcebook.* Oxford University Press, New York.

T. Kono, Y. Obata, Q. Wu, K. Niwa, Y. Ono, Y. Yamamoto, E. S. Park, J.-S. Seo & H. Ogawa (2004). 'Birth of parthenogenetic mice that can develop to adulthood', *Nature* 428, 860–4.

C. Krauthammer (1998), 'Of headless mice . . . and men', *Time*, 19 January.

Available at http://www.time.com/time/magazine/article/0,9171,987687, 00.html.

W. Kristol & E. Cohen (eds) (2002). *The Future Is Now*. Rowman & Littlefield, Lanham, MD.

J. O. de La Mettrie (1994). *Man a Machine*, transl. R. A. Watson & M. Rybalka. Hackett, Indianapolis.

P. Lauritzen (ed.) (2001). *Cloning and the Future of Human Embryo Research*. Oxford University Press, New York.

C. S. Lewis (1943/1973). *The Abolition of Man*. Macmillan, New York.

J. Loeb (1912). *The Mechanistic Conception of Life*. University of Chicago Press, Chicago.

A. Ludovici (1924). *Lysistrata, or Women's Future and Future Women*. Kegan Paul, Trench, Trubner & Co., London.

E. Machery (2010). 'Why I stopped worrying about the definition of life . . . and why you should as well'. *Synthese*, in press. Available at http://www.pitt.edu/%Emachery/papers/Definition%20of%life%20Synthese%20machery.pdf.

A. Mackay (1999). Introduction to 'Crystal Souls: Studies of Inorganic Life'. *Forma* 14, 1-204.

G. C. Meilaender (1998). *Body, Soul and Bioethics*. University of Notre Dame Press, Notre Dame, IN.

A. K. Mellor (1988). *Mary Shelley: Her Life, Her Fiction, Her Monsters*. Routledge, New York.

H. Melville (1856/1986). 'The Bell-Tower'. In *Billy Budd and Other Stories*. Penguin, London. Available at http://www.melville.org/belltowr.htm.

H. N. Milner (1826). *Frankenstein: or, The Man and the Monster!* John Cumberland & Duncombe, London.

N. Mitchison (1975/1995). *Solution Three*. The Feminist Press at the City University of New York, New York.

A. G. Molland (1974). 'Roger Bacon as magician'. *Traditio* 30, 445–60.

M. J. Mulkay (1997). *The Embryo Research Debate: Science and the Politics of Reproduction*. Cambridge University Press, Cambridge.

H. J. Muller (1935). *Out of the Night: A Biologist's View of the Future*. Vanguard, New York.

M. Murphy (2008). 'The natural law tradition in ethics'. In *Stanford Encyclopedia of Philosophy*, ed. E. N. Zalta. Stanford University. Available at http://plato.stanford.edu/entries/natural-law-ethics/.

J. Needham & L. Wang (1956). *Science and Civilization in China*, Vol. II. Cambridge University Press, Cambridge.

M. Negrotti (1999). *The Theory of the Artificial*. Intellect, Exeter.

M. Negrotti (2002). *Naturoids: On the Nature of the Artificial*. World Scientific, Singapore.

B. Nerlich, D. D. Clarke & R. Dingwall (2001). 'Fictions, fantasies, and fears: The literary foundations of the cloning debate'. *Journal of Literary Semantics* 30, 37–52.

B. Nerlich, R. Dingwall & D. D. Clarke (2002). 'The book of life: how completion of the Human Genome Project was revealed to the public'. *Health: An Interdisciplinary Journal for the Social Study of Health, Illness and Medicine* 6, 445–69.

B. Nerlich & R. Dingwall (2003). 'Deciphering the human genome: the semantic and ideological foundations of genetic and genomic discourse'. In *Cognitive Models in Language and Thought: Ideology, Metaphors and Meanings*, eds R. Dirven, R. Frank & M. Pütz, 395–428. Mouton de Gruyter, Berlin.

B. Nerlich, R. Elliott & B. Larson (eds) (2009). *Communicating Biological Sciences*. Ashgate, Farnham.

W. R. Newman (2004). *Promethean Ambitions: Alchemy and the Quest to Perfect Nature*. University of Chiacgo Press, Chicago.

E. Paul (1930). *Chronos, or the Future of the Family*. Kegan Paul, Trench, Trubner & Co., London.

P. J. Pauly (1987). *Controlling Life: Jacques Loeb and the Engineering Ideal in Biology*. Oxford University Press, Oxford.

G. Pence (1998). *Who's Afraid of Human Cloning?* Rowman & Littlefield, Lanham, MD.

G. Pence (2004). *Cloning After Dolly: Who's Still Afraid?* Rowman & Littlefield, Lanham, MD.

T. Peters (2003). *Playing God*. Routledge, New York.

C. J. S. Picart (2002). *The Cinematic Rebirths of Frankenstein*. Praeger, Westport, CT.

M. Piercy (1976). *Woman On The Edge Of Time*. The Women's Press, London.

Plato (2001). *The Republic*, transl. B. Jowett. Agora, Millis, MA. Available at http://classics.mit.edu/Plato/republic.html.

Plato (2005). *Protagoras*, transl. B. Jowett. Digireads, Stilwell, KS. Available at http://classics.mit.edu/Plato/protagoras.html.

E. A. Poe (1908). *Edgar Allan Poe's Tales of Mystery and Imagination*. J. M. Dent, London.

E. A. Poe (1850). 'The man that was used up'. In J. Harrison (ed.) (1965), *The Complete Works of Edgar Allan Poe*. AMS Press, New York. Available at http://xroads.virginia.edu/~hyper/POE/used_up.html.

President's Council on Bioethics (2002). 'Human cloning and human dignity'. Public Affairs, New York.

President's Council on Bioethics, transcript of debate on 2 January 2002. Available at http://bioethics.georgetown.edu/pcbe/transcripts/jan02/jan17session2.html.

L. M. Principe (1998). *The Aspiring Adept: Robert Boyle and His Alchemical Quest*. Princeton University Press, Princeton.

P. Ramsey (1970). *Fabricated Man*. Yale University Press, New Haven.

J. Rifkin (1999). *The Biotech Century*. Phoenix, London.

J. Rock & M. F. Menkin (1944). 'In vitro fertilization and cleavage of human ovarian eggs'. *Science* 100, 105.

D. Rorvik (1971). *Brave New Baby*. Doubleday, New York.

D. Rorvik (1978). *In His Image: The Cloning of a Man*. J. B. Lippincott, Philadelphia.

C. Rosen (2003). 'Why not artificial wombs?'. *New Atlantis* 3 (Fall), 67–76. Available at http://www.thenewatlantis.com/publications/why-not-artificial-wombs.

A. Rosenfeld (1969). 'Challenge to the miracle of life'. *Life* 13 June, 38–51.

J. Rostand (1959). *Can Man Be Modified?*, transl. J. Griffin. Basic Books, New York.

R. Rowland (1992). *Living Laboratories: Women and Reproductive Technologies*. Indiana University Press, Indianapolis.

B. Russell (1931). *Icarus, or the Future of Science*. Kegan Paul, Trench, Trubner & Co., London.

D. Russell (1925). *Hypatia, or Women and Knowledge*. Kegan Paul, Trench, Trubner & Co., London.

W. St Clair (2004). *The Reading Nation in the Romantic Period*. Oxford University Press, Oxford.

J. St John & R. Lovell-Badge (2007). 'Human-animal cytoplasmic hybrid embryos, mitochondria, and an energetic debate'. *Nature Cell Biology* 9, 988–92.

P. Sargent (1976/1981). *Cloned Lives*. Fontana, London.

E. A. Schäfer (1912). 'Inaugural address to the British Association'. *Nature* 90, 7–19.

S. Schaffer (1983), 'Natural philosophy and public spectacle in the eighteenth century', *History of Science* 21, 1–43.

G. G. Scholem (1965). *On the Kabbalah and its Symbolism*, transl. R. Manheim. Routledge & Kegan Paul, London.

E. Schrödinger (1944/2000). *What Is Life?* Folio Society, London.

J. Schummer (2006). 'Historical roots of the "mad scientist": chemists in nineteenth-century literature'. *Ambix* 53(2), 99–127.

J. Schummer (2009). 'The creation of life in a cultural context: From spontaneous generation to synthetic biology'. In *The Ethics of Protocells: Moral and Social Implications of Creating Life in the Laboratory*, eds M. Bedau & E. Parke. MIT Press, Cambridge, MA.

Science and Technology Committee (2007). *Science and Technology Committee Fifth Report* on 'Government proposals for the regulation of hybrid and chimera embryos', 28 March. Available at http://www.publications. parliament.uk/pa/cm200607/cmselect/cmsctech/272/27202.htm.

J. A. Scutt (ed.) (1988). *The Baby Machine*. McCulloch, Victoria.

M. Shelley (1818/1831/1994). *Frankenstein*, 3rd edn. Penguin, London.

M. Shelley (1818/1831/1974). *Frankenstein*, ed. J. Rieger. University of Chicago Press, Chicago.

P. Singer & D. Wells (1984). *The Reproduction Revolution: New Ways of Making Babies*. Oxford University Press, Oxford.

L. M. Silver (1998). *Remaking Eden*. HarperCollins, New York.

L. M. Silver (2006). *Challenging Nature*. HarperCollins, New York.

F. E. Smith (1930). *The World in 2030 AD*. Brewer & Warren, London.

D. L. Spar (2006). *The Baby Business*. Harvard Business School Press, Boston.

S. M. Squier (1994). *Babies in Bottles: Twentieth-Century Visions of Reproductive Technology*. Rutgers University Press, New Brunswick, NJ.

B. M. Stafford (1994). *Artful Science*. MIT Press, Cambridge, MA.

R. L. Stevenson (1979). *The Strange Case of Dr Jekyll and Mr Hyde, and Other Stories*, ed. J. Calder. Penguin, Harmondsworth.

P. Sullivan (1985). 'Medieval automata'. *Romance Studies* 6, 1–20.

M. Teich (ed.) (1992). *A Documentary History of Biochemistry 1770–1940*. Leicester University Press, Leicester.

S. Toulmin & J. Goodfield (1962/1982). *The Architecture of Matter*. University of Chicago Press, Chicago.

C. P. Toumey (1992). 'The moral character of mad scientists: a cultural critique of science', *Science, Technology & Human Values* 17, 411–37.

A. Tudor (1989a). *Monsters and Mad Scientists*. Blackwell, Oxford.

A. Tudor (1989b). 'Seeing the worst side of science'. *Nature* 340, 589–92.

J. Turney (1998). *Frankenstein's Footsteps: Science, Genetics and Popular Culture*. Yale University Press, New Haven.

G. Vajta & M. Gjerris (2006). 'Science and technology of farm animal cloning: state of the art'. *Animal Reproduction Science* 92 (3–4), 211–30.

J. van Dyck (1995). *Manufacturing Babies and Public Consent*. Macmillan, Basingstoke.

A. Vartanian (1960). *La Mettrie's L'homme machine: A Study in the Origins of an Idea*. Princeton University Press, Princeton.

J. M. M. P. A. Villiers de l'Isle-Adam (2001). *Tomorrow's Eve*, transl. R. M. Adams. University of Illinois Press, Champaign, IL.

M. Warner (2002). *Fantastic Metamorphoses, Other Worlds*. Oxford University Press, Oxford.

M. Warnock et al. (1984). 'Report of the Committee of Inquiry into human fertilisation and embryology'. HM Stationery Office, London. Available at http://www.hfea.gov.uk/2068.html.

D. Watt (ed.) (1975). *Aldous Huxley: The Critical Heritage*. Routledge & Kegan Paul, London.

G. Weissmann (2006). 'Stem cells, *in vitro* fertilization, and Jacques Loeb'. *FASEB Journal* 20, 1031–3.

H. G. Wells (1896/1995). *The Island of Doctor Moreau*. In *H. G. Wells: The Science Fiction*, Vol. 1. J. M. Dent, London.

L. L. Whyte (1927). *Archimedes, or the Future of Physics*. Kegan Paul, Trench, Trubner & Co., London.

M. Willis (2006). *Mesmerists, Monsters, and Madness*. Kent State University Press, Kent, OH.

I. Wilmut, K. Campbell & C. Tudge (2000). *The Second Creation*. Headline, London.

E. B. Wilson (1923). *The Physical Basis of Life*. Yale University Press, New Haven.

A. Wood (2009). 'Darwinism, biology and mythology in the To-day and To-morrow series'. *Interdisciplinary Science Reviews* 34, 22–31.

G. Wood (2002). *Living Dolls*. Faber, London.

Index